Assessing the evidence on Indigenous socioeconomic outcomes:
A focus on the 2002 NATSISS

B. H. Hunter (Editor)

ANU
THE AUSTRALIAN NATIONAL UNIVERSITY

E PRESS

Centre for Aboriginal Economic Policy Research
The Australian National University, Canberra

Research Monograph No. 26
2006

ANU
E PRESS

Published by ANU E Press
The Australian National University
Canberra ACT 0200, Australia
Email: anuepress@anu.edu.au
Web: http://epress.anu.edu.au

National Library of Australia
Cataloguing-in-publication entry.

Assessing the evidence on Indigenous socioeconomic
outcomes : a focus on the 2002 NATSISS.

ISBN 1 920942 19 X
ISBN 1 9209426 4 5 (Online document)

1. Aboriginal Australians - Social conditions. 2.Aboriginal Australians - Economic
conditions. I. Hunter, Boyd. II. Australian National University. Centre for
Aboriginal Economic Policy Research. (Series : Research monograph (Australian
National University. Centre for Aboriginal Economic Policy Research) ; no. 26).

362.849915

Cover design by Brendon McKinley

Table of Contents

List of Figures

List of Tables

Foreword

This monograph presents the refereed, and peer-reviewed, edited proceedings of the conference on Indigenous Socioeconomic Outcomes: Assessing Recent Evidence. The conference was organised by the Centre for Aboriginal Economic Policy Research and held at the Shine Dome, the Australian National University (ANU) in Canberra on 11 and 12 August 2005. The conference aimed to present the latest evidence on Indigenous economic and social status, and family and community life, and discuss its implications for government policy. The recently released 2002 National Aboriginal and Torres Strait Islander Social Survey (NATSISS) provided a valuable new source of data on these issues. The conference featured a variety of presentations that provide an evaluation of the strengths and weakness of the NATSISS methodology and the quality of the survey data and existing output.

In the same year that NATSISS was conducted, the Prime Minister wrote to Gary Banks, as Chairman of the Steering Committee for the Review of Government Service Provision (SCRGSP), to ask for a regular report to the Council of Australian Governments (COAG) against key indicators of Indigenous disadvantage. This is now available on a biennial basis as the Productivity Commission Report *Overcoming Indigenous Disadvantage*. In addition, the annual Report on Government Services issued by the SCRGSP now includes a separate compendium of Indigenous statistics drawn from the administrative databases of Australian, State and Territory governments.

Given the growing number of Indigenous statistics collected and reported in Australia, it is more important than ever before to provide critical scrutiny of the data and related analysis. By publishing this report via the ANU E Press, we aim to ensure the timely and wide access to research findings throughout the Indigenous policy community. I recommend this monograph to anyone who wants to understand the strengths and weaknesses of the existing statistical archive for Indigenous Australians.

Jon Altman
Director, CAEPR
June 2006

Preface

The origin of this book arose from a growing personal concern with the quality of assessment of evidence in public debates. The modus operandi of journalism means that it is necessary to simplify evidence of 'experts' who conduct a detailed analysis of statistical data. The exigencies of daily deadlines mean that the reporting of experts' interpretation of the evidence is sometimes incorrect. However, the alleged experts have fewer excuses, and there is a need to ground their analysis in peer review. This monograph attempts to do this by providing readers with the capacity to critically engage with the evidence presented in public forums.

The volume of quality data in Indigenous affairs has grown exponentially in recent years. Unfortunately, our capacity to analyse it has not increased at the same rate. The main issue is that there is not the number of suitably qualified social scientists focusing on Indigenous issues in Australia to analyse all the data. Another issue is that the sheer volume of data allows some researchers to search for statistics that suit their ideological predisposition. Hughes (2005a) provides an example of the art of selective citation of other people's analysis to support arguments that are otherwise based on assertions and hyperbole (see Hunter's [2005] critique of Hughes [2005a] and Hughes and Warin [2005]). Needless to say, this sort of approach violates fundamental principles that most social scientists hold dear. It should be made clear that advocates on all sides of the debate have used this strategy to look for either a statistic or an expert whose conclusions are consistent with their own.

This monograph presents the refereed and peer-reviewed, edited proceedings of the conference on Indigenous Socioeconomic Outcomes: Assessing Recent Evidence. The conference was organised by the Centre for Aboriginal Economic Policy Research and held at the Shine Dome, the Australian National University (ANU) in Canberra on 11 and 12 August 2005. The conference, which was attended by around 200 people from universities, the government sector, non-profit organisations and private corporations and citizens, aimed to present the latest evidence on Indigenous economic and social status, and family and community life, and discuss its implications for government policy.

After the preliminaries, the conference started with a recent history of the political economy of statistical collections and a brief introduction to NATSISS methodology and the quality of survey questions. While the main focus was on analysing the 2002 NATSISS outputs, many presentations provided an assessment of changes in Indigenous social conditions over time and examined how Indigenous people fared vis-à-vis other Australians in other statistical collections. All contributors were invited to reflect on how NATSISS might be improved when it is next undertaken in 2008. Notwithstanding the fact that one of the main rationales for the conference was to evaluate the NATSISS, considerable time was allocated to providing a broad framework for the discussion, with several overviews provided by generalists who presented fresh perspectives, and a panel discussion of Indigenous public intellectuals who debated the broad Indigenous policy context.

The structure of this monograph closely follows the order of presentations in the conference, with two exceptions: those by Bob Gregory and Tim Rowse. Bob Gregory's presentation was the first presented on the second day of the conference, but his chapter has been placed in the groups of chapters that deal with labour market issues, as his focus is almost exclusively on such issues. This change of order is in no way a reflection on the quality of the presentation, as it received the most votes for 'best paper' at the conference by the participants who filled in the conference evaluation. The contribution from Tim Rowse has been placed at the beginning of this monograph because it provides the historical overview of what he calls the Indigenous statistical archive, a precursor to the rest of the chapters in this volume. Rowse's penultimate position on the conference timetable had been a 'historical accident', as he accommodated last minute changes to the schedule. For the remainder of this preface, unless otherwise indicated, the contributors to the volume have a direct affiliation with CAEPR.

The second chapter in this monograph is written by Jon Altman and John Taylor. It provides a recent history of omnibus social surveys of Indigenous Australians and analyses the political economy of Indigenous statistics. The next two chapters are more methodological in nature. Andrew Webster, Alistair Roger, and Dan Black from the Australian Bureau of Statistics (ABS) examined survey design and output issues, while Nicholas Biddle and I examined selected, more general, methodological issues that were relevant to the NATSISS.

Indigenous demography is given due prominence with John Taylor and Yohannes Kinfu critically examining mobility issues within the NATSISS. Yohannes Kinfu then explores the crucial issues of fertility and child survival issues that drive much of the development dynamics in Indigenous Australia.

The important dimensions of Indigenous socioeconomic status are then addressed, starting with Will Sander's evaluation of the utility of housing in NATSISS. I then examine the treatment of income, financial stress, social exclusion issues with a provocative reference to the so-called 'poverty wars'. Ruth Weston and Matthew Gray, from the Australian Institute for Family Studies (AIFS), provide a framework for analysing family and community life. Matthew then teamed up with Bruce Chapman (RSSS, ANU) to provide an evaluation of labour force issues. Bob Gregory's paper then addressed which were the right questions to ask in Indigenous policy, focusing almost exclusively on labour market issues. The important issue of the customary or non-market economy is then examined by Jon Altman, Geoff Buchanan and Nicholas Biddle.

The proceedings of the first day of the conference wrapped up with the panel discussion by prominent Indigenous Australians. Larissa Berhendt (Jumbunna, University of Technology Sydney [UTS]), Tom Calma, and Geoff Scott (Distinguished Professor of Indigenous Policy at UTS) discussed the diverse perspectives on the needs for various data to be collected for Indigenous Australians. In contrast to most other papers in this monograph, this chapter is an edited transcript which includes the questions asked of the panelists at the end of the formal contributions.

The second day of the conference started with the aforementioned paper by Bob Gregory. Jerry Schwab then discussed the contribution of the 2002 NATSISS in education and training-related issues. The next contribution arose from a session on transport and information technology that packed two distinct papers into one (half-hour) slot. While Sarah Holcombe and Peter Radoll (School of Business and Information Management, ANU) (information technology) were constrained to talk for around 15 minutes because of time pressure on the day, these authors provided excellent separate chapters in this monograph on transport and information technology respectively.

Most of the remaining sessions at the conference covered complex social issues facing Indigenous communities. Russell Ross (University of Sydney) provided some fresh insights into Indigenous health. This is the first time he has written on the area, although he has written a considerable number of papers on economic issues facing Indigenous people. Maggie Brady and Tanya Chikritzhs (Curtin University) then provided a critical analysis of the NATSISS data on substance abuse and tobacco. Mick Dodson (National Centre for Indigenous Studies, ANU) and I then examined crime and justice data in NATSISS, and call for a greater consistency in the survey collections for Indigenous and other Australians. Nicolas Peterson (School of Archaeology and Anthropology, ANU) examines cultural issues in the 2002 NATSISS before Inge Kral and Frances Morphy focus on how the information on language might be optimised in a survey context. Bill Arthur provides an overview of Torres Strait Islanders, using the now conventional distinction of those who live in the Strait and those who live on the mainland. Tom Calma (Human Rights and Equal Opportunity Commission [HREOC]) provides a detailed analysis of social justice issues.

In the final chapter, Jon Altman and I provide some concluding remarks and a wrap-up for the monograph, which pays particular attention to summarising the lessons learned from the conference and highlights how future statistical collections might be improved.

The conference also included an informal presentation by Nicholas Biddle which described an easy-to-use, CAEPR do-it-yourself spreadsheet for customised hypothesis tests on NATSISS data. This spreadsheet is now available on the CAEPR web site, and can be used to validate all the significance tests conducted in this monograph. Since this spreadsheet is publicly available, I have encouraged authors to report results of the significance tests in the text but to not necessarily report the statistics themselves. In addition to avoiding any duplication, it hopefully makes the monograph easier to read without compromising the rigour required in a scholarly publication. Readers should note that the spreadsheet only provides approximate (usually conservative) significance tests based on standard errors expressed in levels. For a discussion about the reliability of 2002 NATSIS data and alternative procedures for estimating significance tests, see ABS (2004c: 68–9).

The authors of the monograph chapters were largely left to their own devices when writing their contribution, before the refereeing process for publication. Before the conference, I sent out an e-mail to get the contributors to think about the substantive outstanding issues in their subject areas and how NATSISS and other future data collections might address the needs identified. It will be apparent in what follows that the authors adopted unique approaches to their respective subject matter(s). This is to be encouraged because it is the strength of an edited volume. As is usual in academic publications, it should be noted that the views expressed in this paper are those of the authors and do not reflect those of the institutions to which they are attached.

Finally, in my time at CAEPR I have encountered some suspicion of statistical evidence mostly from the occasional qualitative researcher who seems to think that statistics objectify Indigenous people and only provide extremely limited insights for Indigenous people and policy makers. I would be the first to acknowledge the limitations of statistical analysis, but it is important to honour the participation of all the 10 000 respondents who spent many hours filling in the form, and work out what they were telling the ABS. Quantitative and qualitative research each have different and valid insights to yield, but both need to acknowledge that they are based on the generosity of respondents who entrusted their knowledge and experience to the people who collect the data. Consequently, the main focus of this conference should be to work out what, if anything, the ABS survey data tells us.

The strength of this monograph, as with all edited volumes, is that it can explore different perspectives. While this point might be characterised as being 'post-modern' by some, it is not necessarily so. Rather, it is an acknowledgment of the obvious—that is, there is considerable uncertainty about which theoretical model should be adopted, and hence there are legitimate issues about how to interpret the 2002 NATSISS data.

Having mentioned the word 'post-modern' I want to assure readers that not all of the authors drink Chardonnay! Indeed, I could not find one person who admitted to drinking Chardonnay with an informal straw poll revealing that the most popular drink among authors is a full-bodied red. Furthermore, the award for the best paper at the conference was a bottle of whisky, albeit an Irish whisky. Notwithstanding their drinking habits, none of the authors admitted to disparaging others for their consumption habits.

The diversity of opinions of authors will be obvious to all those who persevere in reading through the entire book. I hope readers will find their perseverance rewarded because the contrasting approaches of the various chapters suggest some research directions for the academy. However, for those who pick and choose which chapters they are interested in, I commend you to the concluding chapter which briefly draws together the disparate themes of the monograph.

Boyd Hunter
Fellow, CAEPR

Acknowledgements

The contributors to the conference were leading analysts, commentators and researchers in Indigenous policy. The presenters came from all over Australia to provide a stimulating, enjoyable, and informed debate. While I am grateful to all contributors, I would particularly like to thank Tom Calma, the Aboriginal and Torres Strait Islander Social Justice Commissioner from the Human Rights and Equal Opportunity Commission, who spent more time on the Shine Dome stage than almost any other presenters. I would also like to give a special mention to the chairs of the respective sessions at the conference who provided rigour, discipline and breadth to the discussion. Four chairs deserve special thanks in this regard: Craig Linkhorn (New Zealand Crown Law Office and Visiting Fellow, CAEPR), Peter McDonald (Demography Program, Research School of Social Sciences), Sandra Pattison (National Centre for Vocational and Educational Research), and Ben Smith (Post-Doctoral Fellow, CAEPR).

The ABS were supportive of the event as they were keen to facilitate responsible and informed use of their data. The main form of that support was extra data provided to facilitate our evaluation of the 2002 NATSISS. While much of the data used in this monograph is available from the official ABS (2004c) publication on 2002 NATSISS (Cat. No.4714.0), some customised cross-tabulations were provided specifically for the conference and this publication. I would like to acknowledge the staff at the ABS's National Centre for Aboriginal and Torres Strait Islander Statistics, who generously gave their time in providing data and clarifying any queries about how the data was collected, coded and outputted.

The final acknowledgment must go to the dedicated and professional Centre for Aboriginal Economic Policy Research (CAEPR) team who were responsible for the organisation of the conference over several months, especially Hilary Bek, Geoff Buchanan, Ruth Nicholls, John Hughes, Vicki Veness and Maria Davern. Without their hard work, the conference would never have been completed successfully. Two anonymous referees and many readers also gave invaluable comments on early drafts of the chapters for this monograph. I would like to take this opportunity to thank ANU E Press for publishing this report in such a professional manner. A final thanks must go to Hilary Bek and Jeneen McLeod for assistance with the copy-editing of a draft of the manuscript for this book.

Abbreviations and acronyms

ABS Australian Bureau of Statistics

ACER Australian Council for Education Research

ACS Australian Construction Services

ACT Australian Capital Territory

ADAC Aboriginal Drug and Alcohol Council

AEDP Aboriginal Employment Development Policy

AEP Aboriginal Education Policy

AGPS Australian Government Publishing Service

AIATSIS Australian Institute of Aboriginal and Torres Strait Islander Studies

AIC Australian Institute of Criminology

AIFS Australian Institute for Family Studies

AIHW Australian Institute of Health and Welfare

ANU Australian National University

ARIA Accessibility/Remoteness Index of Australia

ASGC Australian Standard Geographic Classification

ATSI Aboriginal and Torres Strait Islander

ATSIC Aboriginal and Torres Strait Islander Commission

AVCC Australian Vice-Chancellors Committee

CAEPR Centre for Aboriginal Economic Policy Research

CA Community Area

CAI computer assisted interviewing

CD Collection District

CDEP Community Development Employment Projects

CDHSH Commonwealth Department of Human Services and Health

CESCR Committee on Economic Social and Cultural Rights

CHINS Community Housing and Infrastructure Needs Surveys

CIF Community Information Forms

CIS Centre for Independent Studies

CNOS Canadian National Occupancy Standard

COAG Council of Australian Governments

CPI Consumer Price Index

CURF Confidentialised Unit Record File

DCITA Department of Communications, Information Technology and the Arts

DEWR Department of Employment and Workplace Relations

FaCS Department of Family and Community Services

FATSIL Federation of Aboriginal and Torres Strait Islander Languages

GSS General Social Survey

HILDA Household, Income and Labour Dynamics in Australia

HREOC Human Rights and Equal Opportunity Commission

ICC Indigenous Coordination Centre

ICT Information and Communication Technology

ICVS International Crime Victims Survey

IL Indigenous language

LFS Labour Force Survey

MCATSIA Ministerial Council for Aboriginal and Torres Strait Islander Affairs

MCEETYA Ministerial Council on Education, Employment, Training and Youth Affairs

MURF Main Unit Record File

NATSIS National Aboriginal and Torres Strait Islander Survey

NATSISS National Aboriginal and Torres Strait Islander Social Survey

NCA Non-Community Area

NCSS National Crime and Safety Survey

NCVER National Centre for Vocational and Educational Research

NCSAGIS National Centre for Social Applications of Geographic Information Systems

NDSHS National Drug Strategy Household Survey

NHS National Health Survey

NILF Not in the labour force

NPYWC Ngaanyatharra, Pitjantjatjara, Yankunytjatjara Women's Council

NSCSP National Survey of Community Satisfaction with Policing

NSW New South Wales

NT Northern Territory

OIPC Office of Indigenous Policy Coordination

PAPI pen and paper interview

PNG Papua New Guinea

PSS Personal Safety Survey

QLD Queensland

RCADC Royal Commission into Aboriginal Deaths in Custody

RSE relative standard error

RADL Remote Access Data Laboratory

RSSS Research School of Social Science

SA South Australia

SAE Standard Australian English

SAS Statistical Analysis Software

SCRGSP Steering Committee for the Review of Government Service Provision

SIF Special Indigenous Form

SOIL State of Indigenous Languages

SPSS Statistical Package for the Social Sciences

SRA Shared Responsibility Agreement

TAS Tasmania

TSI Torres Strait Islander

UN United Nations

UNDP United Nations Development Programme

USA United States of America

UTS University of Technology Sydney

VET vocational education and training

VIC Victoria

WA Western Australia

1. Towards a history of Indigenous statistics in Australia

Tim Rowse

In 2003, the Productivity Commission report *Overcoming Indigenous Disadvantage: Key Indicators 2003* demonstrated that Australia now has an extensive (though not time-deep) statistical archive through which we can compare Indigenous and non-Indigenous people. The report points to twelve 'headline indicators', and seven 'strategic areas for action', and it operationalises the 'strategic areas for action' in terms of thirty variables on which Indigenous and non-Indigenous can be compared. [1]

It has not always been possible to make such a wide battery of Indigenous/non-Indigenous comparisons (nor Indigenous/Australia comparisons). In this paper I will trace the steps that Australian governments have taken to recognise an entity that we call the 'Indigenous population' and to state its characteristics in comparative, quantitative terms.

My story is a frankly teleological one, in two senses. First, in looking at the steps taken I will pay particular attention to the steps that led towards what we have now. Second, I will nominate a turning point or watershed: the period 1966–76. In this moment, Australian governments ceased to manage the statistical archive in one way and began to manage it in a new way.

The watershed years

What happened between 1966 and 1976? The 1966 Census was the first Australian census in which the Statistician was prepared to claim 'virtually complete' enumeration of the Aboriginal population; [2] as I will show below, the first comparisons of 'Aboriginal' with Australia, in respect of education and occupation, were enabled by the 1966 Census. However, 1966 was also the last Australian census to make the distinction between 'full blood' Aborigines and Aborigines of mixed descent. So 1966 was not the beginning of comparisons over time.

At first sight, we could understand the repeal, by referendum, of s.127 of the Australian Constitution as a step towards the reformed statistical archive that we now use. However, the referendum was not clearly such a progressive step,

---

[1] I will use 'Indigenous' to refer to the combined figures for those of Aboriginal and/or Torres Strait Islander descent. For much of the period discussed in this chapter, the term 'Aboriginal' was used loosely to refer to those whom I refer to as 'Indigenous Australians'.
[2] Borrie (1975: 468) judged the 1954 Census to be 'the first for which reasonable estimates of the total Aboriginal plus Islander population are available'.

for two reasons. One is that, notwithstanding the words of s.127 ('in reckoning the numbers of the people of the Commonwealth, or of a State or other part of the Commonwealth, aboriginal natives shall not be counted'), all governments had long been collecting data on their Indigenous populations, as they conceived them. The second is that it was open to statistical authorities to interpret the repeal of s.127 as mandating the merging of the enumerated Aborigine with the total population. That is, the repeal of s.127 could be interpreted as a mandate to include, *without distinguishing*, Indigenous Australians. The argument to continue to distinguish Indigenous Australians, and to distinguish them in new ways, had to be made independently of, and even in the face of, the inclusive rhetoric of the 1967 referendum campaign.

In 1971, in a reform that was not evidently related to the repeal of s.127, the census introduced a new classification question that gave respondents the opportunity to identify themselves and their household members as 'Aboriginal' or as 'Torres Strait Islander'. In 1976 we had the second census using this revised question, so 1976 marks the beginning of the possibility of systematic comparison over time of Indigenous with 'all Australia', and Aboriginal with Torres Strait Islander.

I should emphasise the word 'beginning', for in 1976 there was still much to be done to reform those parts of the statistical archive that are supplied by organisations, such as hospitals, that service Indigenous Australians. Borrie complained in 1975 that it was still:

> ...extraordinarily difficult to persuade all the relevant authorities that Aborigines should be distinguished so that separate statistics can be maintained. Nor, as yet, is there any agreement on how the Aboriginal population is to be defined in the collection of statistics for series other than the census...[S]ome officials have argued that it would be offensive or discriminatory to ask people if they were Aborigines, or even to ask their race (Borrie 1975: 460–1).[3]

Notwithstanding Borrie's sense of the difficulties of further reform in Australia's Indigenous statistics, I think that we can see the years 1966 to 1976 as a turning point. The old way that prevailed from the late nineteenth century to 1966 had used the discredited terms of racial science. Under this old way it had been difficult, if not impossible, to compare the Indigenous population with the non-Indigenous population (or with 'all Australians', which is much the same

---

[3] As well, Borrie argued in 1975 that 'substantial changes in racial identification...are occurring'. He expected that 'the group of people identifying as Aborigines and Islanders will become relatively stable, so statistics relating to this group should become increasingly useful as time goes by' (Borrie 1975: 473). In fact, demographers have since noticed that identity change is a persistent dynamic, a source of growth in the Indigenous population. This does not seem to have diminished the utility of the census.

thing). The new era has two features: ethnic identities (Aboriginal and Torres Strait Islander) have replaced racial distinctions; and there are now more socioeconomic and biomedical variables for making distinctions *within* the Indigenous population. These two features of the reformed statistical archive combine to make it possible to compare, across many variables, certain policy-relevant trends in the Indigenous and non-Indigenous populations. Those comparisons have become the basis of an idea of social justice that is new, or newly prominent, in Australian public life: 'practical reconciliation'.

Statistics for protection and assimilation

I do not have the space to review, jurisdiction by jurisdiction, the structures of colonial, State and Territory statistics on Aborigines and Torres Strait Islanders up to 1966. However, in a sense I do not need to, because the various jurisdictions worked to a common model, that I will call here the protection/assimilation model. I derive this name from the commonly acknowledged periodisation of Australian Indigenous public policy. The era of 'protection' was instituted by a wave of colonial, State and Commonwealth law-making that lasted from the 1880s to the end of the First World War. The era of 'assimilation' lasted from the late 1930s until the early 1970s, as each State or Territory amended and then repealed these statutes and strove to bring Indigenous Australians within the same legislative and administrative frameworks as all other Australians, invoking the ideal of an 'Australian way of life' to which all were tending. I will argue that as the sequence of post-World War Two legislative reforms approached its end in the 1960s, many officials and policy intellectuals pointed to the inadequacies of the statistical archive in its protection/assimilation form.

What were the elements of the protection/assimilation statistical archive? By the third quarter of the nineteenth century, most Australian jurisdictions were keeping some record of what they thought to be the absolute size and the sex composition of the Aboriginal population. The various jurisdictions began to record the ages of Aborigines in different years (Victoria from 1871, New South Wales and Western Australia from 1891, Queensland from 1901, and South Australia, Tasmania and the Northern Territory from 1911).[4] From 1860 to 1905, as each colony began to form a specialised administration and statutory regime through which to govern their Indigenous populations, they recorded two other features of the populations: their genetic character (differentiating 'full blood' from 'others') and their relationship to administrative control. There were two variables within what I am calling administrative control, and each jurisdiction made use of at least one of them. The Indigenous population could be acknowledged as subject to enumeration or as living beyond enumeration, a distinction sometimes conveyed by the distinction between 'settled districts'

---

[4] I infer these starting dates from Smith's figure 8.4.1 (Smith 1980: 219).

and regions that were beyond settlement and enumeration. The other 'administrative control' variable has to do with some kind of institutional authority. Thus, some Indigenous people were classified according to whether or not they were 'in employ', or 'under the Act' or living within reserves and government institutions.

When the Commonwealth government began to standardise the Indigenous statistical archive in the 1911 Census, it adopted the Western Australian 1901 Census's version of this classifying practice. That is:

- Aborigines were enumerated if they were accessible to ordinary enumeration procedures
- all those not enumerated were assumed to be 'full-bloods' and their number was estimated
- the general census population included 'half-castes' (but not 'full-bloods')
- the Commonwealth published separate figures on 'full-bloods' and on 'half-castes' (Smith 1980: 27).

When the Commonwealth collaborated with the States in every June from 1924 to 1941 to make an annual count, they were interested in the size, the sex composition, the age composition, the race composition and the relationship to governing authority of their Indigenous populations. This protection/assimilation model of the Indigenous population lasted about eighty years, continuing until two-thirds of the twentieth century had elapsed.

The crisis of the protection/assimilation model

What happened in the 1960s to upset this way of constructing a statistical archive? This is going to be a research question for me and for Len Smith over the next few years, so my answer today is rather limited.

According to Borrie (1975: 455–6), Australia's statisticians had long defended their very limited practice of Indigenous enumeration by pointing to two constraints. One was practical (the difficulties of enumerating 'tribal people'), and the other legal (their interpretation of s.127 of the Constitution). By 1967 both of these obstacles had been overcome. However, Borrie also found among government officials the view that it was difficult to define who is an Aborigine and also 'a vaguely formulated, but nevertheless strongly held view that separate statistics are in some way discriminatory, even if collected in order to make special provision for Aborigines' (Borrie 1975: 456).

The word 'discrimination' is our clue to the crisis in the Indigenous statistical archive. Liberal opinion favouring 'assimilation' was obliged, by the mid-1960s, to consider whether there was a positive sense of 'discrimination'. That is, if Aboriginal people were not doing well and if the state had an obligation to help them do well, might it not be necessary for the state to discriminate in their favour in certain ways? And was it not necessary to 'discriminate' (in the sense of distinguish) Aboriginal from non-Aboriginal in order to know how badly or how well Aborigines were doing, so that 'positive discrimination' could be soundly based? Some historians now interpret the 1967 referendum as an expression of a widespread (though not universal) conviction that it was time for public policy to discriminate in Aborigines' favour, at least until they had 'caught up' with the rest of Australia in certain respects (Taffe 2005).

The idea that Australia was failing in its duty to improve Aborigines' conditions of life was given powerful intellectual expression in the Social Science Research Council of Australia's multi-author project 'Aborigines in Australian Society'. The Rowley Project (as I like to call it, after its Director, Professor C.D. Rowley) produced 14 books between 1970 and 1980. It was a common complaint of the Rowley Project authors that social scientists had insufficient data on Indigenous Australians. Finding the census inadequate, Leonard Broom and Frank Jones remarked of Aboriginal affairs administration that 'the management of a rubbish tip is more carefully monitored' (Broom & Jones 1973: 75). For example, they reported that there were no reliable statistics on Aboriginal mortality other than about 15 years of records for the Northern Territory (Broom & Jones 1973: 63). For W. E. H. Stanner, 'the very absence of more precise information is itself the best evidence of past indifference' (Stanner 1970: vi–ix). Peter Moodie, in his study of Aboriginal health, pointed to official 'caginess' about quantifying 'the Aboriginal problem' (Moodie 1973: 275). However, he feared also that Aborigines themselves might resist distinguishing 'any self-identifying Aborigines' in databases (Moodie 1973: 274). He urged Aborigines to consider that by allowing their 'statistical visibility', Maoris and American Indians had improved their mortality (Moodie 1973: 122).

The argument for Aborigines' 'statistical visibility', carried strongly by social scientists, was heard by the Whitlam government (and possibly it was influential under the Coalition governments in the formulation of the race question in the 1971 Census). In 1975, Borrie welcomed the terms of reference of the Whitlam government's National Population Inquiry 'that it should include the Aboriginal population not only in the total situation, but also as a separate sub-study'. He commented that, 'There is no doubt that a separate study is needed. In every conceivable comparison, the Aborigines and Islanders, whom it is proposed in general to treat in one group, stand in stark contrast to the general Australian society, and also to other "ethnic" groups' (Borrie 1975: 455).

However, social scientists in the Rowley project had had to work with what they could find in the unreformed statistical archive. To illustrate the variety of their responses to its limitations, I will briefly outline what Rowley, Broom and Jones did.

Rowley showed one way that the racial distinctions in the extant statistical archive could be meaningful. He argued that there were two kinds of situation facing Aborigines: 'colonial' Australia and 'settled' Australia. Though his description of the difference between the two situations drew attention to differences in their characteristic histories of colonisation and in their resulting social institutions (Rowley 1970c: 2), he suggested a numeric index for deciding whether a region was 'colonial' or 'settled': the relative proportions of the full-blood and the mixed descent components in the Aboriginal population in each Statistical Division, according to the 1961 Census. We retain much of Rowley's geographic binary when nowadays we compare 'remote' and 'non-remote' Indigenous statistics using the 2002 NATSISS.

In the *Destruction of Aboriginal Society*, Rowley (1970a) used 1961 Census and 1964 State/Territory estimates to show, State by State (but not Tasmania) the varying density (gross number per shire) of the Indigenous population. Differentiating full-blood from half-caste, he calculated their proportions in States and in capital cities, and he compared their distribution between urban and rural in each State/Territory. Distinguishing 'full-blood' from 'mixed-blood', he compared the age structures of each jurisdiction. He projected full-blood and mixed-blood population growth for 1961–81. The analytical value of distinguishing 'half-caste' from 'full-blood' is not clear, as Rowley preferred to account for Aborigines' behaviour by reference to their socialisation, not by reference to their genetic characteristics. However, the statistical archive that he had to work with was shaped by the genetic terms of an earlier era, so we get table after table distinguishing 'full-blood' from 'half-caste'.

Where he could do without the inherited statistical archive, Rowley did not distinguish half-castes and full-bloods. Researching *Outcasts in White Australia* (Rowley 1970b) he conducted two regional surveys (NSW outside Sydney, and Eyre Peninsula) with Aboriginal respondents. Without any reference to genetic characteristics, he presented data on Aborigines' housing quality, school attainments, books in dwellings, institutional background, water supply, bathing facilities, sanitation, garbage disposal, household composition, occupation, previous and current employment, skills and experience, post-primary education and training, and average weekly wage. He looked at their receipt of social security benefits, their ownership of property, their use of hire purchase, and their use of insurance policies. On some of these variables, he tabulated men and women separately.

In a few of his arguments, Rowley compared Aborigines with non-Aborigines. In *Destruction of Aboriginal Society*, he did so only to the extent of comparing the age structures of the full-blood, mixed-blood, and 'all Australians' groups. In *Outcasts in White Australia*, Rowley analysed his data on 'weekly income from all sources' and on 'membership of clubs and organisations' by comparing

Aborigines in rural NSW and Eyre Peninsula with non-Aborigines in the same regions. He obtained from the Commonwealth Bureau of Census and Statistics and from the Western Australian Government statistics on types of offence, leading to conviction or committal, for 1962–64. From NSW, Victoria and South Australia he obtained comparative data on types of offences for which Aborigines and non-Aborigines were arrested and charged in the second six months of 1965.

In *A Blanket a Year* (1973) Leonard Broom and Frank Jones had a rather different agenda, enabled partly because, writing later than Rowley, they had the fruits of the first reforms of the statistical archive. The 1966 Census, as well as being the first to claim complete enumeration, was the first to release data on Aborigines in the same terms as the data for other Australians. Broom and Jones were able to compare the educational and workforce status of Aborigines with non-Aborigines. However, we should note exactly what they meant by 'Aborigines', for in their efforts to compare Aborigines and non-Aborigines, Broom and Jones were still burdened by the racial classifications of the unreformed archive. Their data on 'Aborigines' included only those classified in the 1966 Census as having '50 per cent or more Aboriginal ancestry'. They wrote that 'comparable data for approximately 16 000 to 17 000 identifiable Aborigines of less than 50 per cent have not been released, and they therefore cannot be dealt with here' (Broom & Jones 1973: 13). Broom and Jones were concerned that by delimiting the Aboriginal population to those with 50 per cent or more Aboriginal descent, their comparison 'may exaggerate the…dissimilarity between Aborigines as a whole and the rest of the population' (Broom & Jones 1973: 24). In a section headed 'selection as a factor in Aboriginal health statistics', Moodie also worried about the 'bias' in the construction of Aboriginal health data, though the distortion that he pointed to was in terms of the institutional rather than the genetic ordering of the 'Aboriginal population' (Moodie 1973: 23). That is, data on Aboriginal health tended to be about 'the more closely supervised government settlements, mission settlements, and the larger cattle stations'. These people were both 'in a dependent situation' and enjoyed 'better access to medical and health services' (for which they did not have to pay) than the more independent, but less well serviced 'relatively large fringe-dwelling and metropolitan groups' (Moodie 1973: 23).

Although they issued such caveats about the bias in their 'Aboriginal' categories, Broom and Jones and Moodie can be said to have initiated the research program that is now familiar to us as the comparative study of Indigenous and non-Indigenous labour market status, human capital acquisition and health status. These are among the pioneer works of our contemporary paradigm.

Broom and Jones were pioneers in another way. They constructed a 'total' Aboriginal population when they sought to project the growth in Aboriginal numbers. In this endeavour, they benefited from the Commonwealth

Government's interpretation of its responsibilities after the repeal of s.127 in the May 1967 referendum. The referendum of 1967 led to the undifferentiated inclusion of Aborigines in birth, marriage and death registrations, making it possible for Broom and Jones to estimate plausibly Aboriginal fertility and infant mortality in the only jurisdiction where Aborigines were a large minority—the Northern Territory. By using these registration data and the 1961 and 1966 Census data, in two papers (Jones 1970, 1972) and in chapter four of his book with Broom, Jones gave an account of the distribution, fertility and mortality of *the entire Aboriginal population*—combining the 'dark' and the 'light' segments.

Some features in the new archive

Let me conclude by pointing to some features of the Indigenous statistical archive in its reformed, contemporary condition.

First, there has been a sustained agenda of reform. This agenda has been driven by a strong conviction that social justice demands Indigenous/non-Indigenous comparison across many socioeconomic and health variables. It has been notably effective in the reform of State and Territory registrations of vital events. Len Smith says that agitation on this point began around 1965 (Smith 1982: 16). According to the ABS (2000a: Table 11.6, p.162), these are the periods within which the various Australian governments put an Indigenous identifier into databases relevant to demography and to population health analysis:

- birth notification forms (1984–1996)
- death notification forms (1985–1996)
- medical certificates (cause of death) (1983–1999)
- medical certificates (cause of perinatal death) (1983–1999)
- hospital separations (1979–1997)
- maternal/perinatal collections (1982–1996)
- cancer registrations (1977–1992)
- communicable diseases notification forms (1988–)[5]

On at least one occasion, the reform of administrative data was stimulated by the demands of a focused inquiry with statutory powers. When the Royal Commission into Aboriginal Deaths in Custody (RCADC) began, its staff knew they could draw on the prison censuses that had begun to use an Indigenous identifier in 1982. But what data was there about police custody? The Royal Commission used its prestige and legal powers to persuade all of Australia's Commissioners of Police to allow the research staff to conduct a National Police Custody Survey in August 1988 (Commonwealth of Australia 1991: 191–2). Analysis of these data led to a conclusion that determined much of the Commission's subsequent agenda: that deaths in custody were disproportionately

---

[5] I have no information on the changes in communicable diseases notification forms since 2000.

Aboriginal not only because of factors within custody but also because Aborigines were much more likely than non-Aborigines to be in custody. What accounted for this higher rate of incarceration, the Commission asked? The Commission postulated the related concepts of 'self-esteem' and 'underlying issues' in order to adduce research on education, labour market status, health and other factors (using data from the 1971 to 1986 censuses) to explain Aborigines' disproportionate entry into police or prison custody.

Second, the reformed statistical archive makes some use of the organisational capacities known as 'the Indigenous sector', for example in Community Housing and Infrastructure Needs Surveys (CHINS) of 1992 and 1999 and in a current Mental Health Survey. The 1999 CHINS collected data from 707 Indigenous housing organisations and 1291 'discrete communities' (1089 of which had an identified housing organisation). 'Data … were collected through personal interviews with key members of Indigenous housing organisations and communities who were knowledgeable about housing and infrastructure issues. Such people included community council chairpersons, administrators, coordinators, clerks, housing officers, water and essential service officers. Information regarding health services was generally collected from health clinic administrators.' (ABS 2000b: 62).

Third, there is a continuing concern to improve the methodology of data collection in remote and very remote Australian communities (Martin et al. 2004).

Fourth, we risk becoming data rich and theory poor. Our data are useful only if we have some theoretical framework in which to make sense of them. Perhaps the most developed theoretical model that we have, at the moment, is that which I mentioned in my introduction. The Productivity Commission's 2003 Report operationalises the 'strategic areas for action' in terms of thirty variables on which Indigenous and non-Indigenous can be compared (see Steering Committee for the Review of Government Service Provision [SCRGSP] 2003). The theoretical model at the centre of this strategy highlights the familial and educational conditions of Indigenous childhood that are conducive to (or destructive of) an Indigenous person's later fortunes in the labour market. It would be a useful exercise to compare this theoretical model with predecessors such as Rowley (1966) who drew on North American studies of the pathologies of institutions, and the Royal Commission's, with its focus on the material determinants of low self-esteem. That is the task for another paper. One of the questions that such a paper might address is how 'Indigenous culture' is defined as a quantifiable variable in our (more or less conscious) theoretical models, and what significance, if any, is attributed to it.

Fifth, we are producing more and more data at a time when the government is terminating the existence of at least one of the bodies—the Aboriginal and Torres Strait Islander Commission (ATSIC)—that should be the primary users of the data. I endorse the point, made by a number of authors in this volume, that one of the best reasons for having a rich Indigenous statistical archive is to enable the Indigenous sector to talk back to government about policy and about program effectiveness.

Sixth, in the often angry, frustrated and demoralised public discussion of Indigenous affairs policy, we now have two kinds of language for talking about Indigenous disadvantage. Here I would like to invoke the work of the Canadian philosopher Charles Taylor. In his recent book, *Modern Social Imaginaries* (Taylor 2004a), he says that we moderns have developed the ability to think about social life in two distinct ways, both of which are valid and useful. On the one hand, we can think about society as a series of interactions between variables. There are several papers in this monograph in which that was the idiom for thinking about society. The Productivity Commission Report is, again, a wonderful example of that way of thinking. On the other hand, we can think about society as an interaction between responsible, intentional agents, such as individual people, and organised collectives of people or organisations, including governments. Taylor says that the modern social imaginary—that is, our taken-for-granted ways of thinking about 'society'—is 'bifocal'; we use both mechanistic and agent-centred thinking. We imagine society in mechanistic terms—interactions between variables—and we imagine society humanistically—interactions between thinking, feeling agents who can be held responsible for what they do.

In contemporary Australia, both ways of thinking about society have been recently intensified. On the one hand, due to the reforms of the statistical archive, we have data with which to think about the interactions among an increasing number of variables. On the other hand, we have the language of welfare reform, with its emphasis on 'mutual responsibility' and 'Shared Responsibility Agreements' (SRAs). This language also has a horror of any 'welfare' that is 'passive' and a strong implication that Indigenous Australians may fail in their responsibilities to take up the opportunities that the public and private sector provide. In contemporary Australian public culture, the mechanistic social imaginary that is fortified by our rich statistical archive is in daily juxtaposition with the voluntaristic social imaginary that asks: are governments living up to their responsibilities to Indigenous Australians and are Indigenous Australians taking responsibility for their own advancement?

2. Statistical needs in Indigenous affairs: the role of the 2002 NATSISS

Jon Altman and John Taylor

Over the last two years, Indigenous affairs policy at the national level has shifted direction dramatically: the central tenets of policy have shifted from terms such as self-determination, self-management and national Indigenous representation and advocacy to mainstreaming, mutual obligation, shared responsibility and a whole-of-government approach.

This broad change in direction has been predicated in large measure on a widespread perception that the socioeconomic situation of Indigenous people in Australia has, at worst, been a failure over the past 30 years or, at best, has not improved fast enough.

The new approach has been based on a growing emphasis on what has been termed 'practical reconciliation', or the pursuit of statistical equality between the standard of living of Indigenous and other Australians in the areas of health, housing, education and employment. In the foreword to the latest and influential *Overcoming Indigenous Disadvantage: Key Indicators Report 2005*, Productivity Commission Chairman Gary Banks refers to these issues as follows. Firstly, he notes the new determination by Australian governments to address the root causes of Indigenous disadvantage. Secondly, referring to the Prime Minister's speech at the National Reconciliation Conference, he identifies the shared goal [sic] that Indigenous people can ultimately enjoy the same standard of living as other Australians, for them to be as healthy, as long-living and as able to participate in the social and economic life of the country (SCRGSP 2005: iii).

We are almost certain that no-one would quibble with the need to address the causes of Indigenous disadvantage or the need for improvements in standards of living. A more lively debate might ensue about whether in the last 30 years any government has ever pursued objectives different to these; or whether the *new* approach is likely to be more successful than previous approaches. Arguably, it is difficult enough having conversations about what has worked and what has not in the past, let alone predicting what might be the best approach for the future.

Under such circumstances of policy contestation, one might anticipate that statistics collected by the official agency, the ABS, would play a crucial role in clarifying both the causes of Indigenous disadvantage and trends in Indigenous wellbeing, in absolute and relative terms. That is certainly one expectation that animated the conference 'Indigenous Socioeconomic Outcomes: Assessing Recent

Evidence'. Our focus is on the 2002 NATSISS, the second social survey conducted by the ABS between August 2002 and April 2003. It is based on information collected from 9400 Indigenous Australians aged 15 years and over across all States and Territories of Australia (ABS 2004c: v).

The aims of this chapter are three-fold:

1. To provide some historical background about the emergence of this survey instrument by linking this conference to earlier workshops convened by CAEPR in 1992 and 1996 about the initial 1994 NATSIS.
2. To broadly examine the role and value of the 2002 NATSISS as one element of a broad ABS strategy to enhance the availability of statistics on Indigenous Australians.
3. To explore the longer-term role that the NATSISS instrument, now to be conducted every six years, might play in tracking Indigenous policy performance, informing Indigenous policy development, and in meeting the statistical needs of all stakeholder groups, including Indigenous Australians.

These three aims should be differentiated from the overall aim of the conference, which is to examine the 2002 NATSISS in great detail from a variety of methodological (e.g. coverage), thematic, and conceptual perspectives as the latest evidence about Indigenous socioeconomic outcomes. Then we anticipate some discussions of issues for consideration in developing the 2008 NATSISS. For these purposes, a wide array has been assembled of social scientists and Indigenous policy practitioners who are specialists in disciplinary areas.

A brief history of the 1994 NATSIS and the 2002 NATSISS

Conference participants were provided with two earlier monographs published by CAEPR, *A National Survey of Indigenous Australians: Options and Implications* (Altman 1992) and *The 1994 NATSIS: Findings and Future Prospects* (Altman & Taylor 1996a). We summarise some salient issues from these monographs to provide a historical backdrop to this conference.

The first of these monographs reports the findings from a workshop, titled 'A National Survey of Aboriginal and Islander Populations', that was convened in April 1992. This was held, coincidentally, just a week after the Keating government committed $4.4 million to the ABS to conduct a special national survey as outlined in recommendation 49 in the *Final Report of the Royal Commission into Aboriginal Deaths in Custody* (Commonwealth of Australia 1991). The workshop provided an early opportunity for the ABS to engage with a diversity of academics and Indigenous stakeholder groups about the conduct and potential content of such a national survey.

The impetus for this recommendation in 1991 came from two directions. On one hand, the Royal Commission itself experienced an acute shortage of information about Indigenous Australians in its deliberations. In particular, its conclusions about the underlying factors that precipitated higher Indigenous incarceration were mainly drawn from the 1986 Census (Gray & Tesfaghiorghis 1991). There was a strong view articulated by the Royal Commission that additional, and more timely, information about distinct and diverse social, demographic, health and economic characteristics of the Indigenous population was needed.

On the other hand, there was an equally strong view at the time that information was urgently needed at the just-established ATSIC Regional Council level to facilitate the development of regional plans and to gauge community infrastructure needs. It is interesting to note other policy drivers for a national survey and more statistical information on the situation of Indigenous Australians in 1992, including:

- a need for regional statistics so that Indigenous-specific programs could be better targeted to Indigenous people in greatest need
- debates about whether Indigenous-specific or mainstream programs were more efficient and effective
- a recognition of the need for more concerted and coordinated inter-governmental effort in Indigenous affairs, including the need for legally-binding bilateral agreements between the Commonwealth and the States, and
- a need for more information to inform the reconciliation process and track change in 'Indigenous disadvantage' over time.

The 1992 workshop ended with some recommendations both for the forthcoming 1994 NATSIS and for the collection of Indigenous statistics more generally. In terms of items to be included in the survey, there was a consensus view that a national survey should focus on distinct Indigenous issues not covered in other collections. Even by 1992, the ATSIC Housing and Infrastructure Needs Survey (Australian Construction Services 1993) was addressing many of the topics specified for inclusion by the Royal Commission. And there was a strong view expressed that a one-off survey would be of limited value without commitment to future surveys conducted on a similar and regular basis.

More importantly, it was highlighted that a national survey could not be an omnibus for all statistical needs about Indigenous Australians: there was a need for enhanced use of Indigenous identifiers in regular ABS household surveys, for greater availability of Indigenous identified information in administrative databases, and more information at the State and Territory level. Finally, there was a strong view that, while committing $4.4 million for data collection was welcome, there was also a need for resources to ensure data analysis, publication and dissemination, especially at the regional level, to Indigenous stakeholders.[1]

The second monograph reports the findings from the workshop 'Statistical needs for effective Indigenous policy: findings from the 1994 NATSIS'. This workshop was convened in late August 1996, just on nine years ago, not long after the new Howard government's 1996–97 Budget where it signaled its new direction in Indigenous affairs.

The workshop was convened to assess 1994 NATSIS outputs from academic, bureaucratic, Indigenous and ABS staff perspectives with a special focus on issues to be considered in the development of a future national survey. While the 1996 workshop examined NATSIS from a number of disciplinary perspectives that closely correlated with survey questions, it was not an evaluation of the survey, which had already been conducted by the ABS (1996b) itself (with input from CAEPR & other stakeholders).

In considering the need for a repeat survey, the 1996 workshop, like this conference, sought open, constructive and rigorous scrutiny of the 1994 NATSIS. This is something that is not always easy to achieve, as new and innovative surveys evolve over time into statistical institutions with an identity and a sponsoring agency. Then, as now, the ABS, was an active and lively participant in proceedings.

As for the utility of the 1994 NATSIS, participants were acutely conscious of a fundamental dilemma: the more critical they were of particular questions, the greater the likelihood that questions would be modified or omitted, thus undermining the longer-term comparative utility of a future survey. The second workshop also examined what alternate sources of data were emerging that could be excluded from a repeat of the 1994 NATSIS, again highlighting that a national survey should not provide the means to 'statistically cost shift' from other household surveys where an Indigenous identifier might be justified (examples provided were the Labour Force Survey and the National Health Survey).

It is again interesting to note the political context of the 1996 workshop, when there was a perception that the new government might move to bifurcate its policy approach: targeting Indigenous-specific programs at remote regions and enhancing access to mainstream Commonwealth and State programs in non-remote contexts.

If this was a policy option then, it was only implemented by cutbacks to ATSIC Indigenous-specific programs as a contribution to the then overall 'deficit reduction plan'. It was highlighted in 1996 that the absence of statistics about

---

[1] Expenditure on the 2002 NATSISS was approx $4.5 million over four years: 2000–01 to 2003–04 (that is, from development through enumeration, processing and output). Most expenditure (that is, about $2 million in field costs) was in enumeration year 2002–03.

performance at the program level made agencies like ATSIC vulnerable to cutbacks and, conversely, that comparative statistics would be needed to hold governments accountable for their performance.

In this context, it was highlighted that the political economy of statistics—how statistics influence the distribution of public money—was important.[2] This, in turn, influenced what information was to be collected, how Indigenous participation in the national survey was to be encouraged, and how the empowering information from a survey was to be analysed, accessed (taking into account income disparities and the ABS's user pays approach) and disseminated.

The 1996 workshop raised some important issues. Two that stand out in current policy debates are the need for information at appropriate regional levels, and the links between political and statistical cycles.

In relation to the former, it was noted that ATSIC regional council boundaries were liable to change and, in any case, might have limited utility given that the discretion available to the ATSIC Board of Commissioners to make regional allocations had declined significantly. While ATSIC has disappeared, it is still important to consider the appropriate regional jurisdictions for data collection.

On the latter, it was noted in 1996 that there would be an urgent need to monitor outcome changes in Indigenous affairs as policy takes *new* directions. While this observation may not have been heeded at the time, it certainly resonates with the contemporary situation. The 1996 workshop made three concluding recommendations:

- that there should be a repeat of the 1994 NATSIS, preferably in 1999 after a five-year interval, but that this should be leaner, meaner and sharper
- that careful consideration be given to maintaining some questions in the national survey that are comparable over time, while also allowing for new questions that allow comparability with other data sets, and
- that consideration be given to enhancing existing ABS household surveys to allow the identification of Indigenous participants in a statistically valid way that reflects their different geographic distribution. This is done, for example, with the Labour Force Survey once a year, but little is heard about the statistical outcomes.

Each of these issues could well be revisited in the context of this conference.

---

[2] Having made this observation, one might question whether better statistically-based arguments would have altered the Australian Government's 'surplus distribution plan' in the run-up to the 2004 election.

The 2002 NATSISS in the overall statistical framework

In the not-so-distant past, official processes served to exclude, devalue, and deter full Indigenous statistical representation. By contrast, the contemporary politics of data collection have sought to encourage inclusion via self-identification. This is manifest most recently in the greater involvement of Indigenous personnel in the collection of census and survey data, as well as in ministerial-level agreements for the adoption of a standard self-reported Indigenous status question in administrative collections.

In addition, the ABS has embarked on an ambitious Indigenous household survey and census program (see Table 2.1) that can trace its origin to the need for a government response to the findings of the 1991 Royal Commission into Aboriginal Deaths in Custody (Sims 1992).

As with other Indigenous-specific population surveys, such as the Western Australian Aboriginal Child Health Survey (Zubrick et al. 2004), this response is recognition of the need for non-standard approaches to developing census and survey content and methodology with an emphasis on Indigenous participation and, in some cases, control (Zubrick et al. 2004).

Complementing these enhanced census and survey initiatives, we now also have regular reporting of Indigenous outcomes from administrative data. In 2002, the Prime Minister wrote to Gary Banks as Chairman of the SCRGSP to produce a regular report to the COAG against key indicators of Indigenous disadvantage. This is now available on a biannual basis as the Productivity Commission Report *Overcoming Indigenous Disadvantage*. In addition, the annual Report on Government Services issued by the SCRGSP now includes a separate compendium of Indigenous statistics drawn from the administrative databases of Australian, State and Territory governments.

Table 2.1. ABS Indigenous household surveys program: 1999–2011

Year	Collection	Indigenous sample	Level of geography supported
1999	Housing Survey (a)	850–900 households	National
1999	CHINS	All discrete Indigenous communities (approx. 1300)	Community level
2001	NHS (& Indigenous supplement)	Indigenous sample of 3400 persons	National
2001	CHINS	All discrete Indigenous communities (approx 1300)	Community level
2001	Population Census Indigenous Enumeration Strategy	All persons	Small geographic regions
2002	NATSISS	9 400 persons	States/Territories
2004–05	Indigenous Health Survey	11 000 persons	States/Territories
2006	CHINS	All discrete Indigenous communities (approx. 1300)	Community level
2006	Population census	All persons	Small geographic regions
2008	NATSISS	11 000 persons	States/Territories
2010–11	Indigenous Health Survey	11 000 persons	States/Territories
2011	Population Census Indigenous Enumeration Strategy	All persons	Small geographic regions

Source: Adapted from www.abs.gov.au http:/true/truewww.abs.gov.au/true , *Themes—Indigenous: Directions in Aboriginal and Torres Strait Islander Statistics*, March 2000

While these developments clearly demonstrate rapidly enhanced collection and publication of statistical information, much basic information remains unavailable – not so much because data do not exist, but rather for want of appropriate planning frameworks. Viewed historically, there appears to be a growing gap between the scales at which Indigenous polities organise and plan, on the one hand, and the scales for which statistics are available, on the other. Thus, while we have never been so data-rich, leaving aside questions about the cultural relevance of the data, the information that is generated at great expense tends increasingly to be only useful for national, State and Territory level analysis, and/or broad remote/non-remote distinctions. There are a number of methodological, practical, and ultimately political dimensions to this observation.

First, from a methodological perspective, sample instruments such as the NATSISS and its companion General Social Survey (GSS) are best deployed to inform high-level policy discussion about the broad nature of inter-relationships between social circumstances and outcomes. Does crime impede employment prospects? Is health related to income? Do educated women have fewer children? For reasons of sample size and non-response error, they are less well suited to establishing absolute levels of need and comparisons over time.

Related to this is the fact that major constraints arise in relation to the geographic disaggregation of data to regional or community levels. Although the sample size of the adult population in the 2002 NATSISS was equivalent to that of the

1994 NATSIS, data from the latter were released at the scale of 36 ATSIC regions, whereas the basic geography for 2002 comprises the remote and non-remote areas of the States and the Northern Territory (NT). The basic rationale for this shift is to be found in the trade-off between spatial detail and robustness of results.

Secondly, from a practical perspective, in terms of gathering administrative data, a combination of under-reporting, confidentiality provisions, and the grinding and increasingly guarded nature of bureaucratic processes render even the acquisition of data at State and Territory levels a major achievement (see Taylor & Stanley 2005).

Finally, from a political perspective, the current paradigm for the collection and dissemination of Indigenous statistics is suited to the measurement and reporting of gaps. For the most part, this assists processes of government by bureaucrats in Canberra and the State capitals. This paradigm strongly reflects a deficit model of Indigenous socioeconomic need as measured by standard social indicators, and not a community development model. This approach suffers from all the pitfalls of averaging diverse circumstances leading to questions about the utility of data for Indigenous regional and community organisations and their members.

Worst of all, it lacks local Indigenous context. Demographers, for one, were warned some years ago of an imbalance between their statistical precision and detail, on the one hand, and their casualness of treatment of non-demographic contextual variables on the other (McNicoll 1988: 20 cited in Riedmann 1993: 107). Yet perennial problems about the ability of such an approach to capture cultural difference and the diversity of Indigenous circumstances looms yet again (Morphy 2002; Peterson 1996). As social scientists and policy practitioners, we need perhaps to revisit the original goal of the NATSISS instrument to capture information that is distinctly Indigenous and that can be made available at the regional level.

NATSISS and Indigenous policy and practice

Notwithstanding the above comments, there is no doubt that the 2002 NATSISS is already playing a role in broadly assessing Indigenous policy and practice.

The key task as defined by the SCRGSP is to identify indicators that are of relevance to all governments and Indigenous stakeholders and that can demonstrate the impact of program and policy interventions. In its overview, the SCRGSP (2005: xix) notes that its first report (2003) relied heavily on 2001 Census data, while the 2005 report has relied largely on 2002 NATSISS results.

The SCRGSP is quite cautious in its use of 2002 NATSISS information, suggesting that:

- it is not comparable with earlier 2001 Census outcomes
- it comes from a period before the adoption of the reporting framework by governments, and
- it would not be reflecting outcomes from more recent interventions.

Nevertheless, the SCRGSP uses 1994 NATSIS and 2002 NATSISS information to make some broad observations about what has improved and what has not, although also basing much of its analysis on administrative data.

While this reporting of available information is instructive at a discursive level, it is very unclear how it demonstrates the impacts of program and policy interventions. Indeed, rigorous evaluations of specific policies targeting Indigenous Australians do not appear in the SCRGSP report and are largely missing in Indigenous policy discussions. It is generally recognised that interventions cannot be evaluated by comparing broad undifferentiated social indicator outcomes over time: programs and policies have to be evaluated at the micro target group community or regional level and there is little independent analysis of this kind in Australia today.

At one level, this statistical development is ironic given the government's recent focus on targeted regional COAG trial sites, SRAs and planned Regional Partnership Agreements, none of which have clearly defined evaluation mechanisms and associated data requirements. A notable exception here is the research undertaken at the Thamurrurr COAG trial site by Taylor (2004b) and Taylor and Stanley (2005) to gather baseline information that might inform evaluation. However, the intensity of research effort required to glean information at this level of disaggregation is instructive of the overall problem to which we are alluding—this research is expensive and not readily replicable.

At another level, this is symptomatic of the demise of a legally sanctioned (Indigenous) authorising environment for statistical collection and dissemination at regional levels, a role fulfilled until recently by ATSIC. The availability of ABS published outputs for ATSIC regions, co-signed by the Chair of the ATSIC Board and the Australian Statistician, was path-breaking (ABS 1995, 1996a). Unfortunately, this exercise in geographic disaggregation has not been repeated for the 2002 NATSISS.

In light of the ABS's own formal evaluation of the 1994 NATSIS, this last observation is significant. Among the key issues highlighted were the appropriateness of output mediums and the accessibility of results to Australia's Indigenous people and their organisations (Sarossy 1996). Findings on these matters raised the importance of regional level reporting and recommended that the dissemination strategy for any future survey should ensure that the results are readily available to the Aboriginal and Torres Strait Islander populations. To ensure this, it recommended that the ABS consider delivering basic statistical training in the interpretation of results to Aboriginal and Torres Strait Islander communities (Sarossy 1996: 192). This is now implicit in the ABS Indigenous Community Engagement Strategy implemented in 2004 (see Webster, Rogers & Black, this volume).

To the extent that such recommendations are deemed to be still relevant, and with the demise of the ATSIC regional structure, the question has to be raised about what geographic scale of reporting is now appropriate. And with whom will the ABS engage in order to enhance the use of NATSISS information by Indigenous communities and organisations? Accessibility of survey data has been enhanced for academic and other researchers via the Australian Vice-Chancellors Committee (AVCC) agreement on data access and by the introduction of the Remote Access Data Laboratory (RADL). However, this enhancement is unlikely to benefit most Indigenous stakeholders, given the technical capabilities required and the fact that access to the CURF costs $8000 for non-AVCC authorised users.

Conclusion

Without pre-empting the directions that discussions at this conference might take, we conclude with just a few complex issues of statistics and Indigenous public policy for consideration over the next two days.

Historically, the NATSISS instrument was established at a time when there was a recognised dearth of official statistics to document the socioeconomic situation of Indigenous Australians, though this is not to deny the importance of the five-yearly census—especially for comparison with the general population. In recent years, the ABS has greatly enhanced its Indigenous household survey and census program to the extent that in the 13 years 1999 to 2011, only three—2000, 2007 and 2009—are survey free.

Despite this, there appears to be a growing and worrying mismatch between the broad direction that Indigenous affairs policy is taking on the one hand (focusing effort on partnerships with specific communities and regions), and the availability of non-census derived information. In particular, 2002 NATSISS data are not amenable to disaggregation to the regional level, unlike 1994 NATSIS data. This suggests that recent data might be of less value to Indigenous community and regional end-users and that comparative analysis between 1994 and 2002 will be impossible at the regional level.

This raises important issues of statistical accountability and the politics of statistics. From 1990–2005, there have been checks and balances on the increased ABS activity in this area, provided by its alliance with ATSIC. It is noteworthy that under s.7 of the *Aboriginal and Torres Strait Islander Commission Act 1989*, the ABS and ATSIC had a statutory relationship and that ATSIC was legally required to develop policy proposals to meet Indigenous needs and priorities from the national to the regional level. ATSIC was also required to assist, advise and cooperate with Indigenous communities, organisations and individuals, again to the regional level. Irrespective of ATSIC's capacity to effectively address such complex statutory functions, as a national representative organisation it did provide a degree of institutional authorisation to the Indigenous data collection and analysis activities of the ABS.

Today, we have a new partnership approach in Indigenous affairs, but we lack the statistical framework to assess its effectiveness. This situation might suit governments, but it also has the potential to entrench power and economic inequalities, especially for those Indigenous communities that might want to sign off on agreements in good faith and with the expectation of information to assess their effectiveness. We have already alluded to the emerging problems inherent in the absence of NATSISS data at the regional level. It is also clear that there is an emerging hierarchy of privileged access to 2002 NATSISS information, with the ABS being uniquely positioned in accessing Main Unit Record File (MURF) data at levels of disaggregation that are just not available publicly. [3] Others, like the Productivity Commission, can purchase data, and academics can use the CURF without charge. Indigenous organisations and individuals appear least well placed to access such information. Such issues raise a number of important questions, with which we end:

- Is a survey instrument, like the NATSISS, sufficiently at arms-length from governments of the day?
- Bearing in mind the urgent need to track Indigenous wellbeing over time, what precise role will the 2008 NATSISS play in this process?
- Will 2008 NATSISS outputs, available in 2010, be helpful in adaptively managing the new approach in Indigenous affairs?
- Given that the 2002 NATSISS sample size was too small for regional disaggregation of findings, should the ABS look to augment the sample size in future?
- In the absence of a national Indigenous representative organisation, what institution will be empowered to ensure that official statistical information is collected that will help Indigenous communities and regions plan their futures?

---

[3] See, for example, the analysis in Australian Social Trends 2005 at remote, very remote and discrete Indigenous communities (ABS 2005a: 52–57) that would be impossible for public users of 2002 NATSISS data to undertake.

3. The 2002 NATSISS — the ABS survey methodology and concepts

Andrew Webster, Alistair Rogers and Dan Black

In 2002, the ABS conducted the second national social survey of Australia's Aboriginal and Torres Strait Islander population, the NATSISS. The first national survey was conducted by the ABS in 1994 as part of the Australian Government's response to the 1987–1991 Royal Commission into Aboriginal Deaths in Custody. The Royal Commission had brought to light the urgent need for more and better data about the social circumstances of Aboriginal and Torres Strait Islander people and had recommended a special national survey to cover a range of social, demographic, health and economic topics (recommendation 49).

The 2002 NATSISS was conducted and funded by the ABS as part of its ongoing household surveys program, whereas the 1994 survey was funded by a $4 million Commonwealth grant as a direct response to the Royal Commission (ABS 2004c). In 2000, the ABS established the National Aboriginal and Torres Strait Islander Social Survey as an ongoing six-yearly survey that comprised one element in a much broader Indigenous statistics strategy. Other elements in this strategy include:

- a six-yearly national survey of the health of Indigenous people
- continued improvements to the enumeration of Indigenous people in the population census
- further work to improve the identification of Indigenous people in administrative collections, and
- further improvements to population estimates and projections.

At the time of writing, the 2004–05 National Aboriginal and Torres Strait Islander Health Survey, which had a sample size similar to that of the 2002 NATSISS, has already been enumerated and first results are due to be released early in 2006.

ABS statistical activity

Cultural considerations, the geographical location of Indigenous people and policy requirements influence the statistical activity of the ABS with regard to Australia's Indigenous population. Some of the developments in statistical activity over the past 10 years include:

- improvements in the quality, analysis and availability of information about Indigenous people from the five-yearly population census

- implementation of a regular cycle of household surveys of the Indigenous population, alternating every three years between a survey of social circumstances and a health survey
- the Community Housing and Infrastructure Needs Survey, funded by the Australian Government Department of Family and Community Services, which is now conducted in conjunction with preparations for the five-yearly population census
- progressive implementation of consistent standards of Indigenous identification in administrative data sets
- production (with the Australian Institute of Health and Welfare) of the biennial report *The Health and Welfare of Australia's Aboriginal and Torres Strait Islander Peoples* (ABS cat. no. 4704.0), and
- improved quality and availability of Indigenous population estimates and projections.

As part of its survey activity, ABS undertakes extensive consultation with government agencies, Indigenous people, Indigenous organisations and researchers. This takes the form of bilateral agency consultations, survey reference groups, direct community consultations and topic-focused working groups. In addition, the ABS is currently implementing an initiative to increase its engagement with Aboriginal and Torres Strait Islander communities.

Under its Indigenous Community Engagement Strategy, ABS has recruited State/Territory-based Indigenous engagement managers to provide an ongoing communication channel with discrete communities, community groupsand organisations. Indigenous engagement managers will take a leadership role with Indigenousenumeration in the census and relevant ABS surveys. They are also undertaking dissemination and training activities, such as returning information to communities and Indigenous organisations in their preferred format and in ways that support the use of statistics for their own purposes.

The 2002 NATSISS

The 2002 NATSISS was designed to provide information on the social circumstances of Indigenous people relevant to Indigenous stakeholders and policy researchers. Like the General Social Survey of the total population, the NATSISS is a multi-dimensional survey covering a wide range of areas of social concern, with the capacity to enable analysis of interactions among different topics, including the analysis of multiple disadvantage. The 2002 NATSISS was designed to measure selected changes over time (since 1994) and to allow for comparisons with the circumstances of the non-Indigenous population. To enable this, the survey questionnaire had about a 50 per cent content overlap with the 1994 NATSIS and considerable overlap with the 2002 GSS (ABS 1995, 2003b).

It also included new material, for example, on disability, incarceration and age at first formal charge.

Sample design

The 2002 NATSISS was designed to provide reliable estimates at the national level and for each of the eight Australian States and Territories. In addition, the Torres Strait Islander population was over-sampled in order to produce data for the Torres Strait area and the remainder of Queensland. The sample was spread across the States and Territories in order to produce estimates that would have a relative standard error (RSE) of no greater than 20 per cent for characteristics that are relatively common in the Indigenous population; for example, characteristics that at least 10 per cent of the population would possess.

The 2002 NATSISS incorporated two broad samples that together comprised about 9400 people in 5900 households and covered all areas of Australia. The first was a random sample of about 2100 people in discrete Indigenous communities, predominantly in remote areas. Discrete communities in Queensland, South Australia, Western Australia and the Northern Territory were included in this sample. The second was a random sample of about 7300 people in so-called 'non-community' areas drawn from major cities, regional and remote areas in all States and Territories. The two samples were designed separately, with each involving a multi-stage sampling process. Overall, the 2002 NATSISS sampled one in 30 of the total Indigenous population (see Table 3.1).

Table 3.1. Sample size, 2002 NATSISS

	Non-remote[a]	Remote[b]	Total	Estimate (15 years or over)	Sample fraction
New South Wales	1137	365	1502	83 800	1 in 56
Victoria	806	–	806	17 400	1 in 22
Queensland	1018	847	1865	76 000	1 in 41
South Australia	605	405	1010	15 800	1 in 16
Western Australia	465	1097	1562	39 600	1 in 25
Tasmania	713	23	736	10 900	1 in 15
Northern Territory	168	1380	1548	36 200	1 in 23
Australian Capital Territory	330	–	330	2600	1 in 8
Australia	5242	4117	9359	282 200	1 in 30

a. Comprises major cities, inner regional and outer regional areas.
b. Comprises remote and very remote areas.
Source: The 2002 NATSISS MURF

In 2002, survey content and methodology were specifically designed to take account of the different circumstances of Indigenous people in remote communities and non-remote areas. In the community sample, the standard household survey approaches were modified as a result of pre-testing, to accommodate language and other issues associated with the geographic remoteness of these communities. ABS interviewers were accompanied by local Indigenous facilitators who assisted in the conduct and completion of the interviews. In the community sample, interviewers used a pen and paper interview (PAPI) questionnaire, while in the non-community sample interviews were conducted predominantly using Computer Assisted Interviewing (CAI).

While PAPI was only used in the four jurisdictions (Queensland, South Australia, Western Australia and the Northern Territory) that had households in the community sample, about half of the national sample of 4100 people surveyed in remote/very remote areas were interviewed using CAI and half using the PAPI form. Of the 1600 people surveyed in the remote component, just over two-thirds were interviewed using CAI, whereas of the 2500 people in the very remote component, two-thirds were interviewed using PAPI (see Fig. 3.1). Use of PAPI varied from 30 per cent of the total remote samples in South Australia and Western Australia to nearly 50 per cent in Queensland and almost 100 per cent in the Northern Territory.

While wording was modified in the PAPI questionnaire, most underlying concepts remained consistent with those in the CAI questionnaire. Data items or output categories that were specific to either PAPI or CAI, such as some output associated with disability status, are identified in the main ABS publication and in the documentation for the CURF.

Figure 3.1. Form type by area, 2002 NATSISS

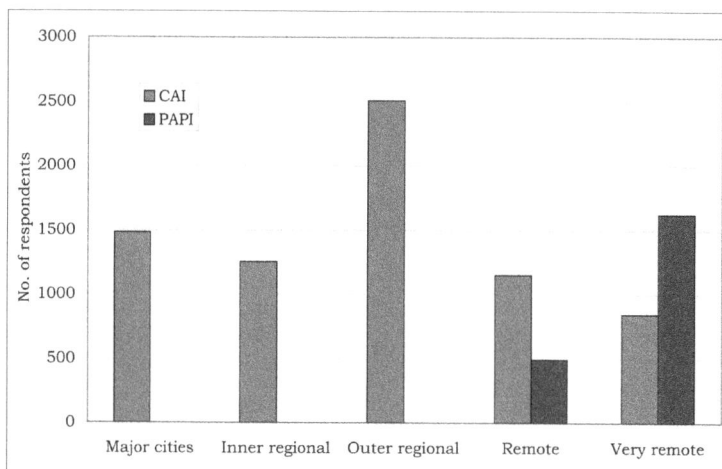

Source: The 2002 NATSISS MURF.

Comparison between 1994 and 2002 survey methodologies

Scope

The 2002 NATSISS collected information from a maximum of three people aged 15 years or over in each selected private dwelling. The 1994 survey collected information from people in non-private dwellings, such as hospitals and prisons, as well as from people in private dwellings. In addition, the 1994 survey collected information about each person, including children, in each selected private dwelling.

The selection, for the 2002 NATSISS, of a sub-set of residents of a dwelling was designed to manage the provider load on any one household. The detailed questionnaire for this survey would otherwise place a burden on large households. Some information was collected from a household spokesperson on the composition of the entire household, including information on the number of children, so that households with children can be identified in output.

The restriction of the number of people sampled from any one household enabled the total 2002 survey sample to be spread across a large number of households. The 1994 survey sampled 17 800 people in total, 9400 of whom were aged 15 years or over and resident in 4000 private dwelling households. The 2002 NATSISS sampled about the same number of people aged 15 years or over in 5900 households. As a result, the 2002 NATSISS was less clustered than the 1994 survey and there were consequent reductions in sampling error.

While people in non-private dwellings may have characteristics that are different from the rest of the population overall, their exclusion from the 2002 NATSISS was judged to have only a minimal effect on the representativeness of results. By restricting the sample to private dwellings only, a larger number of people could be sampled within the survey budget. Also, the size of the population in non-private dwellings is relatively small. Based on census results, about 4 per cent of the Indigenous population may be resident in non-private dwellings at any point in time.

In addition, there is now a much broader range of administrative statistics available on people in prisons and other corrective service institutions than was available in 1994. In conjunction with the corrective service sector, the ABS publishes information on Indigenous people in prison from the annual prisoner census and quarterly movements in the population in corrective service institutions (ABS 2004e). Furthermore, in recognition that the prisoner population is not static, the 2002 NATSISS collected retrospective information from those respondents who had been incarcerated in the previous five years.

In all data comparisons between the 1994 and 2002 surveys that have been published by the ABS, the population of the 1994 survey has been restricted to align exactly with the 2002 population, namely people aged 15 years and over living in private dwellings.

Geography

The 2002 NATSISS was designed to produce reliable estimates for all eight Australian States and Territories, resulting in higher quality estimates at the State and Territory level (particularly for Victoria and the Australian Capital Territory) than was the case for the 1994 survey (e.g. see the employment data in Fig. 3.2). In addition, robust estimates by ABS remoteness categories are available at the national level. While the 2002 survey was not designed to produce other regional data, reasonably robust estimates for some data items for some ATSIC regions can be produced. In contrast, the 1994 sample was spread across the then 35 ATSIC regions and the Torres Strait area to provide some estimates for each region. Under the 1994 strategy, estimates at the regional level were nevertheless associated with relatively high sampling errors.

Figure 3.2. Mainstream employment by State, 1994 and 2002

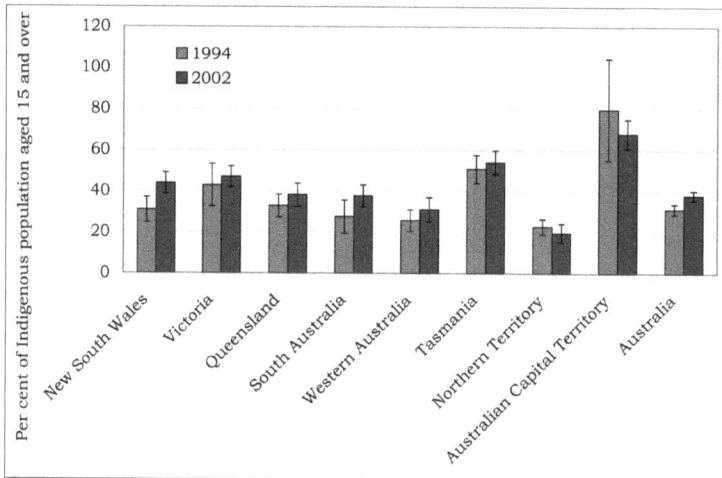

Source: The 1994 NATSIS MURF and 2002 NATSISS MURF.

Content

In designing the 2002 NATSISS questionnaire, ABS was responding to user expectations that the survey achieve both of the following:

(1) Measure the social circumstances specific to Aboriginal and Torres Strait Islander people throughout Australia.

(2) Provide valid measures of comparison with the circumstances of the non-Indigenous population. While there was generally a high degree of overlap between the topics covered in the 2002 NATSISS and those in the 1994 survey, there were some differences in question wording to align the 2002 survey more

closely with the 2002 GSS and with other ABS surveys. As a result, some changes were made to the education and employment modules that were used in 1994.

Statistics for policy

Initial results from the 2002 NATSISS were released in the *National Aboriginal and Torres Strait Islander Social Survey 2002* (ABS 2004c) and a series of web-based State and Territory spreadsheets (www.abs.gov.au). These releases covered topics as diverse as family and culture, health, disability, education, work, income and housing, law and justice, information technology and transport.

In response to the expectations of policymakers and researchers, the ABS has undertaken a range of initiatives to provide users not only with relevant statistics, but also with associated information about the quality of those statistics. In particular, ABS has calculated precise estimates of sampling error for 2002 NATSISS data using a grouped jack-knife methodology based on partitioning the primary sample into 250 cross-sectional groups (Wolter 1985). Estimates of relative standard errors have been released for all data shown in the main summary of findings report and in the web-based output at State and Territory level. The 250 replicate weights used in the jack-knife methodology are available on the CURF. Further, the results of tests of statistical significance are shown in selected tables in the main report and web-based spreadsheets, to inform users of the statistical significance of differences between estimates, for example in the measures of change over time or between the Indigenous and non-Indigenous populations.

Since the Indigenous population has a younger age structure than the non-Indigenous population, use of age-standardisation is sometimes appropriate when comparing estimates from the two populations. In the national report, age-standardisation was used for health, employment, law and justice, and information technology items, as these data items were judged to be associated with age in ways that warranted this statistical technique.

Responding to user demand for unit record data in order to facilitate more flexible and detailed analysis of survey files, the ABS developed the RADL in 2004. This facility allows users to access CURFs over the web. By maintaining the data file within the ABS computer environment and monitoring its use by researchers, the ABS is able to provide a more expanded data file than would be possible if the CURF was released on CD-ROM. While universities have access to CURFs via the RADL under an arrangement between the ABS and the Australian Vice-Chancellors' Committee, other agencies and researchers can purchase a licence to access CURFS via the RADL (i.e. over the web).

The 2002 NATSISS CURF contains a higher degree of disaggregation than the 1994 survey CURF. In particular, the 2002 NATSISS CURF includes geographic items by State/Territory as well as by remoteness area. For more information, see the National Aboriginal and Torres Strait Islander Social Survey CURF Technical Paper (ABS 2005b).

The strength of the 2002 NATSISS data set is that the broad range of socioeconomic variables enables the exploration of associations among different outcomes, including analysis of multiple disadvantage. To this end, the ABS is engaging in a number of collaborative research projects to examine different topics, including self-assessed health, victimisation and cultural issues.

4. Selected methodological issues for analysis of the 2002 NATSISS

Nicholas Biddle and Boyd Hunter

The most commonly used data for information on Indigenous Australians is the Census of Population and Housing. The five-yearly census allows us to generate reasonably reliable social statistics about Indigenous people as a by-product of the introduction, in 1971, of a question which asked whether people identify as Indigenous. However, the census is a blunt instrument that is designed primarily to count the national population rather than to measure and track changes in complex socioeconomic conditions of population sub-groups. Furthermore, census questions are limited in their number and scope by the exigencies and costs involved in collecting information from the entire population. Surveys are more flexible and cost-effective instruments for collecting a wide range of information, even though the resulting data is subject to sampling error.

In 1991 the Royal Commission into Aboriginal Deaths in Custody recommended a national survey of the Indigenous population. It was the dearth of information with which to inform the Royal Commission that resulted in the first NATSIS in 1994. This survey provided the first nationwide inter-censal estimates of Indigenous socioeconomic status.

The 2002 NATSISS is the second major nationwide survey specifically targeted to collect a large range of information on Indigenous Australians. Carried out between August 2002 and April 2003, it collected information from 9359 individuals aged 15 years and over from 5887 households. Some of the information had never been collected before for the Indigenous population, whereas a number of the questions were broadly comparable to the 1994 NATSIS.

The 2002 survey was also conducted more or less concurrently with the 2002 General Social Survey (GSS) which collected information about the total adult Australian population (the Indigenous and non-Indigenous populations are not separately identifiable in the GSS). Many of the data items in the 2002 NATSISS are comparable with the GSS, but the GSS did not collect information in very remote areas and was limited to individuals 18 years and over.

The ABS has a program (or cycle) of Indigenous household surveys, with the next NATSISS survey scheduled for 2008. Other ABS data collections with significant Indigenous components planned before then are the 2004–05 Indigenous Health Survey, the 2006 Community Housing and Infrastructure

Needs Survey (CHINS) and, of course, the 2006 Census of Population and Housing.

The NATSISS survey was designed to 'enable analysis of the interrelationship of social circumstances and outcomes, including the exploration of multiple disadvantage' (ABS 2005c: 1). Information is provided across a range of topics:

- demographic characteristics of the individuals and household and geographic characteristics of the area in which they live
- cultural and language information and the family and community context
- health and disability
- education participation and achievement
- employment
- income, housing and financial stress, and
- information technology, transport and law and justice.

This chapter seeks to outline a number of the methodological issues concerning the 2002 NATSISS, with a particular focus on helping readers of this monograph understand the remainder of the empirical results and analysis presented. The next section will outline the survey methodology, including the scope, sample selection and survey design and implementation. Section 3 will provide an overview of some of the potential issues one might need to take into account in an empirical analysis of the NATSISS, while section 4 will provide a brief critical analysis of the existing ABS outputs from the 2002 NATSISS. Finally, section 5 will summarise and highlight the main implications of this chapter.

Survey methodology

Scope of the survey

The 2002 NATSISS collected information from Indigenous Australians aged 15 years and older who were usual residents in private dwellings at the time of the survey. In line with standard ABS household survey scope, it excludes visitors to the randomly selected private dwellings.[1] The survey was carried out across Australia with the aim of collecting enough information to make conclusions at either the State/Territory level or by the Australian Standard Geographic Classification (ASGC) remoteness classification.[2] The coverage of the 2002 NATSISS data is different to the GSS (which only collected information on people aged 18 and over in private dwellings) as well as the 1994 NATSIS (which collected information from people aged 13 years and over in both private and non-private dwellings).

---

[1] Some information may have been collected from these visitors if their usual residence was also selected as part of the collection.
[2] All geographic definitions of remoteness in this paper are based on the ASGC remoteness structure.

The difference in age structures is reasonably easy to take into account by re-weighting, though it should always be kept in mind during comparative analysis. However, given that the 2002 NATSISS did not collect information on people in non-private dwellings, differential coverage may be more problematic when making comparisons with the 1994 NATSIS or the 2001 Census, or when drawing conclusions about the total Indigenous population. Non-private dwellings include hotels, motels, hostels, hospitals, short-stay caravan parks and—perhaps most importantly—prisons and other correctional facilities. Such dwellings can be identified in both the 1994 NATSIS and the 2001 Census.

According to the ABS (2005c), at 31 December 2002 there were an estimated 19 320 Indigenous people living in non-private dwellings, or about 4 per cent of the entire Indigenous population. The following discussion gives some information from the 1994 NATSIS on people who were usual residents of non-private dwellings. The sample size is 375 adults.

By definition, all prisoners in the 1994 NATSIS data were in non-private dwellings. Biddle and Hunter (2004) provided a profile of Indigenous people in non-private dwellings using the original weights for the 1994 NATSIS. Residents of non-private dwellings were more likely than Indigenous residents in private dwellings to have been arrested in the last five years. They were also concentrated outside capital cities, and were more likely to be male and young. Also, a higher percentage of respondents from non-private dwellings were taken from their natural families. Without access to accurate sampling errors for the 1994 NATSIS data, it is difficult to definitively claim that the differences between private and non-private dwellings are significant. However, it seems reasonable to assert that the people living in private and non-private dwellings are drawn from different populations.

Therefore, when statements are made regarding data from the 2002 NATSISS, it may not always be appropriate to say 'the Indigenous population has a given characteristic' but rather make a more qualified statement about 'the Indigenous population living in private dwellings' or, if referring to law and justice data, 'that portion of the Indigenous population that is not currently in prison or other non-private dwelling'.

Given the substantial growth in the Indigenous population since 1994, it is important that comparisons between the 1994 and 2002 surveys use the re-weighted 1994 data set (when the ABS makes it available as a CURF). Customised cross-tabulations provided by the ABS will use the re-weighted data as a matter of course.

Sample selection and survey design

The overall sample was spread across States/Territories in order to produce estimates that have a relative standard error of no more than 20 per cent for characteristics that are relatively common in the Indigenous population (for example, that at least 10% of the population would possess). However, there were two components to the 2002 NATSISS sample designs. The first (in parts of Queensland, South Australia, Western Australia and the Northern Territory) was based on a sample of discrete Indigenous communities and the outstations associated with them. This is the Community Area (CA) sample. In the remainder of these four States and Territories, as well as in all of New South Wales, Victoria, Tasmania and the Australian Capital Territory, the survey methodology and sample design was somewhat different. The data from these other areas are described as the Non-Community Area (NCA) sample. Around 30 per cent of the sample came from the CAs and 70 per cent from the NCAs. Those in NCAs were interviewed using Computer Assisted Interviewing, whereas those in CAs were interviewed using a pen and paper interview.

The differences between survey questions and survey technique raise one of the most important issues for people analysing 2002 NATSISS data. Before documenting such issues, it is also necessary to briefly discuss how individuals were selected in each of the two types of areas.

The CA sample was obtained from a random selection of discrete Indigenous communities and outstations. The sample frame used to design the survey was based on both 2001 Census counts, and information collected in the 2001 CHINS (ABS 2005c: 4). Once the communities had been selected, a random selection of dwellings was made.

Dwellings—and therefore individuals—in NCAs were selected using a stratified multi-stage area sample based on the 2001 Census. A random selection of dwellings within selected census Collection Districts (CDs) was then screened to assess their usual residents' Indigenous status. An insufficient number of households with Indigenous Australians was initially collected, so additional CDs were sampled during February to April 2003.

Before moving on to the CA and NCA survey design, it is important to make clear the difference between the CA sample and the concept of remote areas. The sampling in non-remote areas (i.e. major cities, inner regional and outer regional areas) was carried out entirely under the NCA methodology. This included 5242 of the surveyed individuals.

In remote areas (which includes the remote and very-remote Accessibility/Remoteness Index of Australia classifications), both CA and NCA sampling methodology was used. Remote areas that were not identified as 'discrete communities' used the same sampling methodology and interviewing techniques as were used in non-remote areas (i.e. the NCA methodology). In remote areas where NCA methods were used, there were 1997 respondents. In discrete communities and outstations, CA sampling was used, with information collected on 2120 individuals. Although the distinction between CAs and remote areas is not entirely clear in published record to date, according to correspondence with the ABS, the majority of those collected under the CA sample were from very remote areas rather than remote areas.

Questionnaire design and output

The questionnaire for the 2002 NATSISS was designed with the assistance of a special advisory group. Although the questions in the CAs and NCAs were broadly similar, there were still differences in what was asked, and the way the data was presented for publication by the ABS. The variables that were affected by such decisions are listed in Table 4.1 and can be classified into three main categories: those that were collected in both CAs and NCAs but that have different output categories; those collected in NCAs only; and those collected—but not released—in remote areas.

Table 4.1. Differences in data collection in CA and NCA areas

Restriction	Variable
Collected in both NCAs and CAs, with different categories outputted in remote areas	Main reason for last move
	Type of stressor in last 12 months
	Type of social activities in last three months
	Presence of neighbourhood/community problems
	Neighbourhood/community problems
	Whether used formal child care in last four weeks
	Type of child care used in last four weeks
	Main reason for not using (more) formal child care in last four weeks
	Type of organisation undertook unpaid voluntary work for in last 12 months
	Disability status
	Tenure type
	Type of major structural problems
	All sources of personal income
	Principal source of personal income
	Where used computer in last 12 months
	Where used internet in last 12 months
	Modes of transport
	Type of legal services used
	Attendance at cultural events in last 12 months*
	Self-assessed health*
	Type of government pension/allowance (auxiliary)*
	Whether working telephone at home*
Collected in NCAs only	Whether has an education restriction
	Whether has an employment restriction
	Disability type
	Multiple job holder
	Cash flow problems
	All types of cash flow problems
	Number of types of cash flow problems
Collected, but not released in remote areas	Whether ever used substances
	Type of substances ever used
	Whether used substances in last 12 months
	Type of substances used in last 12 months

Source: ABS (2004d)
Notes: An * refers to variables where the only difference between the CAI and PAPI samples is the presence or absence of a 'not stated' option.

For those variables in the first part of the table, it is not always clear, from the published record, why the ABS chose different output categories in remote and non-remote areas. After reading the questionnaires and through correspondence with the ABS, it would appear that for these questions, a reduced set of options was available to interviewees. It also seems that this was done mainly because the ABS, on advice from stakeholders and their testing processes, felt that these options would not be relevant to those living in CAs, so the benefits of having a more streamlined survey could be gained without adversely affecting the quality of the survey. However, the ABS needs to be clearer and more transparent about how it came to the conclusion that these particular variables needed different data outputs, and hence potentially limited the ability to use such data items with total confidence.

For example, full information on where computers and internet were used in the last 12 months is not available in remote areas. The missing category was internet/cyber cafes. While it could be argued that there are no internet cafes in CAs, it presumes that community residents are not mobile and have not visited urban settings where such cafes may exist. If the ABS has information that this is the case, then it needs to be made clear. Even though it should be acknowledged that the final survey content for CAs were a product of advice from ABS stakeholders and testing which identified items that were either inappropriate or did not work in the household interview environment in CAs, the ABS does not necessarily provide the level of detail that more sophisticated users may require (e.g. ABS 2004c: 20–26).

The second category of variables in Table 4.1 is those that were collected in NCAs, but not in CAs. Obviously, the structure of the survey prevents comparisons for these variables, but it is worth a brief reflection on why these decisions may have been made. For example, it is unlikely that many respondents held more than one job in CAs, so it does not make much difference that there is no relevant data in such areas. With respect to the lack of cash flow data in CAs, it is arguable that the notions of cash flow and poverty differ in remote and non-remote areas, so will differ between CAs and NCAs (Altman & Hunter 1998). Once again, though, it would be useful for the ABS to publish why such decisions were made without requiring analysts to speculate themselves, with incomplete information.

The disability variables warrant special mention, as quite different variables were constructed for the NCA and CA samples. As a result of field testing and consultative processes, data items and questions for a range of topics, including disability, were modified to take account of language and particular circumstances of people living in very remote communities. In addition, Indigenous stakeholders advised that attempts to measure psychological disabilities in remote communities required development of an appropriate instrument sensitive to the circumstances of people in these areas. The interim instrument developed by the ABS and used with Indigenous stakeholder endorsement in the recent 2004–05 National Aboriginal and Torres Strait Islander Health Survey is the initial response to that requirement. In the longer term, a culturally appropriate social and emotional wellbeing question module is being developed by Indigenous stakeholders.

Full disability was collected in NCAs, and this is comparable to data items in the GSS. A modified set of disability questions that did not include psychological disability was collected in CAs. This question was combined with the relevant options from the NCA sample to create a new variable which can be used across both samples (see ABS 2005c), though not in comparative analysis with the GSS.

While the different nature of the labour markets in remote and non-remote areas may mean that comparative data on education and employment restrictions caused by a disability may not be very meaningful (that is, given the binding constraints evident in the lack of employment prospects outside the CDEP scheme), this issue must still be of concern for the remote areas using NCA methodology.

The last category in Table 4.1 is the substance use variables for which data was collected but not released in remote areas. The relevant questions in the 2002 NATSISS were based on the National Drug Strategy Household Survey (NDSHS) and had a response rate of over 90 per cent. In NCAs, a voluntary self-enumerated form was used to collect this information, whereas in CAs, respondents were required to respond verbally to questions asked by an interviewer. The low prevalence of substance use reported in CAs has been assumed, by the ABS, to be the result of the use of direct questioning in CAs (ABS 2005c). It is further assumed that this led to a significant adverse effect on both the level of response and the quality of responses to questions on substance use. For this reason, information on substance use in remote areas was considered to be unreliable and has not been released.

Interviewing techniques

Not only were some of the questions different in the CAs and NCAs, so too were the interviewing techniques. As indicated above, interviews in NCAs were conducted using a CAI where interviewers use a notebook computer to read the questions and to record the data gathered. If respondents were asked to choose from a range of options, then prompt cards were used. For the substance use questions, a voluntary self-enumerated form was used with a response rate of 90 per cent.

In CAs, the surveying techniques were modified to take into account the cultural and language differences predicted for these areas. Firstly, the interviewing was conducted by more traditional pen and paper interviews. In addition, Community Information Forms (CIFs) were used to collect information about the community from the local council office. In every community, Indigenous facilitators were used to improve the validity of the data. However, not all interviews in CAs were conducted in the presence of facilitators. These facilitators 'explained the purpose of the survey to respondents, introduced the interviewers, assisted in identifying the usual residents of a household and in locating residents who were not at home, and assisted respondents in understanding questions where necessary' (ABS 2005c).

While the *differential* use of facilitators may have introduced potential interviewer bias in the response to some of the questions, accurate records were not kept by the ABS as to when facilitators were used. This means it is not possible to control the analysis of 2002 NATSISS for the effect of the presence of facilitators. However, it is still important to appreciate the relative importance of so-called 'non-sampling error' (which includes interviewer bias when facilitators are and are not present) and sampling error (which is present, to a greater or lesser extent, in all survey data).

Remote/non-remote and CA/NCA

The issue of remote/non-remote and CA/NCA is a confusing one, especially with regard to analysis of the CURF. The biggest issue is that on the CURF, those 1997 individuals in remote areas collected under the NCA methodology cannot be distinguished from those 2120 who were collected under the CA methodology. This is problematic when the questions asked across the two methodologies are quite different. Consider the question used to obtain information on whether a person was a victim of assault. In the NCA sample, the person was asked, 'In the last 12 months, did anyone, including persons you know, use physical force or violence against you?'. In the CA sample, on the other hand, the person was asked, 'In the last year, did anybody start a fight with you or beat you up?'.

There can be arguments made for or against the results from the two questions being comparable or not. However, from a methodological point of view, the point is that it is impossible for analysts using the CURF data to make that decision themselves. To repeat, users of the CURF are unable to identify whether people were surveyed under the NCA or CA sample, only whether they lived in remote or non-remote areas. The ABS does have such information available on the 2002 NATSISS MURF, and it may be possible for analysts to access it through customised tables. Ultimately, it is up to users as to whether they need to pursue such options to investigate data quality issues in greater detail.

Sampling and non-sampling issues

This section looks at a number of issues one must take into account when analysing the 2002 NATSISS or when interpreting results published elsewhere. It begins with a discussion of sampling error and non-sampling error so that analysts appreciate the relevant issues when interpreting the data. The section then benchmarks data from the 2002 NATSISS to other comparable sources. The remainder of the section then looks at a number of non-sampling issues which arise when comparing data items through time, comparing data to other contemporary sources, or interpreting results from ABS publications. These issues are not necessarily all the potential pitfalls involved with analysing the data. Rather, this section uses several examples in an attempt to highlight that analysts need to be aware of the context of the survey before making conclusions based on the data.

Sampling error

Sampling error arises from the fact that samples differ from their populations in that they are usually small sub-sets of the total population. Therefore, survey sample results should be seen only as estimations. Henry (1990) note that sampling errors cannot be calculated for non-probability samples, but they can be determined for probability samples, such as the 2002 NATSISS. First, to determine sample error, look at the sample size. Then, look at the sampling fraction—the percentage of the population that is being surveyed. The more people surveyed, the smaller the error. This error can also be reduced, according to Fox and Tracy (1986), by increasing the representativeness of the sample.

The 2002 NATSISS gathered information from 9359 Indigenous Australians. This was estimated to represent about 1 in 30 Indigenous Australians aged over 15 at the time of the survey (ABS 2005c). Such a large sample allows reasonably accurate inferences to be made about the population as a whole, as is shown by the quite similar results from 2002 NATSISS and other data sources (for which sampling error is not an issue). For example, the number of CDEP participants as recorded by the 2002 NATSISS is almost identical to administrative records (2005c: 27). However, irrespective of the concordance of results with administrative or population data sources, all data in the 2002 NATSISS is subject to sampling error and hence analysts must be aware of the reliability of estimates.

A standard error is the extent to which we expect a calculation from a sample of individuals to differ from what the same calculation would be if we had information for the whole population. Alternatively, it can be thought of as the amount of variation around the estimated value due to the fact that only a sample of people was taken.

Fortunately, sampling error can be quantified with reasonable accuracy. To the credit of the ABS, and in comparison to previous surveys, information is now given in the survey publications and CURF to enable outside users to make their own estimates of the possible sampling error. Therefore, in contrast to much of the existing analysis of the 1994 NATSIS, it is now possible to put some figures around how confident we can be about certain conclusions we make from the data.[3] That is, we are able to calculate the standard error of an estimate.

Standard errors for estimates of means and proportions can be used in two ways: constructing confidence intervals and performing hypothesis tests. Strictly speaking, a confidence interval refers to what one would expect to occur under repeated sampling. So, a 95 per cent confidence interval means that, if a given value was the true population value, then in 95 per cent or 19 out of 20 repeated

---

[3] The ABS is attempting to calculate replicate weights for the 1994 NATSIS CURF. This is a non-trivial exercise that has been facilitated by the development of computing processing since 1994 and driven, in part, by the increasing sophistication of users.

samples, we would expect the estimated value from a sample to lie within the range of the confidence interval. Although it is not strictly true, a 95 per cent confidence interval can also be thought of as the area bounding an estimate within which there is only a 5 per cent chance that the true population percentage is outside this range (but a sample of this size came up with the estimate that it did).

Instead of putting a range around the estimate, we may also want to test whether a certain estimate is different from another value. This value may be one we come up with, or another estimated value from the survey. To see whether this difference is significant in a statistical sense, a researcher can perform a hypothesis test. They would come up with a hypothesis, and then decide whether they can or cannot reject this hypothesis based on the estimated value from the sample, and the associated standard error. Note that one can only say conclusively that the population value is not a given percentage, never equivocally that it is.

There are three main ways in which one can estimate standard errors and hence estimate confidence intervals or perform hypothesis tests. Firstly, the ABS provides an approximation of the standard errors in tabular form which can be used with an Excel spreadsheet created by CAEPR for this conference. Secondly, the ABS provides a spreadsheet for estimates from the initial results publication that gives (relative) standard errors. Users can perform their own calculations with these estimates. Finally, on the CURF there are replicate weights that can be used to construct exact estimates of standard errors which users can then use to perform hypothesis tests. [4] There are formulas for deriving standard errors from replicate weights provided in ABS (2005c: 19).

Benchmarking 2002 NATSISS data against population estimates in the 2001 Census

This section benchmarks some selected results from the 2002 NATSISS using similar data from the 2001 Census to provide an indication of the underlying population estimates. The census estimates are calculated using almost the entire Indigenous population (i.e. less a relatively small undercount), and hence are reasonably reliable (from a sampling point of view). As indicated above, 2002 NATSISS data have sampling error associated with them, and hence the benchmarking exercise uses standard errors to calculate the 95 per cent confidence intervals around the results. These confidence intervals are represented as 'whiskers' around the 2002 NATSISS estimates in the following figures.

Strictly speaking, one should only compare 2002 NATSISS results to private dwellings in the census. However, the non-private dwelling estimates are also

---

[4] These replicate weights have the added benefit of being useful in calculating standard errors for estimates from regression style models that take into account the design of the survey.

presented to allow us to see who is missing from the sample. Figure 4.1 shows the proportion of the population aged 15 and over with various levels of attainment at secondary school in non-remote areas. Figure 4.2 illustrates the same results for remote areas.

Residents in non-private dwellings in both remote and non-remote areas appear to be more heterogeneous than those in private dwellings. That is, non-private dwellings are more likely to contain both people who did not go to school and people who have Year 12 or equivalent. This reflects the fact that non-private dwellings include a diverse range of residences: prisons and hostels that contain a disproportionate number of people who are extremely disadvantaged; and nursing homes and hotels that may include people with better access to resources (e.g. people with later-year secondary qualifications).

Figure 4.1. School attainment in non-remote areas, 2001 and 2002

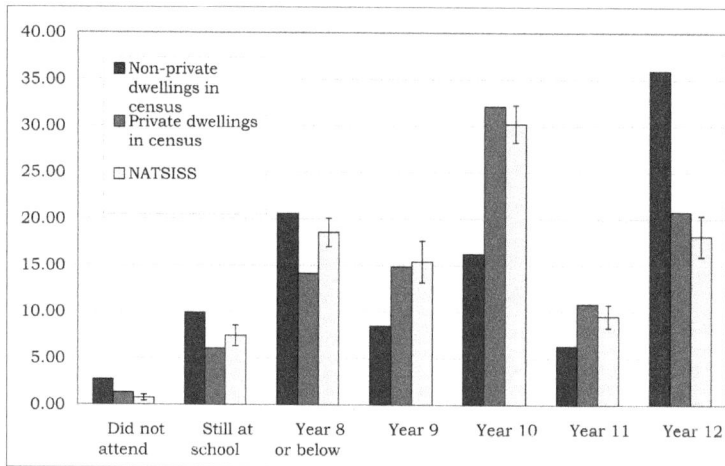

Source: Customised cross-tabulations.

Figure 4.2. School attainment in remote areas, 2001 and 2002

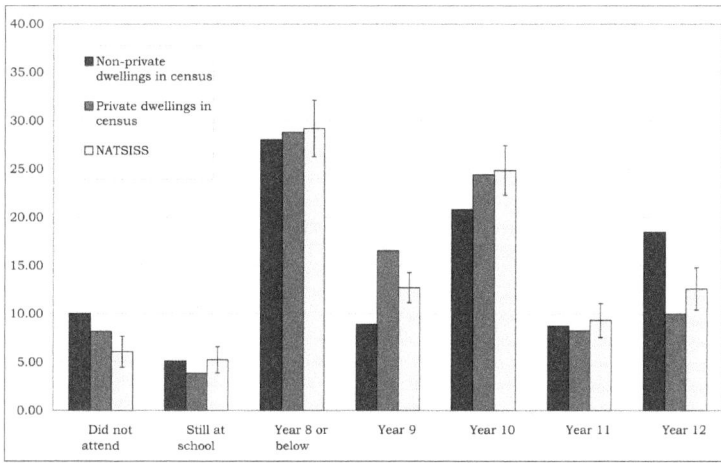

Source: Customised cross-tabulations.

In non-remote areas, it would appear that compared to the census, 2002 NATSISS had slightly lower completion rates for the later years of schooling. That is, there was a significantly lower proportion of Indigenous residents in private dwellings who have completed Year 11 or Year 12, no significant difference between those who completed Year 10 or Year 9, and a significantly higher proportion who were reported to have completed Year 8 or less (all at the 5% level of significance). 2002 NATSISS respondents were significantly more likely to be at school but less likely to have not gone to school at all.

In remote areas, there appeared to be less systematic variation. There is no significant difference between 2002 NATSISS estimates and private dwellings in the census for Still at school, Year 8 or below, Year 10, and Year 11. However, there was a significant difference between 2002 NATSISS and comparable census results for Year 12, Year 9 and for 'Did not go to school'.

Figures 4.3 and 4.4 report labour force outcomes in the 2001 Census and 2002 NATSISS, once again presented for non-remote and then remote areas. Residents of non-private dwellings tend to have lower labour force participation rates than those in private dwellings, despite the apparent heterogeneity noted above. This probably reflects that being in jail and nursing homes tends to constrain an individual's opportunities to work or look for employment, irrespective of whether they completed the later years of secondary school.

Figure 4.3. Labour force status in non-remote areas, 2001 and 2002

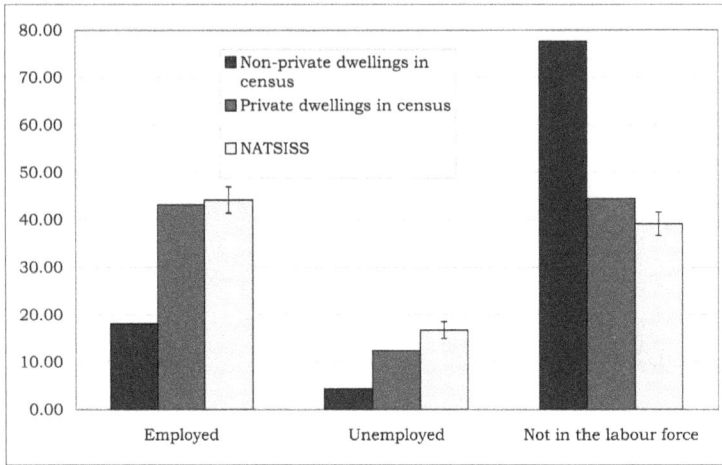

Source: Customised cross-tabulations.

Figure 4.4. Labour force status in remote areas, 2001 and 2002

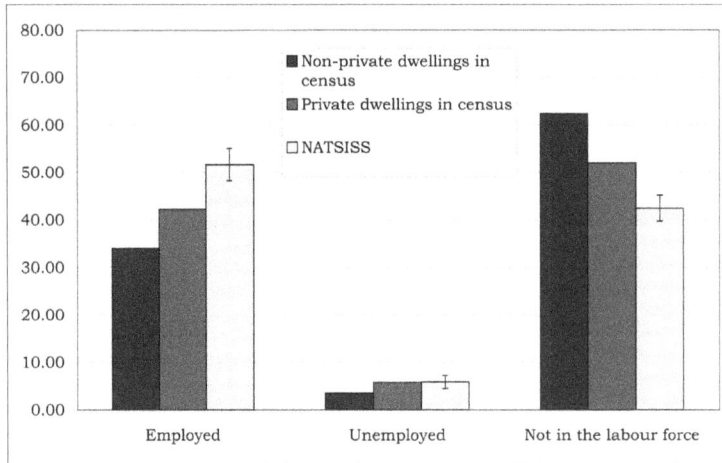

Source: Customised cross-tabulations.

In non-remote areas, the proportion employed in the 2002 NATSISS is not significantly different from the 2001 Census estimates for private dwellings. However, compared to the census, more NATSISS respondents are classified as unemployed, but fewer are classified as being not in the labour force. This could either represent a change in the outcomes over time, seasonality differences caused by the respective collections occurring at different times in the year, or the greater opportunity within a face-to-face interview to clarify what is meant by unemployment.

In remote areas, there appears to have been a shift from respondents being not in the labour force in the census to being employed in the 2002 NATSISS. This apparent shift occurs despite the fact that 2002 NATSISS provides accurate estimates of CDEP scheme employment that is disproportionately concentrated in remote areas. Once again, it is unclear as to whether the difference between 2002 NATSISS and the census is a result of differential timing of the collections (although the magnitudes make that unlikely) or a greater chance of clarification in a face-to-face interview compared to the largely self-enumerated census. What we do know, though, is that the differences we have reported in the labour market and educational status are unlikely to have been caused by sampling error alone.

Notwithstanding this, overall there appears to be less difference between the labour force status estimates from the 2002 NATSISS and the adjacent census than was evident in the 1994 NATSIS. Hunter and Taylor (2001) showed that there was a substantially higher employment and unemployment rates in the earlier survey when compared to the 1996 Census. While participation rates were again higher when survey methodology was used in 2002, unemployment-to-population ratios were less different in 2002 than they were in 1994. Indeed, unemployment ratios in remote areas were not significantly different in the 2002 NATSISS and the census. This may reflect the fact that the estimates of CDEP scheme employment appeared to be accurately measured in the 2002 NATSISS, and consequently there was less scope for mis-classification of CDEP scheme employment as unemployment.

Benchmarking these variables showed a number of differences that are unlikely to have been explained by sampling error alone. Although it is only possible to speculate, these differences could have been caused by:

- the question sequencing in the two data collections
- the use of face-to-face interviewing in the 2002 NATSISS as opposed to self-reporting in the census
- real changes in the year or so between the collections, or
- differences caused by the timing of the census and the survey (i.e. seasonality).

These can all be classified, to a certain extent, as non-sampling error (either in the census or the 2002 NATSISS). Although the differences were mostly small, the remainder of this section looks at three areas where non-sampling error is perhaps a bigger issue.

Non-sampling errors

Non-sampling errors can be defined as errors arising during the course of survey activities rather than resulting from the sampling procedure. [5] Unlike sampling errors, there is no simple and direct method of estimating the size of non-sampling

---

[5] Note that non-sampling errors can be present in both sample surveys and censuses.

errors. In most surveys, it is not practical to measure the possible effect on the statistics of the various potential sources of error arising from things other than the statistical sample. However, there has been a considerable amount of research on the kinds of errors that are likely to arise in different kinds of surveys. By examining the procedures and operations of a specific survey, experienced survey analysts may be able to assess its quality. Rarely will this produce actual error ranges, as for sampling errors. In most cases, the analyst can only state that, for example, the errors are probably relatively small and will not affect most conclusions drawn from the survey, or that the errors may be fairly large and inferences are to be made with caution. In rare instances, researchers may be able to say with some confidence in what direction the error might be.

Non-sampling errors can be classified into two groups: random errors whose effects approximately cancel out if fairly large samples are used; and biases which tend to create errors in the same direction and thus cumulate over the entire sample. With large samples, systematic errors, and resultant biases, are the principal causes for concern about the quality of a survey. For example, if there is an error in the questionnaire design, this could cause problems with the respondent's answers, which in turn, can create processing errors, etc. These types of errors often lead to a bias in the final results and analyses. In contrast to sampling variance and random non-sampling error, bias caused by systematic non-sampling errors cannot be reduced by increasing the sample size.

Non-sampling errors can occur because of problems in coverage, response, non-response, data processing, estimation and analysis. The following discussion is adapted from an excellent exposition on the subject from the Statistics Canada web site (see the non-sampling error section of Statistics Canada 2005).

An error in coverage occurs when there is an omission, duplication or wrongful inclusion of the units in the population or sample. Omissions are referred to as under-coverage, while duplication and wrongful inclusions are called over-coverage. These errors are caused by defects in the survey frame: inaccuracy, incompleteness, duplication, inadequacy and obsolescence. There may be errors in sample selection, or part of the population may be omitted from the sampling frame, or weights to compensate for disproportionate sampling rates may be omitted. Coverage errors may also occur in field procedures (e.g. if a survey is conducted but the interviewer misses several households or people).

Response errors result from data that have been requested, provided, received or recorded incorrectly. The response errors may occur because of inefficiencies with the questionnaire, the interviewer, the respondent or the survey process. Subject matter experts are often in a good position to identify flaws in such aspects of the survey.

Poor questionnaire design is a common aspect of non-sampling error. It is essential that sample survey or census questions are worded carefully in order to avoid introducing bias. If questions are misleading or confusing, the responses may end up being distorted.

As alluded to above, an interviewer and facilitators can influence how a respondent answers the survey questions. This may occur when the interviewer is too friendly or aloof or prompts the respondent. To prevent this, interviewers must be trained to remain neutral throughout the interview. They must also pay close attention to the way they ask each question. If an interviewer changes the way a question is worded, it may impact on the respondent's answer.

Respondents can also provide incorrect answers by their own volition. Faulty recollections (recall bias), tendencies to exaggerate or underplay events, and inclinations to give answers that are more 'socially desirable', are several reasons why a respondent may provide a false answer. Individuals may conceal the truth out of fear or suspicion of the survey process and the institutions sponsoring it (i.e. governments and their agencies). Other respondent errors may arise through a failure to understand the underlying concepts or a basic lack of knowledge about the information requested.

Non-sampling errors can also arise from the survey process. Using proxy responses (taking answers from someone other than the respondent) or a lack of control over the survey procedures are just two ways of increasing the possibility of response errors. Processing errors sometimes emerge during the preparation of the final data files. For example, errors can occur while data are being coded, captured, edited or imputed. Coder bias is usually a result of poor training or incomplete instructions, variance in coder performance (e.g. tiredness or illness), data entry errors, or machine malfunction (some processing errors are caused by errors in computer programs). Sometimes, errors are incorrectly identified during the editing phase. Even when errors are discovered, they can be corrected improperly because of poor imputation procedures.

Non-response errors—another category of non-sampling error—can also result from having not obtained sufficient answers to survey questions. Complete non-response errors occur when the survey fails to measure some of the units in the selected sample. Reasons for this type of error may be that the respondent is unavailable or temporarily absent; the respondent is unable or refuses to participate in the survey; or the dwelling is vacant. If a significant number of people do not respond to a survey, the results may be biased, since the characteristics of the non-respondents may differ from those who have participated. Given the high rates of mobility among Indigenous people, it is difficult to discount these issues in the 2002 NATSISS (and indeed other surveys involving mobile populations, Hunter & Smith 2002). Taylor and Kinfu's chapter on mobility will explore these issues in some detail.

Researchers and policy makers need to make themselves familiar with the discussion of non-sampling error in ABS (2004c: 56–7). The discussion on imputation issues for 2002 NATSISS in ABS (2004c: 61–2) is also relevant: analysts should pay careful attention to the list of variables where imputed data was used. To summarise that discussion, a small amount of missing data was imputed for a range of educational and training variables due to errors in the CAI instrument. While the ABS has been admirably candid about non-sampling errors in their publications, it is necessary to explore a few specific examples, to assist researchers and policy-makers in interpreting 2002 NATSISS data.

Possible non-sampling issues for selected variables in 2002 NATSISS

This section looks at three sets of variables in the 2002 NATSISS for which the possibility of non-sample error is likely to be pronounced. By non-sample error, we mean aspects of the survey methodology that are likely to result in differences between results from the 2002 NATSISS and other published data. The variables examined are: hunting, gathering and fishing; substance use and high-risk drinking; and education variables. This is not an exhaustive list of variables that are potentially problematic—rather, the discussion provides several examples of some of the issues users need to be on top of when interpreting 2002 NATSISS data.

Hunting, gathering and fishing can be an important part of the social life of a person's community, as well as their wider social relations. Although this may be true for many non-Indigenous Australians (especially in rural areas), it is particularly true for Indigenous Australians. However, for many Indigenous Australians—especially in remote areas—hunting, gathering and fishing also provide an important supplement to cash income and are hence an important part of their economic life. This is also true, although perhaps to a lesser extent, in non-remote areas (Altman, Gray & Halasz 2005).

Altman, Buchanan and Biddle's chapter on the customary sector illustrate that the information on the customary economy in the 1994 NATSIS was potentially biased by the structure of the questionnaire. Uncertainty about the definition of voluntary work meant that many people who may have engaged in hunting, fishing and gathering probably did not indicate this—for example, in Kununurra the incidence of both variables was zero, an implausible finding. That is, there was a strong geographic correlation between the incidence of voluntary work and hunting, fishing and gathering. This is probably driven by the response to the voluntary work question, as the anthropological evidence points to the existence of substantial hunting and gathering by the Indigenous people in the area. Consequently, the 1994 measures of hunting, fishing and gathering tend to understate the extent of the customary economy.

The way in which information on hunting, gathering and fishing was collected in the 2002 survey also makes it difficult to obtain a meaningful measure of the extent to which individuals participate in such activities. In 2002, the question was asked as part of a group cultural activity, and refers to 'fishing or hunting in a group'. In the 1994 NATSIS, the questions were asked as part of the category of voluntary work. However, an analysis of the Wallis Lake Catchment (which covers non-remote areas of Forster/Tuncurry) showed that local Indigenous people did not necessarily see hunting, gathering and fishing as being either voluntary work or a group cultural activity (Altman, Gray & Halasz 2005).

The ABS was probably right not to include hunting, gathering and fishing as a question incorporated under voluntary work, but collecting the information as a group cultural activity is not an adequate alternative. Needless to say, given that different questions were asked in 1994 and 2002, and the fact that both were problematic, the change in hunting, fishing and gathering between the two surveys is uninformative, if not meaningless.

The scope for non-sampling error is also apparent in the substance use questions. In NCAs, these questions were asked via a voluntary self-enumerated form. Pilot testing by the ABS concluded that due to English literacy problems in CAs, better information could have been obtained by asking respondents to respond verbally to questions asked by the interviewer. However, the very low prevalence of substance use in CAs led the ABS to conclude that the information obtained was not reliable and, as such, information is only available for NCAs.

The prevalence of high-risk drinking in 2002 NATSISS is substantially lower than was found in both the 2001 NHS (conducted by the ABS) and the 1994 NDSHS, Urban Aboriginal & TSI Peoples Supplement. The NDSHS had 3000 confidential interviews from metropolitan and other urban areas.

The relatively low prevalence of alcohol abuse identified in the chapter by Brady and Chikritzhs may have been caused by the different survey methodology (e.g. populations sampled, sample size, survey method, alcohol questions). Unlike the NDSHS, the alcohol questions in the 2002 NATSISS were neither confidential nor self-completed. Although a one-to-one interview was invited, respondents often answered questions in the presence of other family members and may have been reluctant to give accurate estimates. That is, asking other members to leave the room is an implication in itself. Another issue may be recall problems when questions refer to a longer time period (e.g. 12 months rather the last two weeks).

The important issue arising from the discussion of substance abuse is that non-sampling error is likely to be particularly important when asking sensitive questions about embarrassing issues or illicit activities

There were sequencing errors in the section of the NCA questionnaire on education, and the ABS needed to impute values for these variables based on the responses to other questions and information from the 2001 Census. The first sequencing error affected the 733 respondents aged 20–24 who were not studying full-time. These individuals were not asked whether they were currently studying, nor were they asked the type of education institute they were attending. Those individuals in that age group who were studying part-time would therefore not have been recorded as such. Using data on Abstudy receipts and distributional information from the census, the ABS coded 4 per cent of this sample with missing information as studying part-time.

An additional sequencing error occurred with the 1399 respondents who had used employment support services in the last 12 months. These individuals were sequenced past four questions on vocational training. According to the ABS, imputation for the five vocational training questions was conducted using 'donor records' (ABS 2005c: 19), where information from another person was matched to records based on sex, age and labour force status. These donor records were derived from other individuals in the sample. The ABS should indicate which categories of individuals were used to impute records so that analysts can assess whether they can discount the possibility of any significant bias arising from the imputation procedure.

In general, it would be useful for the ABS to publish the exact imputation techniques for both of these problems, so other researchers using the CURF can test the sensitivity of their results. It may be the case that the methodology used by the ABS is as robust as possible for most purposes, but certain applications or modelling frameworks may require a different technique.

A further issue with regard to the education questions is that we can see a difference that cannot be explained by sampling error alone. This is clear when comparing the data on the proportion of people with non-school qualifications in the 2001 Census with the results from the 2002 NATSISS. This can be shown in Figure 4.5, where once again the 'whiskers' refer to the 95 per cent confidence interval.

Figure 4.5. Non-school qualifications, 2001 and 2002

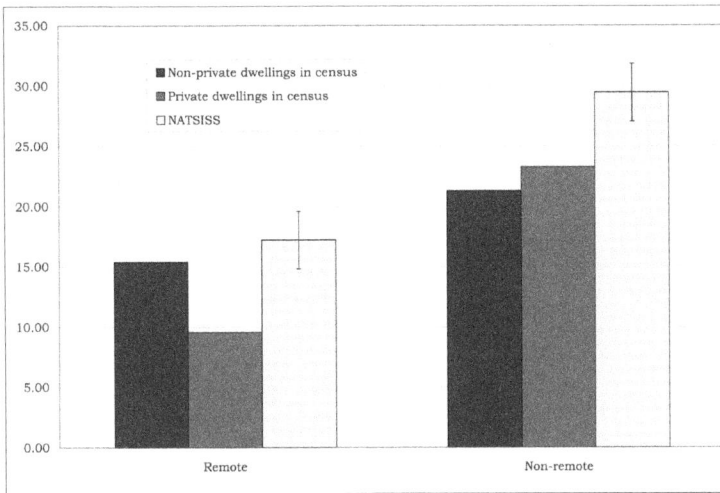

Source: Customised cross-tabulations.

This discrepancy between the 2002 NATSISS and the 2001 Census data is too great to be explained by the year or so gap in between the two data collections. The difference has, however, been noted in ABS (2005c) and is explained as a consequence of census reporting being undertaken by people on behalf of those being enumerated. That is, the non-sampling error in this comparison quite possibly arises from the census methodology.

Outputs from the 2002 NATSISS

There are three broad categories of output from the 2002 NATSISS that individuals are likely to use in their analysis of the data. These are: ABS publications (mainly the initial results publication), customised tables ordered from the ABS, and the CURF. We critically explore the first and the last of these categories as the chapter by Webster, Rogers and Black (in this volume) documents the process of securing customised tables in some detail.

ABS publications using the 2002 NATSISS

In June 2004, the ABS released the initial results from the 2002 NATSISS (ABS 2004c). This document included a 15-page summary of findings as well as 23 tables presenting the data in different ways and making comparisons with other data sources. The summary of findings focuses on selected changes through time (i.e. comparing the 1994 NATSIS and the 2002 NATSISS) as well as making explicit comparisons between the Indigenous population and the non-Indigenous population (as measured by the GSS).

While the depth of data available is impressive, and the provision of standard errors commendable, one criticism of ABS (2004c) could be made. In Table 9 of that publication, the proportion of people with various characteristics is presented, broken down by their equivalised household income. Based on analysis of the non-Indigenous population (ABS 2003c), the ABS outputs data for the second and third decile as a measure of 'low income', arguing that the lowest income decile has characteristics closer to those with higher incomes. ABS (2004c) use this 'low income' group as an indicator for 'poverty' in their discussion of results. This appears to be an implicit endorsement of criticism of income-based measures of poverty, especially the claim that measurement error (i.e. under-reporting) is pronounced for low income earners, especially those who indicate they have an income less than or equal to zero (Tsumori, Saunders & Hughes 2002). One group in which it is particularly difficult to accurately measure income is the self-employed. The recent public debate on the difficulties in identifying a socially accepted minimum standard of living illustrates that this criticism is not totally uncontested (Harding, Lloyd & Greenwell 2001; Saunders 2002; Tsumori, Saunders & Hughes 2002). One reason to be cautious about adopting the ABS definition of 'low income' without question is that self-employment is not prominent in the Indigenous population (Hunter 2004a: chapter 5).

Even if one can make a case for using this 'low income' indicator for the non-Indigenous population, a cursory examination of Table 4.2 shows that this assumption is suspect for the Indigenous population. Compared to those in the second and third decile, those in the first decile are less likely to be employed and own or purchase a home, and are more likely to have fair or poor health. There is no significant difference between the bottom decile and the ABS 'low income' group in terms of qualifications.

Table 4.2. Implications of ABS definition of low income[a]

	1st decile	2nd–3rd decile	4th–10th decile	Not stated
Employed	16.9	30.2	76.0	48.0
Has qualification and/or completed Year 12	17.7	21.7	44.5	27.6
Owns or purchasing home	11.9	15.1	43.8	29.0
Health fair or poor	30.5	26.5	16.6	23.4

a. The difference between the values for the 1st decile and the 2nd-3rd decile is always significant at the 5% level of significance.
Source: Customised cross-tabulations

Other variables not reported in Table 4.2 indicate that the bottom decile respondents are more disadvantaged than the 'low income group', as they are more likely to have not completed Year 12, be unemployed or outside the labour force, have been arrested in the last year, and have transport difficulties. The ABS definition of low income therefore tends to understate the incidence of Indigenous disadvantage, and should not be used.

Another criticism of ABS (2004c) is the lack of clarity regarding the difference between the concept of remote areas and the different sampling methodology employed in CAs. One example is that information on cash flow problems were collected in NCAs, but not in CAs (ABS 2004d: 32). However, in Table 4 of ABS (2004c), the footnote says the data was collected in non-remote areas only. However, this begs the question of what information was collected from the 1997 individuals who were collected as part of the NCA enumeration strategy but whose usual residence is in a remote area. The ABS probably has retained this information in the MURF and people may be able to access it when requesting customised tables. If this is the case, it would be helpful if the ABS made this clear. If the data is not available because of reliability concerns, it would minimise misunderstanding by explaining why this is the case.

There also needs to be more discussion about the underlying models and theory that appear to motivate the layout and choice of variables for some of the tables. For example, if you compare the age breakdowns in tables 10 and 11 from ABS (2004c), quite different age groups are used. While a reader could speculate as to why the different age groups are chosen, the publication needs to provide links to the previous research or theoretical models the ABS used to inform their judgments. In addition, the ABS needs to provide justification for why certain variables were deemed to be amenable to age standardisation in the official publication. The age standardisation of health is a reasonably standard technique, but it is unclear why employment status is better suited to age standardisation than, say, education participation (ABS 2004c: Table 5).

The ABS has also issued a number of publications and related tables for each State and Territory, but these follow the same basic conventions and formats used in ABS (2004c). Webster, Rogers & Black (this volume) give more detail about future planned output from the NATSISS.

The 2002 NATSISS CURF

On 7 June 2005 the ABS released the CURF for the 2002 NATSISS. Containing a unique record for all those households and individuals who were part of the survey, the data set enables the researcher to run their own cross-tabulations (subject to constraints) as well as more detailed statistical analysis of the data. Unlike the 1994 NATSIS and the GSS, the 2002 NATSISS is only available via the RADL. That is, individuals who have access to the data submit Statistical Analysis Software (SAS) or Statistical Package for the Social Sciences (SPSS) programming code which is then checked to make sure confidential information is not released. This is then run by the ABS and the output posted on the user's section of the RADL web site.

There are two separate files available via the RADL. The first has household information and a weight for the 5887 households in the sample. The second has information on each of the 9359 individuals, a household identifier that enables household information to be merged with the data, and a person level weight. These person and household weights can (and should) be used to turn information on the sample into estimates for the population, taking into account a person's or household's chance of selection. Furthermore—and this is a substantial improvement over the 1994 NATSIS—each household and person has 250 replicate weights which can be used to generate standard errors for estimates (as shown in ABS 2005c).

Another major improvement is that the 2002 NATSISS CURF has some information that has not been available to researchers on other ABS data sets. For example, previous data sets that included significant Indigenous populations, including the 1994 NATSIS, 2001 NHS and past censuses, have not included continuous income data. The 2002 NATSISS, on the other hand, contains both continuous individual data and household income data, up to a cut-off beyond which income data is censored.[6] This will enable distributional analysis outside the ABS that has not been possible before. Household rent and mortgage payments are also given as continuous values, although they are also right censored.[7]

There are, however, data items which have been made confidential, restricting the type of analysis that is possible. A number of employment variables are outputted in ranges. These include duration of unemployment and CDEP employment, as well as the number of hours usually worked per week. This latter variable places particular restrictions on analysis of hourly—as opposed to weekly—income.

More importantly, apart from in Queensland, one cannot separately identify Aboriginal Australians from Torres Strait Islanders. This is despite the fact that the Torres Strait Islander population was over-sampled in order to produce reliable (separate) estimates of the characteristics of Torres Strait Islanders living in the Strait and the rest of Queensland (ABS 2005c: 53).

The largest number of—and arguably the most confusing—restrictions on CURF data are imposed on the geographic variables. Apart from the variables mentioned in Table 4.1 that have different categories for CA/NCA or remote/non-remote areas, there are also restrictions on the type of geographic breakdowns that are possible using the 2002 NATSISS CURF. In particular, it is possible to undertake analysis by State/Territory or by remoteness, but not both simultaneously. As an example of the implication of such restrictions, it is possible to examine whether income is higher in remote versus non-remote areas throughout

---

[6] This right censoring of income appears to affect 0.65% of individuals and no households (at least when equivalised household income is used). These estimates of the proportion who are right censored are for those individuals or households who have stated their income (& rent/mortgage in the next endnote). For income, it includes those with negative or nil income, as well as those 'not applicable'.
[7] Censoring of weekly rent and mortgage repayments affects 3.45% and 12.96% of households respectively.

Australia, or higher in NSW versus other States. However, it is not possible to examine whether income is higher in non-remote versus non-remote areas within NSW specifically. While this restriction is understandable, given the imperative to protect the confidentiality of respondents, it does restrict the ability to make meaningful interstate comparisons (that is, by constraining the ability to control for the level of remoteness in the respective States).

If users want to obtain some sort of cross-classification of State/Territory and remoteness structure, the ABS insists they use a specific variable (STXREM) when attempting to do this on the RADL. Users are not permitted to submit programs that include both State/Territory and remoteness structure under any circumstances.

An additional restriction on geographic analysis is that Tasmania and the Australian Capital Territory (ACT) are included as one variable. This is despite the statement that 'the NATSISS was designed to provide reliable estimates at the national level and for each State and Territory' (ABS 2005c) and that ABS (2004c) outputted data for each State separately. Within these two combined localities, there were 736 individuals from Tasmania and 330 from the ACT, which when weighted represented about 10 900 and 2600 individuals respectively. The State governments in these two States have quite different policies, and the presence of the federal government and two large universities in Canberra (among other things) make these two populations quite different. In addition, the geography of these two regions in terms of remoteness and access to resources are quite dissimilar. Thus the joining of these two localities somewhat constrains policy analysis and restricts a modeler's ability to control for geography when analysing the relationship between other variables.

At the moment, one of the biggest constraints for analysing the 2002 NATSISS using the RADL is the requirement to undertake the analysis in only two statistical programs: SAS (version 8.2) or SPSS. While SAS performs most of the standard types of analysis, there are a number of things it either cannot do or that are not done as easily as in other statistical packages (such as the calculation of 'robust' standard errors and the easy output of marginal effects from estimation on the probit model). Perhaps more importantly, SAS is expensive. This may not be as big an issue in large government departments, or even perhaps large research centres, where the fixed costs can be spread across a number of users, but for smaller organisations or individual researchers, the costs can become prohibitive. [8] SPSS is more affordable, but it does not have the range of applications that other programs have. Users may be able to visit a data laboratory at the ABS site where they have been able to install and run the widely-used

---

[8] CAEPR cannot afford to use SAS and only a handful of people in the ANU now use SAS because of cost constraints. Stata and other cost effective statistical packages that allow more sophisticated statistical analyses are now preferred by many users.

statistical program, Stata. However, it is likely that this option will cost the user extra money because the ABS may need to employ a staff member to ensure the user does not compromise confidentiality. The ABS hopes to have a fully operational version of Stata on the RADL in the near future.

Concluding remarks

To summarise the discussion in this paper, there are three main messages that we feel users of the data should keep in mind when making conclusions.

Firstly, researchers and policy-makers need to be aware of the different survey methodology used in CAs and NCAs. Given that no identifiers for CAs are publicly released on the CURF or the RADL, this effectively means that analysts have to be aware that a large proportion of remote data in the 2002 NATSISS was collected from CAs. It may be possible to purchase customised cross-tabulations from the ABS if people are particularly concerned about potential non-sampling error arising from the differing methodology in the two types of areas. [9] Whatever one's position, some caution is warranted when making comparisons between the CAs and NCAs, and possible remote and non-remote areas.

The second issue that should be kept in mind, especially when making comparisons with either the 1994 NATSIS or the 2001 Census, is that 2002 NATSISS only collects information from Indigenous people aged 15 and over living in private dwellings. That is, the statistics collected do not provide information on the entire population of Indigenous Australians. Where other data sources are used, analysts need to ensure they are comparing the same age groups; non-private dwellings need to be excluded; and, in the case of the 1994 NATSIS, the re-weighted data set must be used once it becomes available on the CURF.

The final point is that given that the ABS has released information that enables users to reasonably accurately estimate standard errors, it is no longer acceptable to make statements about differences in means or proportions without taking into account sampling error. Analysts need to make themselves aware of the formula for calculating standard errors and confidence intervals. More sophisticated users will have to work out how to use replicate weights. Given that there is considerable room for programming errors, especially when using arrays in SAS, the ABS should consider releasing draft programming code (particularly, if such code is already available) to minimise the possibility of non-sampling error arising from analysts. Ultimately, though, analyst error is the responsibility of the analyst.

---

[9] CA records are separately identified on the 2002 NATSISS MURF.

5. Differentials and determinants of Indigenous population mobility

John Taylor and Yohannes Kinfu

Of the three components of demographic change, geographic mobility is the most nebulous and difficult to measure, and yet it is the one with potentially the greatest impact on population distribution and composition. Difficulties of measurement arise because a variety of definitions of population movement can be construed, all of which constitute arbitrary functions of the distance and length of time involved in relocating from one place to another. Impacts on distribution arise because migrant numbers in and out of a given place could exceed births and deaths, especially in small geographic areas and at higher stages of demographic transition, while age and sex selectivity of migration places a wide ranging effect on the composition of migrant sending-and-receiving regions.

In illuminating these issues, analysis of population movement for both the Indigenous and total Australian population has generally been informed by data yielded from fixed-period usual residence questions in the national census (that is, usual residence now compared to one and five years ago), as well as by information from case studies focused on particular localities or sub-populations (Bell 1992; Bell & Maher 1995; Gale 1972; Gale & Wundersitz 1982; Kinfu 2005; Taylor & Bell 1996a). Admittedly, between 1970 and 1987, the ABS also conducted an annual survey of internal migration in conjunction with one of its monthly population surveys (without an Indigenous identifier) (ABS 1987). However, since that time, few official nationwide surveys have included a mobility indicator in Australia. Even those that have included a mobility indicator do not have an Indigenous identifier or have limited Indigenous samples (such as the Australian Housing Survey and the February supplement of the Labour Force Survey on labour mobility).

To this extent, the inclusion of a question on mobility status and reasons for movement in the 2002 NATSISS was a rare event and it presents two new opportunities for analysts. Firstly, it provides a new source of data with which to validate previous findings from census analysis and case studies. Secondly, the existence of wide ranging individual and household characteristics data allows for testing of the statistical relationship between these variables and mobility, thereby enabling a wider and more direct exploration of the social and economic determinants of movement than hitherto possible. The purpose of this chapter is, therefore, to make use of the survey data to examine the intensity and correlates of spatial mobility among Indigenous Australians. The distinguishing feature is that the determinants of mobility are explored from a micro rather a macro perspective, while reasons provided by respondents for their movement are also explored—something that has only previously been established indirectly by proximate cause using census data (Kinfu 2005).

Background: survey development and questions

Interestingly, the background to the inclusion of mobility questions in the 2002 NATSISS appears to be exactly the opposite to the experience of the 1994 NATSIS. In consultations before the 1994 survey, the topic of 'location and mobility' was proposed as important for inclusion by a number of government agencies under the broad heading of 'culture'. In particular, it was felt necessary to acquire data on the movement patterns of household members over a 12-month period before the survey, including the number of moves, duration of each move and reasons for each move. This mobility history was deemed by agencies such as ATSIC, Department of Employment, Education and Training (DEET), and the Department of Health, Housing and Community Services to be of assistance in planning the location of appropriate services such as community infrastructure and social programs. However, due to recall problems, less than adequate responses to mobility questions were obtained in the pilot survey and it was determined that an improvement in data quality would require overly-indulgent in-depth probing on an already crowded interview schedule. This led to the mobility questions being omitted, with the exception of a question on the number of moves away from the local area for the treatment of health problems (Taylor 1996: 41).

By contrast, in the preparations for the 2002 survey round, questions on mobility were initially excluded in line with 1994 practice. However, they were then included following strong user demand for inclusion, as mobility was viewed as a cross-cutting issue that impacted on many areas of concern (including health, education, employment and housing). While this view of the policy significance of mobility is widely held and articulated, what are less clear are the precise policy questions to be addressed by an understanding and measurement of movement. In short, what is it about mobility that we want to know and can a sample survey, such as the NATSISS, provide the answer?

The decision to include mobility questions turned out to be easier than deciding how to formulate appropriate questions. At the census, and in the now defunct ABS internal migration survey, people are classified as having moved if the address of their usual place of residence was different from that of one year or five years ago. In the 2002 NATSISS, the definition of movement is far more inclusive and refers to all moves made over the 12 months before the survey. In particular, the NATSISS asked (in non-remote and remote non-community sample areas), 'In the last 12 months, have you *lived* (emphasis added) in any other dwellings?' and (in remote community sample areas), 'In the last year have you *lived* (emphasis added) in any other houses or places?'. In both areas, the survey then went on to elicit how many dwellings, houses/places people had lived in over the course of the year.

Why these differences in the nature of questions deployed on community sample and non-community sample forms arose is not known. However, leaving this aside, the key point to note is that the probability of movement should be considerably higher in the 2002 NATSISS compared to the census, given its reference to all moves as opposed to a single specific move as defined by the fixed-period census 'usual residence' question. This lack of any temporal or spatial reference in framing the NATSISS definition of mobility is unusual, since duration and distance of movement are two key variables for mobility analysis (Bell 2001; Taylor & Bell 2005). As mentioned earlier, the NATSISS also collected information on the main reason for the last move and classified these under four major headings: family, housing, employment and accessibility (to services). The issue of whether these questions provide an adequate and meaningful basis for addressing key policy issues—either independently or in association with other survey variables—will be dealt with in conclusion.

The scale and age pattern of mobility

Overall, 30.9 per cent of respondents to the NATSISS aged 15 years and over indicated that they had moved to live in another dwelling/house/place in the 12 months before the survey, with the true value (at the 95% confidence level) lying somewhere between 28.9 and 32.9 per cent. In light of the comments made above about the inclusive nature of the survey question, this is a surprisingly low proportion. This is especially true when benchmarked against 1996 and 2001 census results that show broadly similar propensities for movement for the same age group (29% and 26%) for the periods 1995–96 and 2000–01 respectively. Equally surprising, given the open-ended nature of the NATSISS mobility question, is that only a quarter of respondents who moved reported making more than one move over the 12 months before the survey.

Like all other demographic processes, change of residence is associated with age and certain life-cycle statuses and transitions (Long 1992). To investigate the age pattern of migration, Figure 5.1 compares age-specific movement rates from the NATSISS with equivalent rates derived from the 1996 and 2001 censuses. Clearly, the age pattern of movement derived from the NATSISS is in broad agreement with census results, in so far as movement comes to a peak among young adults aged 20–24 years and steadily declines thereafter. Overall, these patterns conform with more or less universal observations first made by Rogers, Raquillet & Castro (1978) and Rogers & Castro (1981) for the United States and Europe, and subsequently in the Australian context by Bell (1992, 1995). Basically, for the Australian population as a whole, as in other countries, the peak in the age profile of mobility has been linked to the combined influence of departure from the parental home, the start of tertiary education, entry into the labour force and the establishment of independent living arrangements. To the extent that Indigenous people participate in these same life-course events, the age profile of Indigenous movement is likely to reflect similar influences. In the subsequent section, we examine the gross and net effects of such variables on movement propensity.

Figure 5.1. Age-specific Indigenous movement rates: 1996 Census, 2001 Census and 2002 NATSISS

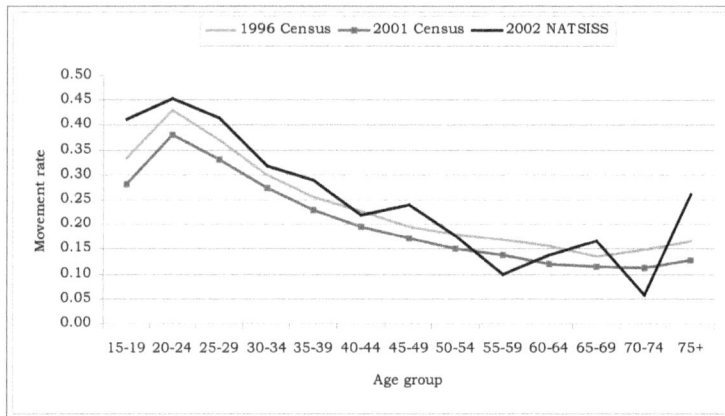

Source: Customised cross-tabulations.

What Figure 5.1 shows is that mobility rates derived from NATSISS are consistently higher than census rates (especially 2001 Census rates) up to age 50, beyond which point the NATSISS pattern becomes relatively erratic, possibly due to the small sample size. Reading these sets of rates together, perhaps the more interesting observation is the consistent reduction in age-specific movement rates between the 1996 and 2001 censuses and the apparent increase from the 2002 NATSISS. This appears to be similar to other systematic shifts in Indigenous demographic indicators revealed by census analysis that are difficult to reconcile with demographic behaviour (Kinfu & Taylor 2005).

Differentials and determinants of mobility

As with age, mobility also varies according to the geographic, socioeconomic and demographic characteristics of individuals and households. One limiting factor of the NATSISS in regard to establishing the geography of mobility and its demographic impact on sending and receiving regions is the lack of any origin/destination data. Instead, we have measures of the propensity to move (or not) at fairly large geographic units, notably States and the Northern Territory and the broad remoteness categories of the ASGC. For the most part, no difference is evident between jurisdictions in the propensity to move, given the spread of the upper and lower bounds of most estimates (see Fig. 5.2). However, stand-out exceptions include Tasmania and the Northern Territory, with the rate of movement in the latter ranging from as low as 15.7 per cent to no more than 20.9 per cent. In effect, the propensity to move among Indigenous people in the Northern Territory is only half the rate in Queensland. As for movement propensity by remoteness category, this reveals that inner regional areas display significantly higher mobility than all other regions except for major cities, while at the other extreme of the remoteness classification, the population in very remote areas displays significantly lower movement than all other regions.

Figure 5.2. Indigenous movement propensities by State and Territory, 2002

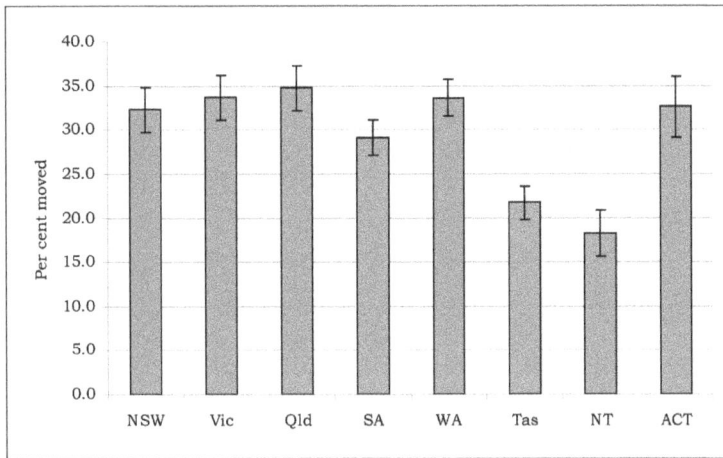

Source: Customised cross-tabulations.

These observations regarding the spatial pattern of movement are interesting for their similarity to consistent findings from census data that display the lowest rates of Indigenous mobility in remote areas of the north and west of the continent, and especially across the Northern Territory (Taylor & Bell 1996b). This impression of relative immobility among Indigenous people in very remote areas and across the Northern Territory is grossly misleading. Numerous case studies attest to the importance of frequent mobility in the daily, periodic and seasonal round of activities associated with Indigenous social and economic life in such places (see Taylor & Bell 2004 for a summary). The gap between these observations and NATSISS results is all the more surprising given the more inclusive nature of the NATSISS mobility question allowing for all moves to be recorded. This differs from the census question which is ill-suited to capturing the frequent short-term circular movements that are common in remote Australia.

While explanation for this anomaly is most likely to be found in respondent error associated with the nature of the NATSISS question and the manner in which it was interpreted, even this is perplexing because previous critiques of census data have pointed to the inappropriateness of applying 'usual residence' concepts for measuring movement among populations that live as much in an area as a single place (Morphy 2002: 44–55). However, the NATSISS does not refer to usual residence, preferring instead to talk of the number of dwelling/houses/places 'lived in' which would seem to be a less ambiguous construct—although it does raise the issue of what 'lived in' actually means.

Along with differentials in spatial mobility, the reported data also show clear differences in the propensity to move across a range of social, economic and cultural characteristics that are selected here for their likely association with movement propensity. As indicated by the age-standardised point estimates for each of these variables in Table 5.1, and by the associated 95 per cent confidence interval, these differences tend to reduce for some characteristics and increase for others once the differences in age composition between the categories are controlled for. In some instances, the standardisation procedure even alters the relative balance of variables. For example, the unadjusted migration rate among married people was lower than for those not married, but higher for the standardised result. Overall, the results reveal that movement propensities vary widely around the national average of 30.9 per cent according to the different variables.

Table 5.1. Social, economic and geographic differentials in movement propensity

Background variable	Reported	Mobility rate %		
		Age-standardised		
		Point estimate	95% lower bound	95% upper bound
Total	30.9	30.9	28.9	32.9
Sex				
Males	30.5	30.3	27.5	33.1
Females	31.2	30.8	28.1	33.5
Marital status				
Married	27.6	34.6	31.7	37.5
Not married	33.7	32.2	29.5	34.9
Labour force status				
Unemployed	45.4	41.1	35.1	47.1
Not in labour force	29.6	30.7	27.6	33.8
Employed	27.6	27.6	24.8	30.4
Public	22.4	24.3	18.5	30.1
Private	29.6	29.3	25.0	33.6
CDEP	28.0	26.9	21.2	32.6
Education				
Post-school	31.5	39.0	33.8	44.2
Yr 11 and 12	37.5	32.6	28.0	37.2
Yr 10	32.0	29.3	25.2	33.4
Yr 9 or below	25.9	28.6	25.3	31.9
Training				
Attended	34.0	30.9	26.9	34.9
Not attended	30.8	29.5	27.3	31.7

Table 5.1. (continued)

Background variable	Reported	Mobility rate %		
		Age-standardised		
		Point estimate	95% lower bound	95% upper bound
Housing tenure				
Owner	20.6	22.1	18.8	25.4
Private rental	52.0	47.1	41.9	52.3
Public rental	32.9	31.6	27.8	35.4
Community	25.6	25.2	21.2	29.2
Place of residence				
On homeland	27.2	27.9	23.6	32.2
Not on homeland	31.9	31.5	28.7	34.3
Does not know or recognise homeland	32.0	30.6	26.9	34.3
Neighbourhood problems				
Has problems	26.5	26.0	23.1	28.9
Does not have problems	34.2	34.1	31.5	36.7
Remoteness				
Major cities	31.5	30.9	27.2	34.6
Inner regional	36.0	36.2	31.3	41.4
Outer regional	30.0	30.8	26.4	35.2
Remote (& very remote)	27.2	27.4	23.6	31.2
Health status				
Excellent	28.8	24.9	20.3	29.5
Very good	32.5	29.9	26.0	33.8
Good	32.7	32.6	29.0	36.2
Fair	30.4	36.8	31.5	42.1
Poor	22.7	30.8	22.9	38.7

Source: Customised cross-tabulations from the 2002 NATSISS RADL

Although this bi-variate analysis is highly suggestive, the net effects of each of these independent variables can only be assessed by a multi-variate analysis. For this purpose, we fit a logistic regression with the dependent variable taking the form of 1 if the respondent lived in any other dwellings (or places) in the 12-month period before the survey, and 0 otherwise (see Table 5.2). In this way, the results indicate the effects of all the selected factors simultaneously on the chances of moving or not. The simplest way of summarising this relationship is to examine what happens to these chances for groups of Indigenous people with different characteristics. These changes in movement probability are best measured relative to a hypothetical reference person and the characteristics of the reference person chosen for this purpose are indicated in the note for Table 5.2.

Table 5.2. Net effects of socioeconomic, spatial and household characteristics on Indigenous mobility: logistic regression results, 2002 NATSISS[a]

Background variables	Odds of mobility
Female	0.999
Not married	1.189[b]
Labour force status/sector	
Unemployed	2.737[b]
Not in labour force	1.841[b]
Private	1.441[b]
CDEP	1.532[b]
Educational attainment	
Post school	1.109[b]
Yr 11 and 12	1.319[b]
Yr 9 or below	0.739[b]
Not attended vocational training	0.753[b]
Housing tenure	
Owner	0.558[b]
Private rental	2.16[b]
Community	0.743[b]
Place of residence	
Lives on homeland	0.895[b]
Does not know or recognise homeland	0.998
Problems identified in neighbourhood	0.663[b]
Remoteness classification	
Major cities	1.051[b]
Inner regional	1.242[b]
Remote/very remote	1.258[b]
Self-reported health status	
Excellent	0.842[b]
Good	1.026[b]
Fair	1.006
Poor	0.686
Model intercept	0.331

a. Reference person is defined as: male, married, employed in the public sector, has Year 10 education, lives in public rental housing, does not live on homeland, neighbourhood has problems, and resident in an outer regional location. The results are not sensitive to the choice of the characteristics of the reference person.
b. Statistically significant at the 5% level.
Source: 2002 NATSISS RADL

These results indicate that marginal labour force status is the biggest predictor of mobility. After removing the effects of other variables, there is a significant underlying pattern of higher mobility among those unemployed or not in the labour force compared to those employed. When this is controlled for by sector of employment, the chances of movement are seen to be greater among private sector and CDEP workers compared to public sector workers. As for educational attainment, a gradient of increased mobility with higher educational attainment is evident. This is strengthened by the related result that non-attendance in a vocational training course has a negative effect on mobility.

Housing tenure is another key variable with much higher chances of movement among those in private rental dwellings, while home ownership and community rental serve to depress mobility. Contrary to expectation, respondents resident in neighbourhoods with perceived social problems are far less likely to move, thereby negating what might have been viewed as a push factor and probably reflecting limitations on residential choice. Other findings related to location appear contradictory. For example, residence on homeland is negatively associated with mobility, while remote and very remote location increases the chances of movement. One possibility is that the latter observation reflects the amalgamation of responses from both non-remote and remote non-community sample forms, whereas the former is based on remote community sample data. Finally, self-assessed health status is interesting, as this deflates mobility at extremes of the range, though no doubt for quite different reasons—one due to choice, and the other due to incapacity.

Reasons for movement

Respondents to the 2002 NATSISS aged 15–64 years were asked to indicate the main reason for their last move. Answers to this question are grouped into four broad categories: housing, family, employment and accessibility (to services), and while this provides for some comparison with findings from the former *ABS Internal Migration Survey*, this is only true for the housing and employment categories. From Table 5.3, it is clear that factors associated with family and housing dominate in the calculus of mobility decision-making. Among the family reasons provided, the single largest sub-category was a desire to be close to family and friends. This is consistent with repeated findings from case studies of Indigenous mobility that stress the importance of kin location in shaping the frequency and pattern of mobility (Gale & Wundersitz 1982; Taylor & Bell 2004: 20–21; Young & Doohan 1989).

Table 5.3. Reasons for last move by age group: 2002 NATSISS

Reason for last move	Age group %				All	Estimated population (no.)
	15–24	25–34	35–44	45–64		
Family	45.5	35.4	30.8	38.2	38.2	32 400
Housing	25.4	39.8	38.3	33.3	33.3	28 400
Employment	12.6	9.2	13.3	10.9	10.9	9300
Accessibility	4.4	2.4	2.6	3.3	3.3	2800
Other	11.7	12.9	15.0	14.0	14.0	11 900
Don't know	0.3	0.4	0.0	0.2	0.2	200

Source: Customised cross-tabulations from the 2002 NATSISS MURF

As for housing, once again the impact of reliance on rental housing has been highlighted in the literature as an important factor in stimulating Indigenous mobility (Gray 2004), although the results of the NATSISS multi-variate analysis suggest that this is especially so for those in private rental accommodation. The overall level of housing reported reasons (33%) is similar to the 29 per cent reported by the general population in 1987, although in the latter case this was predominantly for home purchase (ABS 1987: 13). Of particular note in Table 5.3 is the fact that the estimated numbers moving for employment reasons is relatively low, although this is not much lower than the 14 per cent reported for the total population in the 1987 ABS survey. Likewise, movement to access services appears very limited, which is surprising given the spatial separation of many Indigenous population clusters from basic services such as high schools, hospitals, banks, shops, and government offices (Taylor 2002). Although the category 'other' is equal in size to employment and access reasons combined, it is not clear what this category comprises.

Conclusion

The 2002 NATSISS provides the first survey data on Indigenous residential mobility indicating the propensity to move with reasons for doing so. Brief analysis of these data confirms many of the findings from previous census-based analysis: that the probability of movement peaks among young adults, is similar overall for males and females, is higher for single people and especially high among the unemployed, greater for those in private rental dwellings, and lowest in remote areas. Among the new findings are factors that tend to deflate mobility. These include low educational attainment, home ownership, residence in community housing, residence on their homeland, living in an area with neighbourhood problems, and outer regional location. Also new is the insight that CDEP employment does not dampen movement propensity as previously suggested (Taylor & Bell 1996b). While this much can be gleaned, and while further outcomes and relationships will no doubt be established, even the preliminary findings presented here raise a number of issues regarding the utility and interpretation of NATSISS data on mobility.

First of all is the surprising outcome that the overall propensity to move is only marginally higher than reported in the census, and that in line with the census, movement rates are lowest in the Northern Territory and remote areas generally. So pervasive is the observed fact of residential movement in the daily, fortnightly, seasonal and annual round of Indigenous social and economic life in remote Australia, as elsewhere in the country, that ethnographers have been strained to describe it using evocative terms. These include beats, runs, lines, floaters, visitors, concertina households, or multi-locale relationships (Taylor & Bell 2004). If the 2002 NATSISS fails to reflect the manifest intensity of population movement in remote areas, then questions naturally focus on non-sampling error and, in particular, how the notion 'lived in' might have been presented by interviewers and interpreted by respondents. Having an open-ended question as in the NATSISS is potentially valuable for capturing short-term and repeat movement, but without spatial, temporal, and conceptual structure it can descend into ambiguity.

This brings us to the purpose of a mobility question in a survey such as the NATSISS. In pressing the case, many agencies in submissions to the ABS highlighted the likely impact of mobility on their portfolio areas of concern. If there was ever an expectation that the NATSISS might measure the spatial impact of population movement for the purpose of planning service requirements (the most useful insight from a policy perspective) then this was ill-informed. One assumes, however, that more abstract higher-order questions were in mind, such as those answerable by multi-variate analysis. But again, when we are dealing with movement defined in such a way that it that could be for one night next door as opposed to 11 months across the continent, or from Kintore to Alice Springs instead of Newcastle to Sydney, then structurally—and policy-wise—we are dealing with very different types of mobility that remain totally undifferentiated in the NATSISS. Having said that, further work could be done to isolate sub-groups in the survey population that present potential interest for policy (such as those who indicated employment reasons for movement, or particular difficulties with housing), and their characteristics and behavioural attributes could be explored in more detail. However, much of this is speculative after the event and, as with all such surveys, clear specification of the answers to be addressed by end-users is essential for optimal design.

If we turn to the actual reasons for movement provided by respondents, these conflict (to some degree) with the findings of the multi-variate analysis which highlights structural (economic) factors such as unemployment and private rental housing as key predictors of movement. It is also the case that accessibility to services scores very low as a reason for movement, which is contrary to the experience of many Indigenous people who have to move in order to access basic services such as health care, schooling, training, banking or shopping. The fact is, separating out reasons for mobility is never easy given the multi-faceted purpose and unpredictable nature of many journeys—for example, what may have started out as a shopping expedition to town can turn out to be an extended stay with relatives (Young & Doohan 1989). Nonetheless, it is interesting that Indigenous people choose to emphasise social over economic factors (such as employment), while many housing reasons (such as overcrowding and wanting a bigger house) can be viewed as both social and economic. This emphasis on social factors is consistent with much of the ethnographic research on mobility that underscores the role of kinship networks in directing population flows, even if the primary purpose might be job search or housing (Gale 1972; Gale & Wundersitz 1982; Gray 2004).

6. Aboriginal[1] child mortality in Australia: Recent levels and covariates

Yohannes Kinfu

The subject of infant and child survival continues to be of great interest to policy makers, health planners, social and bio-medical researchers as well as communities and interest groups around the world. Overall, Australia is one of the healthiest and safest countries for young children to live with an infant mortality rate of less than six per thousand live births and an under-five mortality rate of some seven per thousand children (ABS 2004b). However, the same cannot be said with confidence for Australia's Indigenous population, where morbidity and mortality among young children remain excessive. Available evidence suggests that, at current rates, as many as one in fifty Aboriginal children will die before reaching age one, while an additional eight per thousand will have no chance of reaching their fifth birthday (Kinfu & Taylor 2005).

While the broad profiles of such excess mortality have been known for some time, far less is known about the precise levels and determinants of high mortality in this population. For instance little, if anything, is known about how socioeconomic factors influence child mortality or how child mortality risks vary according to household, spatial and residential patterns across the country (Gray 1988), information that is vital for public policy making. This dearth of information in part reflects the paucity of relevant health and mortality data on Indigenous Australians. Identification of Indigenous origin in death certificates in Australia did not begin in most jurisdictions until the mid-1980s, and was not introduced until much later in Queensland (1996), while hospital separation records in the country continue to suffer from inconsistency in the way they collect information on Indigenous status. The situation is no different with regard to the census, which provides the denominator for most health and mortality related studies: despite an early introduction, in 1971, of a direct question on Indigenous status, no two successive censuses in the country have yet produced consistent counts of the Indigenous population. The problem is further compounded by the fact that none of these data sources, with the exception of the census, provide detailed information on the social, economic and other background characteristics of the study population, which limit their utility for the study of differentials and determinants of mortality and ill-health in this population.

---

[1] The term Aboriginal is used in this chapter in a broader sense and includes both Aboriginal and Torres Strait Islander Australians.

The 2002 NATSISS obtained, for the first time since the 1986 Census, data on the number and survival status of children born to women in and past reproductive age groups. While the 1994 NATSIS, which preceded the current survey, had obtained information on the number of children ever born, no data were collected regarding the survival status of children. The existence of both the number and survival status of children ever born in the recent survey, therefore, provides a rare opportunity for analysing child mortality among Indigenous Australians. Specifically, this set of data enables the estimation of independent child mortality measures and provides an opportunity to isolate the risk factors associated with child mortality in the population. Such analyses will not only help to identify underprivileged groups who experience higher mortality levels in the population, they will also improve our understanding of the determinants of mortality and their inter-relationships. This will form a basis on which proper policy measures for reducing mortality may be developed, selected and improved.

The balance of the chapter is set out as follows. The next section describes the source data, assesses its quality, and identifies issues of concern for the intended analysis. This is then followed by an examination of recent levels and trends in Aboriginal child mortality and the factors associated with it. The last part summarises key findings of the study and recommends strategies for future demographic data collection on Aboriginal Australians.

Data quality issues

The starting point for any demographic analyses is a formal evaluation of the quality of the source data. One way of evaluating the quality of the child mortality data collected in the 2002 NATSISS is by computing the average number of children ever born and the proportion of children who have died by the age group of women. Unless there is significant and abrupt change in the scale and patterns of mortality and fertility schedules, both the average number of children ever born and the proportion of children who have died are expected to increase with the age of women. Fig. 6.1 and 6.2 provide the average number of children ever born and the proportion of children who have died by age group of women and jurisdiction obtained from the 2002 NATSISS, respectively.

Figure 6.1. Average number of children born per woman: Aboriginal Australians, 2002

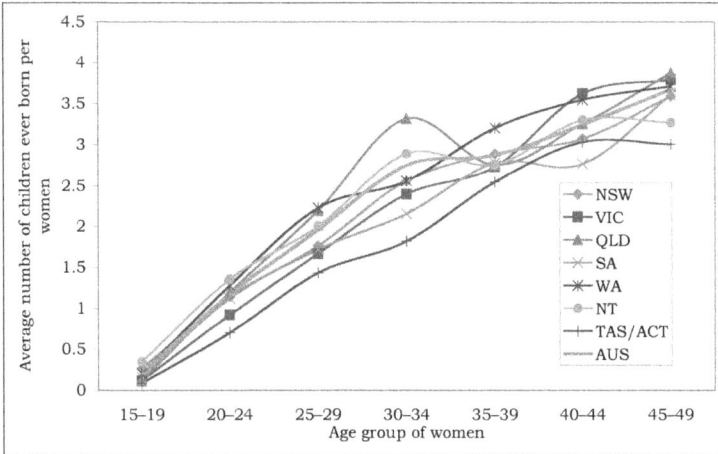

Source: Author's calculations from the 2002 NATSISS RADL.

Figure 6.2. Proportion of children who have died by age of women: Aboriginal Australians, 2002

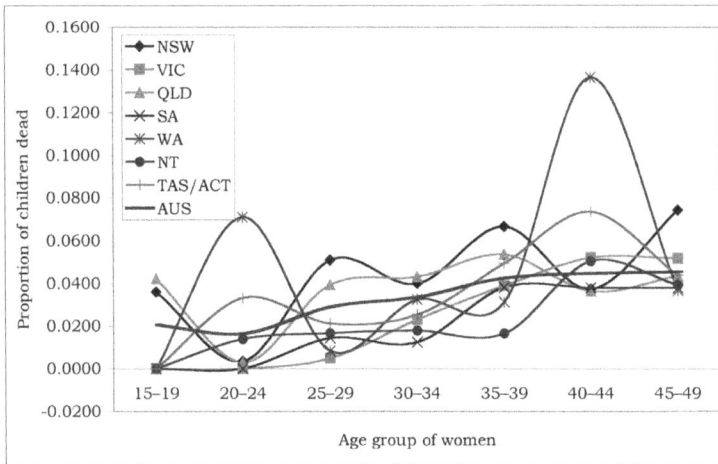

Source: Author's calculations from the 2002 NATSISS RADL.

As can be seen from Fig. 6.1, while for most States/Territories and age groups the average number of children increases with age, for some jurisdictions the reported number of children for some older women often falls short of that reported for younger women. For instance, for the age group 35–39 years in Queensland, 40–44 years in South Australia and 45–49 years in the Northern Territory, Tasmania and the ACT, the average number of children per woman reported in these age groups is lower than that reported for the immediate younger age groups in the respective jurisdictions. The data on the proportion of children who have died (reported in Fig. 6.2) reveals even more anomalies.

With the exception of the data for Victoria and for Australia as a whole, the proportion of children who have died, which is expected to increase linearly with age, depicts a more erratic pattern in all jurisdictions. While part of this unexpected picture could be attributed to sampling, the fact that the problem is also manifested in such jurisdictions as Queensland and New South Wales—where sample size is less of an issue—may point a more structural non-sampling-related quality problem in the data. At any rate, what these suggest is that some care needs to be exercised in using the data for mortality analysis beyond the national level. For this reason, the analyses in the subsequent section will focus on the national level.

Mortality estimates

Table 6.1 presents indirect estimates of early age mortality for the Aboriginal population as a whole. These are based on data on number of children ever born and children surviving collected from all women aged 25–39 years in the 2002 NATSISS survey (Preston & Palloni 1986; United Nations (UN) 1983). As mortality measures obtained through indirect procedures are sensitive to the assumptions on mortality patterns underlying the models—and, in some cases, to methods used in the estimation process—the mortality estimates were generated using three different estimation techniques—the Trussell method, the Preston–Palloni method, and the Palloni–Heligman method—and nine model mortality patterns. These patterns included four from the Coale–Demeny model life table system (north, east, south and west) and five from the UN pattern (Latin American, Chilean, South Asian, Far Eastern and general). For comparison purposes, also presented in the table are probabilities of dying extracted from existing life tables for Indigenous and non-Indigenous Australians.

Table 6.1. Estimates of early age mortality: Aboriginal Australians, comparison of existing estimates with 2002 NATSISS

	Probability of dying before reaching age x: (per thousand live births)		
	$_3q_0$	$_5q_0$	$_{10}q_0$
Panel A: Indirect estimates based on 2002 NATSISS			
Trussell method			
West	27.0	32.0	42.0
East	27.0	32.0	42.0
North	25.0	31.0	42.0
South	27.0	32.0	42.0
Preston–Palloni method			
West	28.0	32.0	44.0
East	28.0	33.0	44.0
North	26.0	31.0	44.0
South	28.0	33.0	44.0
Palloni–Heligman method			
General	28.0	33.0	44.0
Latin American	28.0	33.0	44.0
Chilean	29.0	34.0	43.0
South Asian	28.0	34.0	44.0
Far East	28.0	33.0	43.0
Average	27.5[a]	32.5[a]	43.2[a]
Panel B: Estimates based on registration data [1996–2001]			
ABS [Indigenous][b]	15.0	16.4	18.4
Kinfu & Taylor [Indigen-ous][c]	26.9	27.7	29.2
Kinfu & Taylor [Non-Indi-genous][c]	6.3	6.7	7.4

a. The reference period for the indirectly estimated $_3q_0$ values is approximately 1998, while the indirectly estimated $_5q_0$ and $_{10}q_0$ estimates approximately correspond to the period 1995 and 1992, respectively.
b. From the latest experimental Indigenous life table (ABS 2003a: 81–82).
c. From Kinfu & Taylor (2005). Note that the reference period for all three estimates from the registration data (i.e. $_3q_0$, $_5q_0$ and $_{10}q_0$) is around 1999.
Source: Author's calculations from the 2002 NATSISS RADL

The results in Table 6.1 show that the estimated probabilities of dying, generated using alternative assumptions and estimation procedures, fall within a narrow range, which provide confidence on these estimates. These estimates also display a good deal of consistency with the estimates from conventional life tables that are derived from registration data. The probabilities of dying obtained by averaging the results from the different methods and model life table families indicate that the probability of dying before reaching age three was 27.5 children per thousand around 1998, which is fairly comparable with the estimate based on registration data reported by Kinfu and Taylor for the period 1996–2001 (2002, 2005). The estimated under-five mortality rate stood at around 33 per thousand children in 1995, while the probability of dying before age 10 was around 43 per thousand children in 1992. Both of these estimates are higher than the respective estimates for 1999 obtained by Kinfu and Taylor from the registration data, providing some evidence of possible recent decline in child mortality in the population. However, despite this promising trend, there are still significant differences in survival between Indigenous and non-Indigenous children. As can be seen from the table, early age mortality among Indigenous children is three to four times higher than their non-Indigenous counterparts. To reduce—and eventually eliminate—this inequality, it is important to identify the most vulnerable group of the population or isolate the factors to which Aboriginal child mortality is highly responsive. This is the subject of the next section.

Covariates of child mortality

A regression model is used to analyse the determinants of excess child mortality among Aboriginal Australians. The independent variables examined in this study are:

- individual characteristics describing the educational and employment status of the mother
- the geographical context of residence (degree of urbanisation and location)
- household level characteristics reflecting the family's emotional and material wellbeing (home ownership, marital status of the mother, presence of stressor, smoking and alcohol consumption)
- the characteristics of the dwelling and the neighbourhood, reflecting the family's material and social living conditions (quality of sanitary facilities and existence of problems in the neighbourhood), and
- cultural factors (Indigenous composition of the household, difficulty in communicating with service providers, whether they recognise and live on their homeland).

The age of the mother is also introduced as a control variable to indirectly capture the effect of duration of exposure on mortality of children. In the regression, with the exception of maternal age (which is treated as continuous), for each of the remaining explanatory variables, one of the categories is taken as a reference category.

The model was fitted using the Poisson regression technique with the number of children who have died as the dependent variable and the number of live births to the mother as the exposure variable. This method is appropriate for the available data given the fact that the dependent variable has only non-negative integers (Cameron & Trivedi 2001; Gurmu & Trivedi 1996; Long 1997). In the course of the analysis, other variants of the count data model (such as the negative binomial regression model and the zero and hurdle variants of Poisson and negative binomial regression) were also tested, but as these did not produce a statistically superior outcome, the results are not reproduced here.

Table 6.2 presents the effects of socioeconomic and other background variables on child mortality estimated using the Poisson model. While interpreting these results, it is important to note that as death is a biological process, factors affecting child mortality in the most direct manner are bio-medical in nature; background variables such as those identified in the present study impact on child mortality only in an indirect manner. Another important limitation of the present analysis is that almost all the explanatory variables are contemporary variables relating to the time of the survey and these are used to 'predict' an event (the loss of a child) that happened in some distant past. In addition, some of the variables (such as drinking, smoking, labour force status, marital status and presence of stress) may indeed have been shaped by the negative shock to one's life that can occur through the death of a child. Ideally, these processes entail a dynamic analysis, but this is limited in the present study due to the nature of the data. However, from the point of view of future data collection, one way of addressing this problem in a cross-sectional survey such as the NATSISS is by collecting birth history data. This enables tracing the timing of the death of children which could then be linked with the characteristics of the mother at the time of the event.

The multi-variate regression presented in Table 6.2 shows that better environmental quality and home ownership have positive and statistically significant association with the risk of dying in childhood. The result shows that the odds of child mortality increase by almost 53 per cent for children in community housing and by some 72 per cent for children in private rentals. These differentials, which persist when the effects of other factors are controlled, underscore the importance of a family's economic standing in determining the probability of survival of its children. Moreover, as both size and quality of housing facilities are often correlated with household income/wealth level, those in privately-owned premises are likely to enjoy better and well maintained facilities, and hence be able to minimise or eliminate the chance of environmental exposure of their children to infectious agents. This is also confirmed by the result of the regression result which shows a statistically significant association between child mortality and adequacy of lavatory facilities. These results are consistent with the UN (1985) study which showed that, in general, old housing and deficient sanitary conditions constitute risk factors for child survival.

Table 6.2. Estimated effects of socioeconomic, spatial and household characteristics on Aboriginal child mortality: results of Poisson regression model, 2002 NATSISS

Determinants of child mortality	Regression coefficients[a]	Incidence rate ratios[a]	P values
Age (continuous)	0.320		0.000
Marital status			
Married	[R]	[R]	
Not married	0.301	1.352	0.088
Household composition			
All Indigenous	[R]	[R]	
Mixed household	-0.071	0.931	0.730
Difficulty with service providers			
Does not have difficulty	[R]	[R]	
Has difficulty	0.116	1.123	0.082
Attachment to homeland			
Identifies with homeland	[R]	[R]	
Does not know or identify with homeland	0.148	1.160	0.442
Residence on homeland			
Lives on homeland	[R]	[R]	
Does not know homeland or does not live on homeland	0.244	1.277	0.034
Place of residence			
Major cities	[R]	[R]	
Inner regional	0.330	1.390	0.182
Outer regional	0.195	1.215	0.357
Remote or very remote	0.091	1.095	0.694
Labour force status			
Employed	[R]	[R]	
Unemployed	0.441	1.555	0.162
Not in the labour force	-0.089	0.915	0.609
Educational status			
Diploma or higher	[R]	[R]	
Year 11 and 12	-0.203	0.816	0.543
Year 10	-0.313	0.731	0.231
9 years or below	-0.099	0.906	0.668
Tenure status			
Owner	[R]	[R]	
Private rental	0.543	1.721	0.041
Public or community rental	0.423	1.527	0.049
Toilet facility			
Adequate	[R]	[R]	
Inadequate	0.588	1.801	0.081
Neighbourhood problem			
Neighbourhood does not have problem	[R]	[R]	
Neighbourhood has problem	0.478	1.614	0.002
Household stress			
Stressor reported	[R]	[R]	
Stressor not reported	-0.465	0.628	0.050

Table 6.2. (continued)

Determinants of child mortality	Regression coefficients[a]	Incidence rate ratios[a]	P values
Smoking status			
Ever smoked or currently smoking	[R]	[R]	
Never smoked	-0.131	0.877	0.425
Alcohol consumption			
Never	[R]	[R]	
Low risk	0.128	1.136	0.461
Medium risk	0.139	1.149	0.565
High risk	0.534	1.706	0.032
Constant	-5.844		0.000
Number of observations		3 798	

a. The model includes children ever-born as an exposure variable. R refers to reference category.
Source: Author's calculations from the 2002 NATSISS MURF conducted at the ABS

Neighbourhood problems, existence of stress and high-risk drinking problems among parents increase the likelihood of child mortality. The odds of child mortality are 61 percent higher for children born and raised in a neighbourhood that has problems, while a high-risk drinking problem and the existence of stress within a household each increase the odds of child mortality by some 70 per cent. Similarly, children of never-married mothers and mothers who have difficulty in dealing with service providers show respectively a 35 and 12 per cent elevated risk of mortality compared to their counterparts who live with both parents or live with a mother who has no difficulty in accessing services. A strong and statistically significant association is also evident between maternal age, child mortality and degree of attachment to and residence on their homeland.

However, maternal education, place of residence and labour force participation, which are known to have a strong association with mortality in the literature do not show significant association with child mortality in the present analysis. Other factors that were also found to be not significant were composition of household and smoking behaviour of the mother. These findings may suggest that the relationships between residential pattern and socio-occupational status and child mortality are not direct but operate through other variables.

Concluding remarks

Using the 2002 NATSISS, this study has attempted to generate an alternative estimate of child mortality and identify its determinants among Indigenous Australians. The results of the analysis showed that while there is some evidence of a decline in child mortality in recent years, mortality among young Indigenous Australians still remains three to four times higher than their non-Indigenous counterparts. It is observed that home ownership, quality of dwelling, better neighbourhood environment and absence of stress in the household have a positive influence on child health. On the other hand, other maternal factors such as high-risk drinking behaviour and lone parenthood increase the probability of child mortality, while degree of urbanisation has no effect. However, part of the explanation for the latter may lie in data quality which, as demonstrated in the paper, appears to be highly volatile and deviates sufficiently from known patterns to cast doubt on its quality, particularly at the level of jurisdiction. There is, therefore, a need to explore the causes of these problems and find ways to avoid the shortcomings in future data collections. As the 2002 NATSISS data also only permit analysis of child mortality, it is important to explore the feasibility of collecting information on adult mortality so that the results from these data could be used to counter-check existing estimates that are based on conventional methods. The NATSISS data also provide no information on current fertility, which is a useful input for population projection. This information can be obtained by asking all women in the reproductive age group one simple question on whether or not they had a live birth in the 12 months preceding the survey.

7. Understanding housing outcomes for Indigenous Australians: what can the 2002 NATSISS add?

Will Sanders

Almost a decade ago, when looking at the 1994 NATSIS, I was able to be very complimentary about what that survey could add to our understanding of housing outcomes for Indigenous Australians. That was largely because, in the housing area, the 1994 NATSIS had improved on previous census collections in two important ways: it had developed a better tenure breakdown, distinguishing between community and private rental dwellings, and it had developed an accessible and useful geographic breakdown into capital cities, other urban, and rural or remote areas. Using the 1994 NATSIS, I was able to show just how different the community and private rental tenures were for Indigenous households and, perhaps more importantly, just how different tenure profiles for these households were in rural and remote areas compared to the cities and other urban areas (Sanders 1996).

Today, it is a little more difficult to be glowingly complimentary about what the 2002 NATSISS can add to our understanding of housing outcomes for Indigenous Australians. This is not because the Indigenous survey processes have, in any way, gone backwards, but because other data collection processes, most notably the five-yearly censuses, have moved a long way forwards. Recently, I was able to undertake an analysis of housing outcomes for Indigenous households in five geographic categories of Australia ranging from major cities to very remote areas based entirely on publicly available data from the 2001 Census (Sanders 2005). I could show, even more clearly than I had been able to with the 1994 NATSISS, how different were the tenure profiles of remote areas compared to the cities and regional areas, and again how different were community and private rental. This makes it harder to argue that the 2002 NATSISS now adds so significantly to our understanding of Indigenous housing outcomes. However, it does still add some degree of understanding, and it can be used to confirm and reinforce census-based analysis.

In this paper I will reflect on the housing data from the 2002 NATSISS, drawing figures primarily from the RADL, but occasionally also quoting figures directly from ABS (2004c). The RADL contains two files, one giving characteristics of an estimated 165 674 dwellings containing Indigenous households, and the other giving characteristics of an estimated 282 205 Indigenous adults. When analysing housing characteristics, the predominant practice is to report figures relating to dwellings containing Indigenous households. However, it is also possible, and can be informative, to report by 'persons'. In what follows, I often give both sets of figures, noting in the process that the figures by 'persons' tend to accentuate the differences between tenures and geographic areas. In the first section of the paper, I look at housing tenure profiles and household size by remoteness. In the second section of the paper, I look at a range of affordability and adequacy issues analysed primarily by tenure, but also by remoteness. In the third section of the paper, I look briefly at comparisons with non-Indigenous Australians and changes over time. I conclude by reiterating that almost all of this analysis can also nowadays be done from the five-yearly census data.

Many of the tables that follow divide Indigenous households or people into remote and non-remote areas. This two-fold geography is a sub-set of the ABS's five-fold remoteness classification ranging from cities to very remote areas. The two-fold categorisation combines remote and very remote areas, on the one hand, and cities and inner and outer regional areas on the other. While not as refined as the five-fold categorisation available in the census, this two-fold categorisation is still robust enough to show major geographic differences in housing characteristics among Indigenous households and people.

Housing tenure profiles and household size by remoteness

Table 7.1 presents tenure profiles of dwellings containing Indigenous households in remote and non-remote areas in 2002. In the non-remote areas, three tenures—buying, government and private rental—each account for approximately a quarter of households, with ownership and other less significant tenures making up the last quarter of households. In the remote areas, by contrast, buying, private rental and ownership all fall away to single figure percentages, while government rental also falls away, but not quite so dramatically, to 17.1 per cent. What rises in remote areas in their place is one of the very minor tenures in non-remote areas—community rental. This tenure accounts for 49.7 per cent of dwellings containing Indigenous households in remote areas, but only 6.8 per cent in non-remote areas. Community rental is hence the dominant tenure in remote areas, while being fairly insignificant in non-remote areas. This contrast suggests just how different the housing tenure system is for Indigenous people in remote areas compared to elsewhere.

Table 7.1. Tenure of dwellings containing Indigenous households by remoteness, 2002

	Remote	Non-remote	Total
Number 000s	*29.2*	*136.5*	*165.7*
Owned %	5.9	11.8	10.8
Buying %	7.7	23.2	20.5
Government rental %	17.1	24.0	22.7
Community rental %	49.7	6.8	14.4
Private rental %	4.9	26.3	22.5
Employer and other rental %	7.7	5.9	6.2
Other %	7.0	2.0	2.9

Source: Customised cross-tabulations from the 2002 NATSISS RADL

Some explanation of community rental may be of assistance at this point for those not familiar with Indigenous housing policy. Community rental is the product of a thirty-year government effort to improve Indigenous housing conditions at government capital expense by providing between 500 and 1000 dwellings per year through grant funding to Indigenous community organisations. While these dwellings have been government-funded at the point of construction or capital expenditure, the ongoing ownership and management of them has been vested in Indigenous community organisations. These organisations have then been encouraged to charge income-related rents to cover recurrent housing costs, such as repairs and maintenance, and this has led to the label 'community rental'. The closest cousin of community rental housing within the housing tenure system is government rental, which is also provided at government capital expense, and which also charges income-related rents in the hope of covering recurrent housing costs. However, government rental in most instances is restricted to cities and smaller urban areas. It does not often extend into remote, non-urban places, where significant numbers of Indigenous people live. In these areas, community rental generally takes over the role from government rental of meeting the housing needs of those who do not seem able to do so through market processes. Sometimes government rental and community rental are together referred to as social housing.

Table 7.2, which reports housing tenure by Indigenous people aged 15 years or over (hereafter referred to as adults), accentuates the tenure difference between remote and non-remote areas shown in Table 7.1. While the figures for non-remote areas differ by only one or two per cent for any tenure category from Table 7.1, for community rental in remote areas the figure increases by almost 14 per cent so that 63.6 per cent of Indigenous adults in these areas live in community rental dwellings. The dominance of community rental among Indigenous people in remote areas thus becomes all the more stark when reported by numbers of people, rather than households. The key to this shift is different household sizes in remote and non-remote areas.

Table 7.2. Housing tenure of Indigenous persons aged 15 years or over by remoteness, 2002

	Remote	Non-remote	Total
Number 000s	*77.1*	*205.1*	*282.2*
Owned %	3.8	12.0	9.8
Buying %	4.9	22.4	17.6
Government rental %	13.2	25.4	22.1
Community rental %	63.6	8.5	23.5
Private rental %	3.4	24.3	18.6
Employer and other rental %	5.4	5.6	5.6
Other %	5.8	1.9	3.0

Source: Customised cross-tabulations from the 2002 NATSISS RADL

Table 7.3 gives percentage distributions of household size by dwellings containing Indigenous households in remote and non-remote areas. It shows that whereas 21.2 per cent of these dwellings in remote areas have seven or more members of the household, only 5.4 per cent do in non-remote areas. Table 7.4 accentuates this difference by showing that 39.3 per cent of Indigenous adults in remote areas live in households of seven or more people, compared to 8.7 per cent in non-remote areas. Conversely, the proportions of Indigenous households containing four or fewer people, and of Indigenous people living in such households, are much higher in non-remote areas. This means tenure differences between remote and non-remote areas are accentuated when reported by Indigenous people, rather than households, due to larger Indigenous households in remote areas.

Table 7.3. Household size of dwelling containing Indigenous households by remoteness, 2002

	Remote	Non-remote	Total
Number 000s	*29.2*	*136.5*	*165.7*
1–2 persons %	29.2	39.6	37.8
3–4 persons %	29.2	38.6	37.0
5–6 persons %	20.3	16.3	17.0
7 or more persons %	21.2	5.4	8.2

Source: Customised cross-tabulations from the 2002 NATSISS RADL

Table 7.4. Household size of Indigenous persons aged 15 years or over by remoteness, 2002

	Remote	Non-remote	Total
Number 000s	*77.1*	*205.1*	*282.2*
1–2 persons %	14.9	30.8	26.5
3–4 persons %	24.6	40.7	36.3
5–6 persons %	21.2	19.8	20.2
7 or more persons %	39.3	8.7	17.1

Source: Customised cross-tabulations from the 2002 NATSISS RADL

Table 7.5 presents one final set of figures relating to housing tenure profiles and remoteness. These figures are for Torres Strait Islanders living in Torres Strait, the rest of Queensland and Australia (excluding Torres Strait). The figures in this table are drawn from ABS (2004c) rather than the RADL, and are based on the number of people responding to the 2002 NATSISS (not the number of households). To ensure that the estimates are reasonably reliable, the table is restricted to just two tenure categories: owned (including buying) and rental. Islanders in the Strait show the usual heavy reliance on rental of Indigenous people in remote areas and the concomitant low levels of home buying or ownership, at just 6.2 per cent, though ABS considers this statistically unreliable.[1] Islanders in the rest of Queensland show higher levels of buying and ownership and those in the rest of Australia, even higher still. We should note in passing here that were this third geographic category to be defined exclusively, as Australia excluding Queensland, (rather than overlapping with the second category and hence including Islanders in the rest of Queensland again), then clearly the level of home buying and ownership among this group of Islanders would be significantly higher again.

Table 7.5. Housing tenure of Torres Strait Islander persons aged 15 years and over by area, 2002

	Torres Strait	Rest of Queensland	Australia (excluding Torres Strait)
Number 000s	*3.6*	*13.0*	*26.2*
Owned %	6.2**	20.9*	34.9
Rental %	84.6*	79.1	63.8

Source: ABS (2004c: Table 23)

Affordability and adequacy issues by tenure and remoteness

Back in 1996, using the 1994 NATSISS, the second set of issues I was able to explore was the affordability and adequacy of different housing tenures including, at times, owning and buying, as well as community, government and private rental. This exploration bore out the usefulness—indeed, the indispensability—of the distinction between community and private rental, as these are not only very different tenures in terms of their geographic distribution but also in terms of their affordability and adequacy characteristics.

Table 7.6 confirms, yet again, the vast differences in rents between community and private rental, with mean weekly rentals of $66 and $152 respectively. Government rental, with a mean rent of $87 per week, is a little more expensive than community rental but considerably less expensive than private rental.

---

[1] One asterisk indicates that the estimate should be used with caution because it has a relative standard error of between 25 and 50%. Two asterisks indicate that the estimate has relative standard error of greater than 50% and the ABS generally considers it unreliable. Clearly estimating even a two-fold distinction for a population group of only 3600 is pushing the NATSISS to its methodological limits.

Table 7.6. Weekly rents (in 2002 $) of dwellings containing Indigenous households by tenure, 2002

	Community rental	Government rental	Private rental
Mean	66	87	152
Median	60	80	145

Source: Customised cross-tabulations from the 2002 NATSISS RADL

Table 7.7 suggests there is, however, a downside to the inexpensiveness of community rental in terms of dwelling adequacy. It presents data from the 2002 NATSISS showing community rental dwellings as being reported most often (at 54.7%) as having structural problems and least often (at 57.6%) as having repairs and maintenance carried out in the last twelve months. Government and private rental were reported to be substantially better on both these adequacy measures and dwellings that were owned or being bought were better still.

A third measure of housing adequacy has been constructed by the ABS by comparing numbers of people and bedrooms in a dwelling against the Canadian National Occupancy Standard (CNOS).[2] On this measure too, community rental comes out as the least adequate tenure, with 34.4 per cent of its dwellings requiring additional bedrooms, compared to 13.8 per cent in government rental, 11.6 per cent in private rental and single figure percentages in dwellings that are owned or being bought. To a large extent, these figures for requiring additional bedrooms are a reflection of household size and, in the fourth line of Table 7.7, I give the mean household size for dwellings containing Indigenous households by tenure. Community rental, at 4.6, has a distinctly larger mean household size than the other tenures, none of which have means above 3.6 people per household.

The fifth line of Table 7.7 provides a housing affordability measure, which is perhaps a little more sophisticated than just rent paid. This is the proportion of households paying more than 25 per cent of their income in rent or mortgage. On this measure, community rental (16.0%) is again a bit more affordable than government rental (22.2%) and a lot more affordable than private rental (50.2%). Also, buying is actually more affordable than all three rental tenures, although buyers may of course have non-mortgage costs, like rates, that renters do not. Furthermore, affordability for buyers can depend greatly on where they are in the repayment cycle, so comparisons between buyers and rental tenures are not straightforward. The most important figure to emerge from this fifth line of Table 7.7 is, I would argue, that 50.2 per cent of Indigenous households in private rental pay more than 25 per cent of their household income in rent. Whereas community rental appears to be an inexpensive tenure with some adequacy issues for Indigenous households, private rental appears to be a tenure category with some quite significant affordability issues.

[2] The CNOS is explained at explanatory notes 49 and 50 in ABS (2004c).

From the sixth line, Table 7.7 presents more fine-grained housing adequacy measures from the 2002 NATSISS relating to household facilities; such as whether a dwelling has a working stove, washing machine or toilet, and adequate kitchen bench and cupboard space. Although the proportions of dwellings containing Indigenous households which have these facilities are generally quite high, there are instances where they drop away. In all instances, bar one, it is community rental dwellings that most frequently do not have these facilities. The three instances where figures drop below 90 per cent in this part of Table 7.7 are all in community rental dwellings; having a working fridge (89.9%), a working washing machine (81.1%), and having adequate kitchen cupboard and bench space (76.5%). The last line of Table 7.7 shows major differences between tenures in having a working telephone in the dwelling, with less than half of community rental dwellings doing so.

Table 7.7. Adequacy and affordability characteristics of dwellings containing Indigenous households by tenure, 2002

	Community rental	Government rental	Private rental	Buying	Owned
Number 000s	*23.8*	*37.7*	*37.3*	*33.9*	*17.8*
Has structural problems %	54.7	41.8	33.5	22.5	22.3
Repairs and maintenance carried out %	57.6	64.5	61.6	72.0	63.1
Requires additional bedrooms %	34.4	13.8	11.6	5.6	6.1
Mean household size	4.6	3.4	3.1	3.6	3.0
Pays > 1/4 of household income in rent/mortgage %	16.0	22.2	50.2	14.1	–
Has working stove/oven %	91.1	99.1	98.3	99.8	97.3
Has kitchen sink %	97.7	100.0	99.1	99.6	98.7
Has adequate kitchen cupboard/ bench space %	76.5	81.2	89.1	91.9	94.9
Has working refrigerator %	89.9	96.8	97.6	99.9	99.9
Has working washing machine %	81.1	90.4	92.6	98.1	98.3
Has working bath/shower %	97.9	99.2	99.5	100.0	99.5
Has laundry tub %	96.7	96.9	94.7	98.1	97.5
Has working toilet %	97.1	99.2	99.4	99.8	99.8
Has working telephone %	47.1	72.7	80.9	95.6	94.3

Source: Customised cross-tabulations from the 2002 NATSISS RADL

Table 7.8 reports the top five of these adequacy and affordability measures by Indigenous adults living in different tenures. This reporting tends to accentuate the differences between community rental and the other four tenures compared to reporting by dwelling. For example, the percentage for dwellings requiring additional bedrooms in community rental increases from 34.4 per cent when reported by dwelling to 51.0 per cent when reported by 'persons', while the comparable shift for private rental is from 11.6 to 18.2 per cent. These shifts need not concern us greatly here, as reporting housing adequacy and affordability characteristics as if they attach to people is sometimes a bit strained. Also, in the 2002 NATSISS, this information was generally obtained from a single household spokesperson, so is probably best reported on a household basis.

Table 7.8. Housing adequacy and affordability characteristics of Indigenous persons aged 15 years or over by tenure, 2002

	Community rental	Government rental	Private rental	Buying	Owned
Number 000s	*66.4*	*62.2*	*52.4*	*49.6*	*27.5*
Has structural problems %	62.2	42.6	34.5	21.4	23.7
Repairs and maintenance carried out %	56.4	66.2	60.9	72.9	64.7
Requires additional bedrooms %	51.0	20.5	18.2	9.8	10.3
Mean household size	5.9	4.0	3.5	3.9	3.4
Pays > 1/4 of household income in rent/mortgage%	8.6	17.2	44.9	11.4	-

Source: Customised cross-tabulations from the 2002 NATSISS RADL

Of more substantial interest is that these housing adequacy and affordability measures from the 2002 NATSISS can also be reported by remote and non-remote areas, as in Table 7.9. Given the very different geographic distribution of the tenures, it is not surprising that this division also shows substantial adequacy and affordability differences. Table 7.9 shows that 49.9 per cent of dwellings containing Indigenous households in remote areas have structural problems compared to 31.9 per cent in non-remote areas. And 32.9 per cent in remote areas require additional bedrooms compared to 9.7 per cent in non-remote areas. Conversely, only 10.4 per cent of Indigenous households in remote areas pay more than 25 per cent of their household income in rent or mortgage payments compared to 26.4 per cent in non-remote areas. Further down Table 7.9 we can also see, for example, that 89.6 per cent of dwellings containing Indigenous households in remote areas have a working refrigerator, compared to 98.4 per cent in non-remote areas, while 50.3 per cent in remote areas have a working telephone compared to 83.4 per cent in non-remote areas. Table 7.10, like Table 7.8 for tenures, reports the top five of these adequacy and affordability measures by Indigenous persons, rather than dwellings.

Table 7.9. Adequacy and affordability characteristics of dwellings containing Indigenous households by remoteness, 2002

	Remote	Non-remote
Number 000s	*29.2*	*136.5*
Has structural problems %	49.9	31.9
Repairs and maintenance carried out %	51.8	65.9
Requires additional bedrooms %	32.9	9.7
Mean household size	4.4	3.3
Pays > 1/4 of household income in rent/mortgage %	10.4	26.4
Has working stove/oven %	89.0	99.1
Has kitchen sink %	96.5	99.4
Has adequate kitchen cupboard/bench space %	75.9	88.4
Has working refrigerator %	89.6	98.4
Has working washing machine %	80.9	94.3
Has working bath/shower %	96.1	99.6
Has laundry tub %	93.5	96.9
Has working toilet %	94.9	99.7
Has working telephone %	50.3	83.4

Source: Customised cross-tabulations from the 2002 NATSISS RADL

Table 7.10. Housing adequacy and affordability characteristics of Indigenous persons aged 15 years or over by remoteness, 2002

	Remote	Non-remote
Number 000s	*77.1*	*205.1*
Has structural problems %	58.4	32.5
Repairs and maintenance carried out %	51.9	66.7
Requires additional bedrooms %	50.3	14.9
Mean household size	5.8	3.7
Pays > 1/4 of household income in rent/mortgage%	7.1	19.9

Source: Customised cross-tabulations from the 2002 NATSISS RADL

Tables 7.7 to 7.10 all show that housing adequacy issues are much more prevalent in remote areas and community rental, while affordability issues are much more prevalent in non-remote areas and private rental. This contrast is now a well established theme in discussions of Indigenous housing need based on census analysis (Neutze, Sanders & Jones 2000). It should also be noted, in this regard, that the measures in lines three to five of tables 7.7 to 7.10 can be—and have, in recent times, increasingly been—constructed from census data (Jones 1994, 1999; NCSAGIS 2003). If these measures are seen as sufficient for understanding the basic adequacy and affordability characteristics of Indigenous housing, by tenure and remoteness, then the census tells us essentially what we need to know.

Comparisons with non-Indigenous Australians and over time

ABS (2004c) draws on the 2002 GSS to construct comparisons between the tenure profiles of Indigenous and non-Indigenous people aged 18 years and over. The findings are reproduced in Table 7.11, although I have left out the figures combining remote and non-remote Indigenous people on purpose. These seem to me to combine two such disparate groups (in terms of their housing tenure characteristics) that they actually obscure, rather than clarify, the situation. Perhaps more importantly, the non-indigenous estimates only refer to non-remote areas—that is, where the GSS was conducted.

Table 7.11 usefully reminds us that, whatever the differences between the housing characteristics of Indigenous people in remote and non-remote areas, the differences between Indigenous people in non-remote areas and non-Indigenous people are just as great. Only 12.4 per cent of Indigenous people aged 18 years or over in non-remote areas live in houses that are owned by an occupant, compared to 38.5 per cent of non-Indigenous people. And the percentages for living in houses that are being bought are 21.0 and 34.6 respectively. Conversely, Indigenous people in non-remote areas rely far more heavily than do non-Indigenous people on both government rental (24.4 per cent compared to 3.8 per cent) and other—mainly private—rental (29.8 per cent compared to 19.9 per cent). Meanwhile, the housing tenure characteristics of Indigenous people in remote areas are so different from those of non-Indigenous people as to not really bare comparison at all.

Table 7.11. Housing tenure characteristics of Indigenous and non-Indigenous persons aged 18 years or over, 2002

	Indigenous remote	Indigenous non-remote	Non-Indigenous non-remote
Number 000s	*69*	*182*	*14 354*
Owned %	4.0	12.4	38.5
Buying %	4.6	21.0	34.6
Government rental %	12.6	24.4	3.8
Community rental %	64.3	9.3	0.6
Other renter %	8.4	29.8	19.9

Source: ABS (2004c: Table 4)

Table 7.12 is again taken from ABS (2004c) and gives changes in the housing tenure characteristics of Indigenous adults from the 1994 NATSISS to the 2002 NATSISS. The biggest change is an 11.2 per cent drop in Indigenous people in government rental, from 33.3 to 22.1 per cent. This seems a surprisingly large drop, which needs further investigation. It probably reflects a quickly growing Aboriginal population at a time when the public housing stock is only growing very slowly or, relative to total housing stock, is perhaps in decline. However, it may also reflect some specific developments in Indigenous housing policy in particular States and Territories, where under recent inter-governmental agreements it is possible that some Indigenous-specific public rental housing is being reclassified as community rental, or at least new investment in Indigenous-specific housing is moving in this direction. This seems to me something that will be worthwhile looking into when the 2006 Census becomes available and we have a much larger set of data spanning 10 years and three time points which distinguishes between government, community and private rental. However, it can be noted in passing, in a preliminary fashion, that when measured by dwelling numbers, the proportion of Indigenous households in public housing does appear to have been dropping since 1986 (Sanders 1996: 108).

The other major changes over time evident in Table 7.12 are a 6 per cent increase in people in dwellings that are being bought, a 5.6 per cent increase in people in community rental dwellings and 4.6 per cent increase in people in other, mainly private, rental dwellings. I think all these changes in tenure percentages should be treated with some caution, not least because they could partly be explained by a significant drop in the proportion of people for whom no housing tenure information was obtained from the 1994 NATSISS (7.6%) compared to the 2002 NATSISS (3.8%). These 'not stated' figures are not given in ABS (2004c), but are calculated in the last line of Table 7.12 by simply subtracting from 100 the percentage sum of the tenures given. Again, I would argue that the 2006 Census should give us some greater ability to look more reliably at tenure changes among Indigenous people and households, at least back to 1996.

Table 7.12. Housing tenure characteristics of Indigenous persons aged 15 years or over, 1994 and 2002

	1994	2002
Number 000s	*214.6*	*282.2*
Owned %	10.9	9.7
Buying % [a]	10.8	16.8
Government rental % [a]	33.3	22.1
Community rental % [a]	18.7	24.3
Other rental % [a]	18.7	23.3
Housing tenure not stated? % [a]	7.6	3.8

a. The difference between 1994 and 2002 is statistically significant at the 5% level.
Source: ABS (2004c: Table 6)

Conclusion

One recurring theme of this paper is that most housing analysis that can be done from the Indigenous-specific surveys can also nowadays be done from the five-yearly national census. Indeed, the census often gives a better tenure and geographic breakdown than surveys can. What the Indigenous-specific surveys have added, in comparison to the census, is some greater detail relating to housing adequacy measures. However, basic adequacy measures can also be derived from the census. So the answer to the question posed in the subtitle, 'What can the

2002 NATSISS add?' is, in fact, 'not all that much'. The 2002 NATSISS confirms census analysis rather than taking it much further. Indeed, in the last section of the paper above, I have suggested that the findings from the 1994 and 2002 Indigenous-specific surveys relating to changes in housing tenure over time among Indigenous people should be treated with some caution and that it may be best to wait until the 2006 Census to work on this topic. Censuses are clearly the pre-eminent data sources for studying Indigenous housing outcomes, but the 2002 NATSISS can offer some insights on adequacy issues.

8. Revisiting the poverty war: income status and financial stress among Indigenous Australians

Boyd Hunter

Australia is at war! First there were the history wars, as Henry Reynolds, Keith Windshuttle and others fought over the technical detail and interpretation of Australia's colonial history. Then came the war on terror, which followed the events of 11 September 2001. One of the latest 'wars' is the poverty war. Note that this is not the 'war on poverty' that LBJ talked about in the 1960s, but rather a battle for the hearts and minds of the Australian public (and media). Professor Peter Saunders has documented the Poverty wars that started with a coordinated series of skirmishes by the Centre for Independent Studies (CIS) against a report written for the Smith Family (Saunders 2005: 6–8).[1] The debate started with some rather technical details of measurement, but quickly became bound up in questions of cause and response revealing stark differences in philosophy about choice, freedom, responsibility and the role of government. As always, the first casualty in war was the truth—or rather, public debate. The ferocity of the public debate caused the Smith family, one of the major non-profit welfare organisations in Australia, to stop using the word poverty and led them to disengage from poverty research. Saunders recommends that poverty research gets less technical and grounds itself in the lived experience of poverty and the social exclusion that perpetrates poverty in the long run (Saunders 2005). While I fully endorse this sentiment, it would largely entail qualitative research that has not yet been done, and is not possible to do using the 2002 NATSISS. However, the latest survey does provide direct information on financial stress and social capital for the first time. This provides an unambiguous advance in our knowledge of the processes underlying Indigenous disadvantage.

Income questions in the 2002 NATSISS tend to be asked in a reasonably similar way to other ABS surveys. The advent of financial stress questions in NATSISS and the GSS provide a broad indicator of how Indigenous and other Australian households are coping with their respective income statuses. Respondents were asked whether they could raise $2000 within a week for something important. Note that the main social capital variables are dealt with in Ruth Weston and

---

[1] That is, the Peter Saunders who happens to be an ARC Professorial Fellow, and had been a Director of the Social Policy Research Centre for many years. He is not to be confused with the Peter Saunders from the CIS.

Matthew Gray (in this volume), with this chapter focusing largely on income characteristics and poverty issues.

The 1994 NATSISS included an impressive and diverse range of income data by source, as it asked respondents to indicate a separate amount for income from wages and salaries (for main job and second job), business income, government payments (that is, from a list of government pensions and a separate amount provided for family payments and rent assistance). [2] Unfortunately, this rich source of information was not used by many researchers. One reason is that the publicly available data was coded into broad income ranges, so did not provide much distributional information. Another reason was that it was not entirely clear how robust the income data was. For example, was it really possible for individual respondents to accurately identify the separate amounts received from those sources when there were considerable flows between labour force states, and reasonably large flows between the various elements of the welfare system?

The official output for income for the 2002 NATSISS has been more modest than was attempted in the 1994 NATSISS (ABS 2004c). In contrast to the earlier survey, there has been no attempt to break down the amount of income from various sources in publications or the CURF. [3] This is not necessarily a bad thing, as the analysts will not be tempted to overstate the amount of information contained in the income data. Indeed, the 2002 survey data has the major advantage that it is now provided in continuous form, and hence can be used for a more robust distributional analysis and a more informative analysis of poverty. [4]

The structure of this chapter is as follows. The next section elaborates on how the poverty war debate appears to have influenced the way the 2002 NATSISS data was presented in the official ABS publication (2004c) and reflects on the utility of their classification of 'low-income groups', which was used as a synonym for poverty in that publication. The discussion in this section will also reflect on the limitations of how the survey was conducted, with specific reference to the problems with how the questions on income and financial stress were asked. The third and fourth sections then examine new insights provided by NATSISS data with respect to income and financial stress. The concluding section reflects on future directions for research to consolidate the analysis in this chapter.

---

[2] While the income questions were asked slightly differently in CAs and NCAs, these minor differences are unlikely to matter much. Cash flow problems, and main source of income were not asked and not outputted for CAs. Note that information was collected and collated on the amount of income from separate sources in both CAs and non-CAs, but this data was not included in the CURF.
[3] Note that there is still information on whether a person received income from various sources but there is no attempt to disaggregate the data.
[4] Biddle & Hunter (in this volume) show that censoring of income and housing costs data in the 2002 NATSISS is not an important issue, so it represents the best opportunity to date to get informative insights into Indigenous poverty.

Poverty wars and the ABS low income category

The ABS is not a direct combatant in the poverty war. They are more like Switzerland, surrounded by belligerents. They are not doing the fighting but their policy appears to be affected by the war. Biddle and Hunter (in this volume) criticise the ABS's measure of 'low income' as misleading, and this section briefly re-visits that discussion and places it in the context of the poverty war.

Based on analysis of the non-Indigenous population, the ABS (2004c) outputs data for the second and third decile as a measure of 'low-income', arguing that the lowest income decile has characteristic closer to those with higher incomes. ABS (2004c) uses this 'low income' group as a synonym for poverty, which appears to be an implicit endorsement of CIS criticism of income-based measures in the poverty wars, especially the claim that measurement error (or, rather, under-reporting) is pronounced for low-income earners—particularly those who indicate they have an income less than or equal to zero. One reason to be cautious about adopting the ABS definition of 'low income' without question is that the self-employed—a group that are often associated with measurement error in their income status—are not prominent in the Indigenous population.

Even if it were true that income is not measured properly for the non-Indigenous population with very low income, Biddle and Hunter (in this volume) show that this assumption is suspect for the Indigenous population. Compared to those in the second and third decile, those in the first decile are significantly less likely to be employed and own or purchase a home, and significantly more likely to have fair/poor health. Other variables not reported here indicate that the bottom decile respondents are also more disadvantaged, as they are more likely to:

- have not completed Year 12
- be unemployed or outside the labour force
- have been arrested in the last year, and
- have transport difficulties.

The ABS definition of low income tends to understate the incidence of Indigenous disadvantage, and it should not be used.

In contrast to the 1994 NATSISS, there were no specific questions for self-employed people in the 2002 NATSISS. However, data was collected on whether there was income from 'profit or loss from own unincorporated business or partnership, rental property, dividends or interest'. Consequently, if one was particularly concerned about the reliability of income for the self-employed, one could eliminate people with some income from a business or partnership. Eliminating people with income from rental properties, dividends or interests might also be considered by analysts. While this strategy directly addresses the major criticism raised by the CIS in the poverty wars debate, it is not possible to implement using the CURF, as such data is not reported in remote areas.

Notwithstanding this, the lack of any solid evidence that income is being measured incorrectly in the low income category means that one should not be overly concerned about such bias in the Indigenous data. However, any comparisons with non-Indigenous populations in the GSS should focus on non-remote areas in the NATSISS and eliminate potentially problematic respondents such as the self-employed to test the sensitivity of the analysis and to maximise the validity of Indigenous/non-Indigenous comparisons.

Revisiting Indigenous income status and poverty

Altman & Hunter (1998) discuss how poverty among Indigenous people conceptually depends upon household composition, non-monetary income (produce from hunting and gathering activities), fundamental cross-cultural issues, and differences in consumer price indexes (CPIs) for various areas. One obvious issue for the financial stress variable is that $2000 will have a different value for Indigenous groups in various parts of the country. Such issues are likely to be particularly pronounced for remote areas where hunting and gathering activities are prominent and the cost of living is likely to be high. [5] Consequently, some caution needs to be exercised when comparing Indigenous financial stress in various parts of the country.

The National Academy of Sciences Panel on Poverty and Family Assistance in the United States of America (USA) undertook a major study measuring poverty in the 1990s. The main report concluded that, '[al]though the empirical evidence helps determine the limits of what makes sense, there is no objective procedure for measuring the different needs for different family types' (Citro & Michael 1995).

While there is no objective basis for identifying the precise needs of families, it is still necessary to attempt to control for the different costs of households of differing size and composition. This is usually done by adjusting income using an 'equivalence scale'; for example, it is more expensive for two adults and three children to live in a household than a couple without dependants. Equivalence scales try to estimate how much more expensive it is for various households to live at a given standard of living, and consequently they vary with the number of people in the households and the mix of adults and dependants of various ages. The ABS calculated the 'equivalent income' by dividing the raw income by the Organization for Economic Cooperation and Development (OECD) scales, ranked this equivalised income for the whole population, and grouped households into quintiles (that is, groups of 20 percentiles of respondents, see ABS 2004c). However, the OECD scale is just one of many equivalence scales and I will discuss the range of feasible scales that could be used in the penultimate section.

---

[5] There is no regular, reliable data for CPIs outside the capital cities.

Income status and financial stress

It will not surprise anyone that Indigenous people are twice as likely as other Australians to be in the lowest income quintile, and almost four times less likely to be in the highest quintile (see Table 8.1). Indigenous people are more likely than other Australians to rely on income from government pensions and allowances and be unable to raise $2000 within a week for something important. This is consistent with Indigenous employment disadvantage and the fact that the mean income of Indigenous households is around 60 per cent of that of other households (for discussion of the former, see Chapman & Gray in this volume).

Table 8.1. Income summary by Indigenous status in non-remote areas, 2002[a][b]

	Indigenous	Non-Indigenous
	%	%
Equivalised household income		
Lowest quintile	41.7	19.3
Second quintile	28.3	18.6
Third quintile	14.4	19.0
Fourth quintile	9.2	19.9
Highest quintile	6.4	23.1
Main source of income CDEP	10.9	N/A
Main source other wages	30.6	56.9
Main source government pensions and allowances	51.7	27.1
Unable to raise $2000 within a week for something important	54.3	13.6
Mean equivalised household income ($)	394	665

a. The population for this table is persons aged 18 and over.
b. All the differences between Indigenous and non-Indigenous statistics in this table are significant at the 5% level.
Source: Table 4 in ABS (2004c)

Table 8.2. Income summary by remoteness, 2002

	Remote	Non- remote	Total
	%	%	%
Equivalised household income			
Lowest quintile equivalised income	40.5	43.2	42.5
Second quintile[b]	37.4	25.3	28.3
Third quintile	12.8	14.5	14.0
Fourth quintile[b]	5.9	10.3	9.2
Highest quintile[b]	3.5	6.7	5.9
Unable to raise $2000 within a week for something important[b]	73.0	47.3	54.3
Mean equivalised gross household income ($)	350	399	387

a. The population for this table is Indigenous persons aged 15 and over.
b. The differences between remote and non-remote areas are statistically significant at the 5% level.
Source: Table 1 in ABS (2004c)

The OECD equivalence scale does not take into account differences in the cost of living across areas, so it is particularly important to look at the income distributions and other characteristics in various areas. Table 8.2 documents the income differences in remote and non-remote areas and illustrates that the overall distributions are not very different: there is a similar proportion of residents in all quintiles, except the second, in which remote areas have 12 percentage points more. The bottom line, in the bottom line, is that the average equivalised income in remote areas is lower, albeit only slightly lower. However, the proportion of people who cannot raise $2000 in a week for something important is much higher than in non-remote areas (73.0% and 47.3% respectively). This may indicate that the cost of living is placing more financial stress in remote areas or that Indigenous people in remote areas do not have access to social networks from which they might borrow the money (i.e. a form of social capital). The latter possibility can probably be discounted, since Indigenous people often are said to have a substantial amount of bonding social capital (Hunter 2004b). While Indigenous people in remote areas tend to have large social networks, the 'problem of collective action' may mean that it is difficult to coordinate the large number of competing claims on group resources (Olson 1965).

Before moving onto poverty measures, I want to describe financial stress by labour force status and remoteness. Figure 8.1 presents the incidence of being able to raise $2000 within a week in the bars. As in the rest of the monograph, the whiskers denote the 95 per cent confidence intervals, and as a rough rule if whiskers do not overlap, there is a significant difference between the estimates. As noted above, financial stress is higher in remote Australia. It is interesting to note that even non-CDEP employees in 'very remote' Australia have close to 60 per cent of workers suffering financial stress. This may reflect the methodology, with CAs being predominantly in such areas, but the size of the difference is remarkable and in my opinion unlikely to be explained by non-sampling error.

Figure 8.1. Financial stress by labour forces status and remoteness, 2002

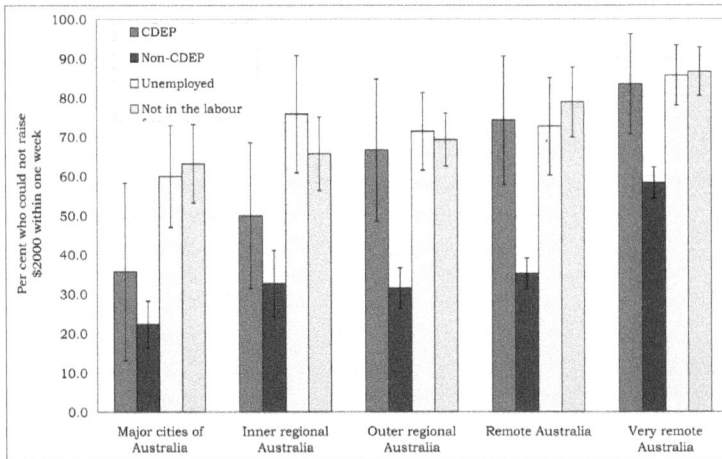

Source: Customised cross-tabulations from the 2002 NATSISS MURF.

Also note that the level of financial stress is high for both unemployed and NILF categories in all areas. CDEP appears to have less financial stress in more urban environments but the reliability of these estimates is particularly low. That is, overall financial stress is high for all groups that are likely to be reliant on some form of government payment, and the best way out of financial stress is to have a job outside the CDEP scheme.

Table 8.3. Indigenous persons aged 15 years or over, selected income characteristics by remoteness, Australia, 2002

	Remote	Non-remote	Australian total
	%	%	%
Main source of personal income			
CDEP[a]	28.2	3.6	10.3
Other wages or salary[a]	18.1	33.2	29.0
Government pensions and allowances[a]	44.2	52.7	50.4
Access to money			
Has a bank account[a]	87.0	96.9	94.2
Method(s) of accessing money[a]			
Over the counter at a bank	21.9	20.4	20.8
EFTPOS/ATM[a]	75.7	87.4	84.2
Internet banking[a]	1.1	6.1	4.7
Phone banking[a]	4.0	9.3	7.9
Over the counter at post office	5.1	4.4	4.6
Other method	2.2	2.3	2.3

a. The difference between remote and non-remote statistics is significant at the 5% level.
Source: Table 18 in ABS (2004c)

It is possible to explore the differences between remote and non-remote areas in more detail by looking at selected income characteristics (table 8.3). Remote areas are more likely to report their main source of income as the CDEP scheme, but are less likely than other areas to rely for most of their income wages from other jobs or government pensions and allowances. The reason for this is simply that CDEP income is identified as a separate source of income. If one argues that CDEP is in some sense a government payment, albeit with a work requirement, then the reliance on government payments in remote areas is substantially higher than in other areas. Remote areas are substantially less likely than other areas to have had government pensions and allowances as their main source of income in the last two years, again presumably because of the flow of people through CDEP schemes.

How do Indigenous people access their money? People in remote areas are significantly less likely to have a bank account, access money over the counter or engage in internet banking. While the incidence of internet banking is much lower than for the rest of Australians in both areas, the particularly low level of internet banking in remote areas is due to the lack of access to personal computers and the internet in such areas (see Radoll in this volume).

Table 8.4 explores several factors that are likely to be correlated with equivalised income. In the first CAEPR Working Paper, 'Three Nations Not One', I described how Indigenous respondents to the 1994 NATSISS only had a weak correlation between such factors and equivalent income. That study was hindered by the grouped nature of much of the income data for the 1994 survey, which meant that my calculations were only rough approximations.

Table 8.4. Indigenous persons aged 15 years or over, mean weekly equivalised gross household income quintiles by selected characteristics, Australia, 2002

	Income quintiles			
	Lowest	Second	Third	Fourth and fifth
	%	%	%	%
Self-assessed health status				
Excellent/very good	38.0	42.8	49.2	56.7
Fair/poor	29.7	23.0	16.5	12.4
Has a disability or long-term health condition	43.5	33.3	30.5	25.3
Risk behaviour/characteristics				
Current daily smoker	57.1	48.1	42.5	32.2
Risky/high-risk alcohol consumption in last 12 months	13.8	14.4	15.5	14.0
Educational attainment				
Has a non-school qualification	17.8	23.4	38.1	43.0
Does not have a non-school qualification				
Completed Year 12	5.9	11.1	12.1	19.4
Completed Year 10 or Year 11	30.5	31.1	28.4	23.5
Completed Year 9 or below	45.7	34.5	21.4	14.1
Total with no non-school qualification	82.2	76.6	61.9	57.0
Labour force status				
Employed				
CDEP	9.3	17.8	11.5	3.4
non-CDEP	8.9	33.6	61.5	84.8
Total employed	18.1	51.4	73.0	88.2
Unemployed	20.8	11.2	7.9	4.7
Not in the labour force	61.0	37.3	19.1	7.1
Financial stress				
Unable to raise $2000 within a week for something important	72.6	57.7	33.6	14.3
Law and justice				
Arrested by police in last 5 years	20.6	16.2	11.7	5.6
Incarcerated in last 5 years	8.5	7.9	7.4	1.5
Victim of physical or threatened violence in last 12 months	29.3	22.7	18.8	18.4

Source: Table 9 in ABS (2004c)

The 2002 data, which permits a more robust estimate of equivalised income than was possible in 1994 (at least using the CURF), shows that there is a significant difference between long-term health in the first quintile and the other quintiles. However, the incidence of such conditions is still very high in the top income group, with over one quarter having a long-term health condition. This level of illness is unprecedented in other high income groups in Australia. Similarly, almost one third of the fourth and fifth quintiles were current daily smokers. All of the variables in this table exhibit a significant difference between the bottom quintile and the fourth and fifth quintiles except risky alcohol consumption, which is slightly higher in the top income group. Indigenous people in the bottom quintile are significantly less likely than the top income group to have an educational qualification, be employed outside the CDEP scheme, or participate in the labour force. However, they are more likely to be in financial stress, and to have been arrested, incarcerated or a victim of physical or threatened violence.

Does the fact that 15.6 per cent of Indigenous people are in the top two quintiles of Australian household income (see Table 8.1)—and that these people appear to be somewhat better off than other Indigenous people—mean there is now an Indigenous 'middle class'? While there may be some signs of an Indigenous middle class emerging, the rates of social ills and socioeconomic disadvantage are too high for the 'rich' groups in Table 8.4. Obviously, addressing income levels is not the only relevant policy. In spite of the conclusion reached, it is clear that one of the main correlates of high income status is securing a non-CDEP scheme job, and addressing Indigenous labour market disadvantage is likely to be the most effective policy emphasis.

A brief digression on validating the top-coding assumptions when using grouped income data for Indigenous people

NATSISS allows us to accurately benchmark the top-coding assumptions routinely used in analysis of grouped data in the respective censuses. In the absence of other data, researchers and policy-makers tend to assume that the average income in the top category is some multiple of the bottom end of the income range, usually a factor of around 1.5. The mean income for the 2002 NATSISS respondents in the top category used in the 2001 Census can be calculated to verify the accuracy of this assumption for Indigenous Australians.[6] The appropriate multiple of the bottom income range is 1.19 for remote areas and 1.32 for non-remote areas. Therefore, the usual assumptions were far too high for Indigenous people, especially in remote areas, which means previous estimates of average income will tend to understate Indigenous income disadvantage.

[6] The income category was adjusted for the weighted average of CPIs in Australian capital cities for the relevant period.

Indigenous and other Australian poverty in 2002

The basic measure of poverty is the headcount measure which can be used to illustrate the disadvantage of Indigenous Australians (that even the researchers at the CIS acknowledge). Table 8.5 presents the headcount measures calculated using a poverty line based on one half of the median equivalised income within the GSS. This was considered to be a reasonably robust poverty line by both sides of the poverty war debate. In one-family households, Indigenous people are twice as likely to be under the poverty line. In households with more than one family living in the dwelling, Indigenous people are over 14 times more likely to be classified as poor. Clearly, household structure is an important 'driver' of Indigenous poverty. The equivalence scales used to calculate equivalent income is defined on household structure, so the poverty rates are in a sense 'driven' by household structure.

Table 8.5. Proportion with less than 50 per cent of median equivalised household income

	Indigenous	All Australian
One family household	41.8	17.29
More than 1 family household	42.5	3.0
No dependants	41.2	23.6
1 dependant in household	39.1	12.6
2–4 dependants in household	44.7	16.1
5 + dependants in household	64.9	53.9
Major cities	39.3	17.9
Inner regional	50.2	23.7
Outer regional	46.1	27.0
Remote/very remote	43.3	NA

Source: Customised cross-tabulations from the 2002 NATSISS and GSS RADL

Having a large number of dependants is another recipe for being in poverty for both Indigenous and non-Indigenous people (i.e. having five or more dependants). Having said this, poverty is also particularly pronounced in Indigenous households with only one dependant.

The variation in poverty by geography is particularly interesting. While poverty is lowest for Indigenous and other Australians in major cities, it is particularly high for Indigenous people living in inner regional areas. Therefore, remote areas are actually less poor than in regional and provincial areas. Despite the lack of an estimate for remote areas from the GSS, poverty for all Australians tends to increase as accessibility to services becomes more difficult.

The tension between the poverty estimates in Table 8.5 and the financial stress data in Figure 8.1 and Table 8.2 is worth noting. Financial stress is higher in remote areas, especially very remote areas. It is particularly noteworthy that non-CDEP scheme employment in very remote areas is also associated with high levels of financial stress. Given that this group is likely to be among the better educated (and most acculturated), this result is unlikely to be caused solely by non-sampling error. The tension between this direct question on financial stress and the standard measures of poverty means that we should question the latter, which are indirectly measured using a series of assumptions.

The theory of poverty is related to utility functions which are used to derive weights for particular households based on cost structures implicit in expenditure surveys (see Hunter, Kennedy & Biddle 2004). Underlying cost structures (and indeed preferences) are not necessarily the same for all sub-population groups, and therefore the imposition of the OECD equivalence scales on NATISS data can—and should—be contested. That is, the following question must be asked: how valid are the OECD equivalence scales when measuring Indigenous poverty?

In technical terms, equivalence scales estimate the 'economies of scale' of living in houses of various sizes and composition. The feasible range of equivalent incomes is bound by raw income and per capita income, with the latter representing an equivalence scale where the cost of each extra person is the same of those currently living in the house. At the other extreme, the equivalence scale underlying raw income assumes that each extra person living in a household cost nothing (i.e. perfect economies of scale). The OECD scale is somewhere between these two extreme assumptions.

Table 8.6 documents the headcount measure of poverty estimated by the proportion of people in households that are classified as poor using the various equivalence scales. Where there is only one person in the household, the actual equivalent income should not change when moving from raw income to per capita income, but the rest of the distribution does change which, in turn, results in a change to the poverty line. Consequently, per capita income pushes such households up the income distribution and they appear less poor. For households with two or more people in them, both the poverty line and the household's equivalent income changes.

Table 8.6. Scoping the feasible range of equivalence scales: proportion with less than 50 per cent of median household income, 2002

	Per capita income	Equivalised household income	Raw household income
Indigenous			
1 person in household	15.21	70.08	81.55
2–4 persons in household	35.36	37.78	40.11
5 + persons in household	65.09	47.64	10.45
All Australian			
1 person in household	6.44	44.79	62.44
2–4 persons in household	11.92	17.06	20.68
5 + persons in household	22.98	13.73	5.38

Source: Author's calculations based on the 2002 NATSISS and GSS RADL

It is particularly interesting to note that Indigenous households with between two and four people in them have a tight range of poverty estimates, so we can be reasonably confident that these poverty estimates are reliable irrespective of the equivalence scale used. However, the story is different for large households with over five people in them, as poverty estimates vary wildly for both Indigenous and other Australians. This is not a criticism of the ABS because this has been observed earlier by many authors (Hunter, Kennedy & Biddle 2004; Whiteford 1985). The main point is that measured poverty in Table 8.6 is particularly sensitive to the choice of equivalence scales for large households, which Indigenous people frequently live in.

Concluding remarks

This chapter has attempted to illustrate that technical details do matter in poverty measurement because there is an implicit equivalence scale underlying Australia's welfare system. The choice of an appropriate equivalence scale(s) for Indigenous Australians needs to be scrutinised and researched so that Indigenous poverty and disadvantage can be addressed adequately.

Where to for future research? The RADL offers exciting potential for research, especially given the more useful continuous income data and financial stress indicators. For my part, I have not published much serious econometric research on predictors of income using the 1994 NATSISS because the grouped nature of that data obscured too much information (one possible exception was some incidental analysis that was conducted to test sensitivity of other analysis in Borland & Hunter 2000). The addition of continuous income data on the RADL for the 2002 NATSISS means that more conventional labour market analysis of Indigenous wages is warranted.

Another potential area for research is that the simultaneous collection of GSS and NATSISS data offers the opportunity to examine the nature of Indigenous disadvantage vis-à-vis other Australians. It would be particularly interesting to pool the NATSISS and GSS data, but that could only be done within the confines of the ABS. One important impediment to the power of any such analysis is that there does not appear to be an Indigenous identifier on the GSS RADL, which means it is not possible to directly calculate non-Indigenous estimates. The inability to compare Indigenous estimates from the NATSISS to non-Indigenous estimates in the GSS is a major weakness of the set-up of the GSS CURF and severely circumscribes the usefulness of both data sets for policy-makers.

9. Family and community life

Ruth Weston and Matthew Gray

Well functioning families are vital to the wellbeing of individuals, their immediate communities, and broader societal groups. The ability of families to function well depends not only on their individual members, but also their physical and social contexts, including communities and wider organisations.

The 2002 NATSISS survey provides information on family and community life for the Indigenous population. It is one of the few nationally representative surveys of the Indigenous population that provides this kind of information and thus makes an important contribution to our understanding of these important aspects of the life of Indigenous Australians.

Collecting information in surveys on family and community life is always a challenge but is particularly difficult for some sections of the Indigenous population. This is in part because of the lack of congruence between the kinship terminology and concepts of local kin systems and those of the mainstream Anglo-Celtic system for many 'tradition-orientated' Indigenous people. [1] It is also a result of many Indigenous households and families being larger and structurally more complex than for the non-Indigenous population. There are also differences in the dynamics underlying the structure and composition of households in mainstream and local Indigenous societies.

At the outset, it is important to keep in mind that virtually all measures examined in the 2002 NATSISS that relate to living standards or wellbeing are highly relevant to family and community life or wellbeing. Personal heath status and health risk behaviours are prime examples, since family and community wellbeing is strongly linked with the wellbeing of individual family members.

Housing quality is another example of a variable that has a big impact on family and community wellbeing. As Taylor & Kinfu (in this volume) indicated, family needs represented one of the most commonly mentioned reasons for moving that respondents provided in the 2002 NATSISS. The fact that 52 per cent of respondents lived in houses needing 'more bedrooms' reveals the inadequate living conditions experienced by the majority of Indigenous families (see Sanders in this volume).

Despite the relevance of so many measures to family and community life, this chapter restricts attention to the more direct measures of family and community

[1] We have borrowed the term 'tradition-orientated' from Morphy (2004a: 3) who used the term to refer to Indigenous people who live in discrete remote communities on or near their traditional country, with limited access to the economic mainstream.

contained in the survey. In this chapter we first discuss the reasons for collecting data on family and community life. This is followed by an outline of the measures of family and community life used in the 2002 NATSISS. Examples of the utility of two of these measures (child care use and the incidence of 'stolen generation' experiences) are then discussed. The final section discusses aspects of family and community life that may be valuable additions to future social surveys of the Indigenous population.

Why study family and community?

As a basic unit of society, families have the key responsibility of caring for their members. This includes helping children and adults alike to be—or become—healthy, well adjusted and productive members of society and supporting elderly, infirm or disabled members. Such functions are complex and multi-faceted and involve the meeting of basic needs of family members, as well as the transmission of pro-social values. While there is no universally accepted definition of healthy functioning, it tends to be linked to meeting needs:

- of a physiological nature
- for educational/cognitive development (e.g. achievement, competence, mastery, independence)
- relating to psychosocial wellbeing (e.g. development of a sense of acceptance, belonging, trust and love, self-esteem, concerns beyond the self), and
- that have a spiritual dimension, including the development of a sense of purpose and meaning in life.

Communities are higher-order systems whose responsibilities also lie in promoting the wellbeing of individuals within them, both directly and indirectly, through supporting families to fulfill their responsibilities and providing opportunities for all members to participate in community life. Fulfilment of these responsibilities is important not only for the wellbeing of individuals and their families but also for national and international wellbeing. At the same time, family and community wellbeing depends on the wellbeing and contributions of their members, as well as those of higher order systems.

Clearly, then, research into family and community life is important for the identification of potential or existing resources and deficits or challenges confronting these social units, their individual members, and broader social systems. Ultimately, such research is important for the development and monitoring of the effectiveness of policies that are directed towards ensuring that individuals live in safe, supportive environments that enable them to reach their potential, adopt health-promoting lifestyles, participate in community life, and develop or maintain a sense of purpose and meaning in life.

Family and community life domains

Given the close interdependence between families and their communities, some measures in the 2002 NATSISS can be treated as indicators of either family or community life. The choice seems arbitrary at times, so we have listed in this section the measures covering either or both of these two domains.

Household and family type

The 2002 NATSISS provides detailed information on household type, family type and social marital status. Information on all the people living in the household was collected from a responsible adult. The survey excluded visitors to the dwelling, and those who stayed in the dwelling the previous night were defined as visitors if they would be staying for less than one month. An important feature of many Indigenous households is that there is a significant amount of mobility through the household, resulting in very complex and dynamic household structures (Morphy 2004a; Smith 2000b). The 2002 NATSISS survey provides virtually no information about such dynamics surrounding household composition. It is difficult (and perhaps not feasible) to collect this information using a cross-sectional survey for households that have a high turnover of people.

The categories and terms used to describe kin relationships are those that apply to the standard Anglo-Celtic system. Although the standard Anglo-Celtic system will be clearly understood and relevant for much of the Indigenous population, many traditionally-oriented Aboriginal people have kinship systems which differ markedly in their structure to the Anglo-Celtic system. For many of these respondents, the 2002 NATSISS questions very likely resulted in incoherent and uninterpretable data (Morphy 2004b).

While the complex familial structures of Indigenous societies are most pronounced in 'traditionally-oriented' communities, Smith (2000b) has shown that they persist in 'settled' Australia. Martin et al. (2004) conclude that, when this household information is used to construct measures of family type, the resulting 'family types' do not coincide with those found in many Indigenous communities (Martin et al. 2004: 95). A further issue is that the 2002 NATSISS survey does not provide any information on linked households, yet linkages between households represent an important feature of Indigenous family and community life.

The main point to be taken from this discussion is that care needs to be taken when interpreting the household composition, family type and social marital status information from the 2002 NATSISS given that, for a proportion of the sample, this information will have little relationship to the family circumstances in which the respondent lives. A detailed discussion of these issues is provided by Martin et al. (2004).

Information on the relationships amongst people in the family or household is obtained by asking the reference person (the person providing information on all household members) the relationship of everybody else in the household to themselves. Although the reference person model works well for simpler household and family structures, it only provides a very partial and potentially misleading picture for more complex family arrangements (particularly multi-generational families) which are so common in Indigenous Australia.

One possible way of improving the quality of this information is through using a household grid to collect information on the relationships amongst members of a household or family (see Brandon 2004). Relationship grids, which obtain information on the relationship of every household member to every other household member (i.e. not just the household reference person), are used in the Household, Income and Labour Dynamics in Australia (HILDA) survey.

Although the household grid has many advantages, it can be quite time-consuming to collect for larger complex household structures. Its inclusion would therefore require the omission of other questions. Furthermore, the grid would involve the use of kin relationship concepts that, as noted above, appear to be inappropriate for *some* Indigenous people.

An important issue in studying Indigenous families concerns 'mixed families' and 'mixed households'. These are families or households in which not all members are of Aboriginal and Torres Strait Islander origin. One strength of the 2002 NATSISS survey is that it allows mixed families to be identified and outcomes for Indigenous families without any non-Indigenous members to be compared to those of families comprised solely of Indigenous people.

Fertility and child survival

Female respondents were asked to indicate how many children they had given birth to, how many were living with them, and how many were living elsewhere. These measures enable an estimate of the number of children who had not survived, although there will be some error in this derived variable (see Kinfu in this volume). While a direct question on child mortality can be extremely stressful for those who have experienced this event, it is noteworthy that HILDA introduced such a question in Wave 5.

Removal from natural family

Given that questions on removal from family could be highly stressful for respondents, interviewers first asked respondents whether it was 'alright' to ask questions on this issue. In total, 4 per cent of respondents indicated a preference to skip these questions (9% in remote areas and 3% in non-remote areas). All other respondents were asked, firstly, whether they had been taken away from their natural family by a mission, the government or welfare, and secondly, whether any of their relatives had had such an experience. Those who indicated that one or more relatives had been removed from their natural family were asked to indicate which relative(s) experienced this. Once again, the terms used to indicate kin relationships were those applicable to the standard Anglo-Celtic kinship system (e.g. parents, aunts, uncles, brothers or sisters, children). The resulting data must therefore be interpreted with caution.

Child care

Although there are some differences in child care questions between the non-remote and remote area questionnaires, the questions on child care use are broadly comparable. These questions were restricted to respondents who had the main caring responsibility for any child living in the household who was aged 12 years or less. In both remote and non-remote areas, access was measured in terms of experiences in the previous four weeks and included questions on the use of formal and informal child care, unmet desire for using any (or additional) formal care, the main reason for not having wanted any (or any additional) formal care, and the main reason for any unmet desire to use such care during the four-week period.

More specifically, respondents who were interviewed using the non-remote questionnaire were asked whether they had used formal child care in the previous four weeks and, if so, the different types of formal care they had used. [2] Respondents were also asked whether they needed any—or any additional—formal child care during this four-week period. Those who answered in the affirmative were asked to indicate the main reason for not having used such (additional) care. Finally, those who had not wanted to make any—or any more—use of formal child care in the previous four weeks were asked to indicate the main reason for not having wanted such (additional) care.

An important difference between the child care data collected using the remote and non-remote questionnaires is that, in remote areas, the respondent was asked whether there was a child care service in the community and, if there was not, whether they would use a service were it available. Those in remote areas which offered such a service were asked questions about their use of this service in their previous four weeks, aspirations regarding usage, and reasons for not having wanted to use the service or for having experienced unmet aspirations regarding service use. (These questions were essentially the same as those asked of respondents in non-remote areas about the use of formal child care.)

A key difference between the data collected in the 2002 NATSISS on child care and many other surveys is that the 2002 NATSISS questions are based on the person primarily responsible for the child(ren) in the household (i.e. an adult)

---

[2] The categories are: before and/or after school care, long day care, family day care, occasional care, pre-school, kindergarten (excluding NSW and ACT), or other formal care (excluding vacation care).

whereas as other surveys provide information on the use of child care for each child (or the study child). Other relevant studies are the ABS Child Care Surveys and the HILDA survey. Thus, caution needs to be exercised in comparing the information on child care from NATSISS with estimates from other sources which are often child-based.

Further discussion of some of these issues appears later in this chapter.

Support in time of crisis

This question tapped the respondents' perceptions of their ability to ask for support from people outside their household in times of crisis and the sources of any such support. The sources included individual acquaintances (e.g. friend, neighbour, family member, work colleague), as well as organisations, professionals and local council or other government services. [3]

It is important to note that, while some potential sources of support are more 'approachable' than others, some people are more confident than others in requesting assistance. Furthermore, some people may be prepared to approach family members and friends, but consider professionals or organisations as 'out of bounds', while the opposite may apply to other people. In other words, reports on support should not be treated as objective measures of the social environment, but rather as perceptions that are likely to be shaped not only by the existence and characteristics of potential sources of support but also by characteristics of the respondents themselves. Nevertheless, a sense of social support is an important aspect of personal wellbeing, and has obvious flow-on effects for the family and community.

Stressors experienced

Respondents were asked about whether they, or a close family member or friend, had experienced various stressful events over the previous 12 months. For respondents in non-remote areas, the events were subdivided into three groups:

- health issues (serious illness—including mental illness, accident, death of family member or close friend, or serious disability)
- relationship breakdown, employment problems and 'risky' behaviour (alcohol or drug-related activities, witness to violence, abuse or violent crime, trouble with the police or a gambling problem), and
- imprisonment, overcrowding at home, pressure to fulfill cultural responsibilities, and discrimination or racism.

---

[3] Respondents in non-remote areas where shown a list of types of support (for example, emotional support, provide emergency accommodation, advise on what to do), while those in remote areas were simply asked whether they could ask somebody who does not live with them for help if they were having 'serious problems'.

The nature and ordering of some items differed slightly for those in remote and non-remote areas. The main question – whether the issues have been a problem for the respondents, or for their family or close friends – does not fit well with some of the actual problems listed for those in remote areas (for example, 'member of *your* family sent to jail or in jail') (emphasis added). In non-remote areas, respondents were asked about 'member of family sent to jail/currently in jail'. In other words, such respondents could include such events in the lives of their close friends' families. This adds to the difficulty of comparing the experiences of respondents in remote and non-remote areas.

It is also important to point out that, given that 'one person's cup is another's poison', the population of stressful events is huge, and any sample from this population is likely to be an inadequate representation of potential stressors in a person's life. Furthermore, non-events can be extremely stressful but there is no attempt to measure these (e.g. failure to obtain the expected promotion, failure to see one's child achieve some strong ambition, failure to establish an intimate relationship with a much admired potential suitor, and so on).

Another difficulty with this measure is that it relates not only to personal experience of events that are typically seen as stressful, but also to the exposure of family or friends to such experiences. While difficulties faced by other people can be personally stressful, it would have been useful to be able to identify whether the experience applied to the respondent, a close family member or friend). It would also have been useful to identify the stressfulness of such events for the respondents. For instance, it appears that, compared with men, women tend to be more emotionally involved in the lives of those around them, more reactive to the moods and experiences of other family members and close friends, and more prone to mention interpersonal difficulties, including family-related concerns, in response to questions about the problems in their lives (see Cross & Madson 1997; Larson & Richards 1994; Thoits 1995). Under these circumstances, the questions about disruptive events experienced by close family and friends may tend to have a greater impact on women than men. It may also have a greater impact on some cultures than others.

It would have been extremely useful to compare the experiences of Indigenous respondents regarding the events listed with those of the non-Indigenous population. As is discussed in chapter 4 of this volume, the GSS is designed to be comparable to parts of the 2002 NATSISS. Unfortunately, while the GSS asks about stressors, the questions in the latter survey focus on stressors experienced by respondents or 'anyone else' close to him or her rather than 'close family member or friend' as is asked in the 2002 NATSISS. This difference may contribute to any systematic variation in reporting that may appear. Nevertheless, it will be possible to compare differences in reports within the Indigenous population—for example, men versus women, those with lower versus higher educational attainment, and, where the stressful events described are identical in the two questionnaires, those in remote versus non-remote areas.

Neighbourhood problems

Respondents were asked about the existence of a series of neighbourhood problems, mainly covering property theft or damage, assault/violence, and neighbourhood conflict.[4] These measures refer to respondents' perceptions and should not be interpreted as objective measures of problems in the neighbourhood. They are relevant to a personal sense of safety and security and views about the safety of family members and others living in the locality—issues that are clearly important aspects of individual, family and community wellbeing.

It would be also useful to include perceptions of neighbourhood wellbeing (as well as ill-being), for example, beliefs about the extent to which people in the neighbourhood are trustworthy, vigilant about each other's wellbeing and property, and generally willing to help each other out.

Voluntary work

Voluntary work represents an important indicator of engagement with society as well as a contribution to community life. It is worth noting, however, that the question focuses on work with organisations and does not capture more informal activities, such as helping an elderly neighbour or friend.

The question taps the type of organisations and number of different organisations to which respondents contribute on a voluntary basis. It should be noted that some respondents may contribute a great deal of time to one organisation or to several organisations of the same type (e.g. welfare/community), while others may contribute time to several organisations. Caution needs to be taken that those who work voluntarily for several organisations are not seen as spending more time in voluntary activities than those whose activities target one or more organisations of the same type.

These two issues outlined above point to the fact that the breadth and amount of voluntary community work are not tapped in this questionnaire.

Two illustrations of the value of the 2002 NATSISS data

In this section, the value of the 2002 NATSISS data for two areas of family life is illustrated. The first is use of child care and the second, removal from natural family.

---

[4] One item in the list is 'Level of personal safety day or night'. This does not seem to fit well with the others that refer to specific problems (theft, gangs, vandals, assault, etc) Perhaps it should be rephrased (for example, 'Concerns about personal safety day or night').

Child care

As discussed above, the 2002 NATSISS survey contains questions on the use of child care in the previous four weeks by respondents who had the main responsibility for children in the household aged 12 years or under. There is relatively little data available on use of child care by the Indigenous population and how it compares to that of the non-Indigenous population.[5] Thus the 2002 NATSISS survey is a valuable new source of information on use of child care by the Indigenous population.

The use of child care by Indigenous people with primary responsibility for children (described as primary carers) by employment status and region of residence (remote compared to non-remote) is outlined in Table 9.1. Some comparisons with the use of child care by the total Australian population (i.e. predominantly non-Indigenous) are made. Comparable estimates for the total Australian population in non-remote areas of Australia were constructed using the HILDA Survey.[6] Although the HILDA estimates are for non-remote areas, given that only a small proportion of Australian children live in these areas, there would be relatively little difference between the non-remote and the total Australian estimates.

Of the Indigenous primary carers, child care was used by a lower proportion who lived in remote rather than non-remote areas (56.7% versus 69.9%). It is interesting to note that Indigenous use of child care in non-remote areas is greater than non-Indigenous use, with 55.6 per cent of the non-Indigenous population using child care.

While similar proportions of the Indigenous population in remote and non-remote areas used informal care (39.7% and 40.9% respectively), those in remote areas were less likely than their counterparts in non-remote areas to have used formal services (15.9% versus 28.9%).

Differences are apparent between the non-remote Indigenous and non-Indigenous populations in their patterns of use of informal care. While for both these populations, the rate of use of formal care was around 28.8 per cent, the non-remote Indigenous population was more likely than the non-Indigenous population to have used informal care exclusively (40.9% versus 26.7%).

---

[5] The Australian Government Department of Family and Community Services (FaCS) Census of Child Care Services provides important comparative information regarding the use of child care by the Indigenous and non-Indigenous population. However, this Census restricts attention to services that are approved and funded by the Australian Government and does not obtain the breadth of socioeconomic information derived in the 2002 NATSISS.

[6] The ABS periodically conducts surveys on use of child care. However, there are several difficulties in using data from the ABS Child Care survey. First, the Child Care Survey covers children aged 12 years, whereas the 2002 NATSISS survey focuses on children aged under 13 years. Second, the ABS Child Care survey is child-based rather than carer-based. That is, information is collected about the child care use of each child in the family. While a carer-based data set can be developed, this is a time-consuming and complicated exercise.

The lower use of child care by Indigenous primary carers in remote than non-remote areas applied to both those who were employed and those who were not employed. However, the difference was particularly marked for those who were employed: 63.1 per cent of employed Indigenous primary carers in remote areas and 80.8 per cent of their counterparts in non-remote areas used child care. This is probably a consequence of the higher rates of part-time CDEP employment in remote areas (Altman, Gray & Levitus 2005). Amongst the total Australian population in non-remote areas, 70.2 per cent of employed primary carers used child care. The pattern of use of formal and informal care differed, with the Indigenous carers being substantially more likely to use informal care compared with the non-Indigenous carers.

There was a large difference in the use of child care by non-employed Indigenous and non-Indigenous primary carers. For example, in non-remote areas, 64.2 per cent of non-employed Indigenous carers used child care compared with just 37.9 per cent of the non-employed non-Indigenous carers. This difference is largely due to a higher rate of use of informal care by Indigenous than non-Indigenous populations (39.9% versus 16.9%). This is a reflection of the extensive kin-based networks that many Indigenous people have.

Table 9.1. Use of child care by persons with primary responsibility for children according to employment status, Indigenous and non-Indigenous Australians, 2002

| | Indigenous (NATSISS 2002) | | Australian population (HILDA 2002) |
| | Remote | Non-remote | Non-remote |
	%	%	%
Primary carer employed			
Used child care	63.1	80.8	70.2
Formal	19.5	37.9	35.4
Informal only	42.4	42.9	34.8
Did not use child care	36.9	19.2	29.8
Primary carer not employed			
Used child care	50.4	64.2	37.9
Formal	12.4	24.4	20.9
Informal only	37.1	39.9	16.9
Did not use child care	49.6	35.8	62.1
Total			
Used child care	56.7	69.9	55.6
Formal	15.9	28.9	28.8
Informal only	39.7	40.9	26.7
Did not use child care	43.3	30.1	44.4

a. For the Indigenous population, the estimates include persons with primary responsibility for children aged 12 years or under who did not state the kind of child care used. The figures for formal care may include persons who also used informal child care. The total proportions who used child care were derived by subtracting the proportions who did not use child care from 100. For remote areas, the latter estimates differ from those derived by summing the proportions who used either formal or informal care.
Source: ABS (2004c: Table 17) and HILDA Wave 2 (details of data release version)

Lack of access to formal child care is often discussed as an issue for remote areas of Australia. It is interesting that, according to the NATSISS 2002, the majority of people in remote areas who had primary responsibility for children indicated that they had access to child care if needed (69.4%). In other words, just under one-third (29.6%) reported that they did not have access (ABS 2004c).

Removal from natural family

Estimates based on the NATSIS 1994 and NATSISS 2002 data sets of the proportion of the Indigenous population who had been taken away their family are very similar. Both surveys suggested that 8 per cent of the population aged 15 or more years (at the time of each survey) had been removed. Furthermore, the 1994 survey suggested that 10 per cent aged 25 years or more had been removed. This proportion is the same as that derived in the 2002 survey for those aged 35 years or more (who would have represented roughly the same cohort). [7]

To measure the number of Indigenous people potentially affected by the removal of children from their families, the 2002 NATSISS asked Indigenous people aged 15 years or over whether they or any of their relatives had been removed from their natural families. As noted above, about 8 per cent of Indigenous people reported that they themselves had been removed (see Table 9.2).

Perhaps the most significant point to be taken from these figures is that, even though a relatively small proportion of the Indigenous population were themselves removed from their natural family, about one-third of the Indigenous population had a relative removed. Indeed, 38 per cent indicated that they and/or at least one of their relatives had been taken from their family (ABS 2004c: 6).

When interpreting the data from the question on removal of relatives from natural family it is important to note that the question had a high rate of 'not known' and 'not stated' responses (20%) (ABS 2004c: 58). This high rate of non-response is not surprising given the sensitivity of this issue to some families. It is probable that the respondents not wanting to discuss this issue disproportionately had relatives removed and so the estimates may be under-estimates.

[7] Statistics from ABS (2004c).

Table 9.2. Removal from natural family

	Remote	Non-remote	Total
	%	%	%
Removal of person from natural family			
Person removed	6.0	9.4	8.4
Person has not been removed	85.0	88.0	87.2
Didn't want to answer	9.1	2.6	4.4
Removal of relative(s) from natural family			
Relative(s) removed	28.1	38.5	35.6
Relatives have not been removed	52.8	41.2	44.4
Didn't know	10.0	17.3	15.3
Didn't want to answer	9.2	3.0	4.7

Source: ABS (2004c: Table 12)

Concluding comments

Family and community life is multi-dimensional and complex. This makes it difficult to design questionnaires that can adequately capture the different dimensions of family and community life.

Overall, the NATSISS 2002 survey does a good job of measuring a range of aspects of family and community life given that these domains are only two of the many domains that a general social survey of the Indigenous population needs to cover. In this chapter we have outlined the measures of family and community life included in the survey and have attempted to highlight some of the issues which need to be taken into account when analysing the data generated by these questions.

Second, the measures focus on the individual, with no information gathered on the quality of relationships, parenting behaviour, family functioning, and so on. Given the crucial importance of such issues for wellbeing, some measures on these issues should be considered for future surveys. The Longitudinal Survey of Australian Children may provide a useful source of questions on some of these issues.

Third, the measures of household structure and composition are problematic for a proportion of the Indigenous population, given the complex and multi-generational nature of many households. The use of a household grid to gather information on family composition is worth considering.

10. Labour market issues

Matthew Gray and Bruce Chapman

The continuing low employment rates and general labour market disadvantage of Indigenous Australians have been well documented (Altman & Nieuwenhuysen 1979; Daly 1995; Hunter 2004a). However, our understanding of the reasons for this labour market disadvantage is constrained by the limited data available for the Indigenous population. This lack of understanding hampers the development of labour market and related policies to improve labour market outcomes for Indigenous Australians.

Before the collection of the 2002 NATSISS, the main source of data on Indigenous labour force status—and the only sources of data that could be used to reliably measure change—have been the five-yearly censuses from 1971 to 2001.[1] While the census data can provide valuable information on trends in labour force status, working hours, occupation and industry, there is very limited information on other important labour market topics such as the duration of unemployment, difficulties experienced in finding employment, and the identification of discouraged workers. Furthermore, the census has very limited or no data on a range of economic, demographic, social and cultural factors which are likely to be important in explaining labour market outcomes.

The only other nationally representative data on Indigenous Australians is the 1994 NATSIS.[2] Although the 1994 NATSIS provides data on a much wider range of topics than the census, these data are now over a decade old, and the 2002 NATSISS provides a valuable *new* source of information on labour market issues.

The 2002 NATSISS collects similar information to the 1994 NATSIS on labour market issues, so it represents a valuable and timely addition to data sets with information on Indigenous labour market outcomes. In broad terms, the information on key labour market variables is comparable between the 1994 NATSIS and 2002 NATSISS, allowing for changes over time to be assessed.[3]

---

[1] However, it should be noted that the ABS has recently released some experimental estimates from the LFS (ABS 2006).

[2] There are a number of data sets which contain limited information on labour market issues and which have a sufficient Indigenous sample to allow meaningful analysis. Examples are the 1995 and 2002 National Health Surveys collected by the ABS. There are also surveys of specific groups of Indigenous people, such as the longitudinal survey of Aboriginal and Torres Strait Islander job-seekers collected by the then Department of Employment, Workplace Relations and Small Business.

[3] There are issues relating to the sampling which need to be taken into account when making comparisons between these two surveys. They are discussed in detail by Biddle & Hunter in this volume.

The purpose of this chapter is to provide an overview of the labour market information available in the 2002 NATSISS and to describe some of the key strengths and limitations of the data. In order to illustrate the value of the 2002 NATSISS, three examples are offered of highly useful types of data that are available.

While no data set is ideal, we consider in some detail an important limitation of the 2002 NATSISS data. This relates to the omission of key variables from the data set, specifically labour market experience and the length of time spent with the current employer. We use an alternative data set with information on labour market experience to illustrate the potential significance of its omission for statistical analysis of both wages and joblessness. The value of our method is that it can be applied to illustrate the significance or otherwise of the omission of other variables from the 2002 NATSISS.

Other chapters in this volume (Biddle & Hunter; Webster, Rogers & Black) and a number of ABS publications provide a detailed overview of the 2002 NATSISS, including sampling, exclusions issues and non-sampling matters. In this chapter, discussion of these issues is limited to those that are specifically related to the labour market data in NATSISS.

Strengths of the 2002 NATSISS

There are three main areas in which the 2002 NATSISS data has advantages over the census for the analysis of labour market issues.

First, the 2002 NATSISS accurately identifies CDEP scheme employment, which is not the case in the census.[4] This is a major limitation of the census because the CDEP scheme represents a crucial difference between Indigenous and mainstream labour market experiences. Under the scheme, funding is allocated to CDEP organisations for remuneration for participants at a level similar to, or a little higher than, income support payments, with the finances being enhanced with administrative and capital support. It is thus used as a means to provide employment, training and enterprise support to Indigenous participants (see Altman, Gray & Levitus 2005 for a detailed discussion of the CDEP scheme). To illustrate how important the scheme is, we note that in 2002, employment in CDEP accounted for over one-quarter of the total employment of Indigenous Australians, with around 13 per cent of the Indigenous working-age population being employed in the scheme.

---

[4] CDEP employment is not reliably identified by the census because the census form does not include CDEP employment as a separate category, although in the 1996 and 2001 censuses a different census form (the Special Indigenous Form, or SIF) was used in some discrete Indigenous communities. The SIF has a separate category for CDEP employment. Although this has improved identification of CDEP employment, many Indigenous people participating in the CDEP scheme are not enumerated using the SIF. Administrative data on CDEP participants for the time of the 2001 Census indicates that 30 474 Indigenous people worked in the scheme, whereas the census identifies only 17 800 participants (Hunter 2004a: 5).

The importance of identifying CDEP employment for different areas of Australia is illustrated in Table 10.1, which shows Indigenous labour force status by region using the 2002 NATSISS. [5] In non-remote areas, just 4.7 per cent of the Indigenous working-age population was employed in the CDEP scheme. In these areas, failure to take account of CDEP employment is likely to have a relatively small effect. But in remote and very remote areas, 16.9 per cent and 42.2 per cent respectively of the working age population was employed in the scheme (see Table 10.2).

Table 10.1. Indigenous labour force status by region, 2002

	Non-remote	Remote	Very remote
	%	%	%
Employed			
CDEP employed	4.7	16.9	42.2
Mainstream employed	41.2	31.7	14.9
Total in the labour force	63.3	58.7	61.6
Population (no.)[a]	196 300	23 100	49 850

a. Table population is Indigenous persons aged 15–64 years.
Note: The remote areas in this chapter, in contrast to most other chapters in this monograph, refers to remote areas that are not classified as very remote by ARIA. That is, tables are not derived from ABS (2004c), which generally provides aggregate results for all remote areas.
Source: Customised cross-tabulations from the 2002 NATSISS (derived from Altman, Gray & Levitus 2005: Table 1)

Using the 2002 NATSISS, it is possible to estimate the effects of CDEP employment on a range of important outcomes, such as income and working hours. With the 1994 NATSIS, it also allows for analysis of trends in labour force status (including non-CDEP employment) to be identified with more confidence than has been previously possible using census data combined with administrative data. It is also possible to analyse changes in the determinants of mainstream employment at an individual level (although not for the same individual, which would require longitudinal data).

Further, the 2002 NATSISS can also be used to estimate the associations between CDEP employment and a range of social, health and cultural variables.

The second major advantage of the 2002 NATSISS is that, for the first time, analysis of labour market issues is possible in very remote areas of Australia. The ability to do this is highly valuable because the labour market context of very remote areas (and, to a lesser extent, remote areas) is very different from those in the rest of Australia, for reasons now discussed.

First, Indigenous people in very remote areas are often living in communities in which the majority of the population is Indigenous. Second, these communities are in sparsely populated regions of Australia which are extremely distant from

---

[5] It is possible to analyse labour force status for the following geographic categories: non-remote, remote and very remote. In Table 10.1, we have aggregated areas into the categories of 'non-remote', 'remote' and 'very remote' in order to simplify the analysis and allow us to highlight the major issues.

markets, both geographically and culturally. Third, these regions were colonised relatively late, with some parts of Arnhem Land and central Australia as recently as during the last 50 years. This has meant that customary (kin-based) systems and practices remain robust and there is ongoing contestation between mainstream Australian and Indigenous world views.

Furthermore, according to conventional economic and social indicators, there is a growing disparity between Indigenous people living in remote areas and both Indigenous and non-Indigenous Australians living in non-remote areas (ABS 2004a, 2004c). There is evidence that some discrete Indigenous communities in remote Australia are in economic and social crisis.

The different labour market context in respective regions is illustrated clearly by the fact that in non-remote areas the mainstream employment rate is 41.2 per cent, in remote areas 31.7 per cent, and in very remote areas just 14.9 per cent (see Table 10.1). Human capital and demographic characteristics also differ dramatically across regions. For example, education levels are much lower in remote and very remote areas than in non-remote areas, and the proportion of the population speaking an Indigenous language is much higher in remote areas than in non-remote areas. These factors are bound to influence the nature, variance and quality of Indigenous labour market experiences and it is a real bonus that the information is part of the 2002 NATSISS.

A third advantage of the 2002 NATSISS is that it contains information on a wide range of somewhat unusual social, demographic, cultural, and economic variables which are potentially important for understanding labour market outcomes. Examples include health status, speaking an Aboriginal language, having used an employment service, access to transport, and having been arrested. Note that many of these variables are not available from the census.

Table 10.2. Labour market data collected in the 2002 NATSISS

Labour force status	Employment support
Duration of unemployment	Whether used employment support services
Hours usually worked in all jobs	Whether needed employment support services
Full-time/part-time status	Reasons did not use employment support services
Employment sector	**Income**
Precariousness – job security in next 12 months	Level of income
Whether work allows for cultural responsibilities	Personal gross weekly income
	Household gross weekly income
CDEP:	**Source of income**
Whether CDEP participant	All sources of personal income
Duration on CDEP	Main source of personal income
Considers CDEP participation to be a job	**Government pension/allowance**
Barriers to employment	Type of government pension/allowance (primary)
Whether had difficulties finding work	Type of government pension/allowance (auxiliary)
All difficulties finding work	Government support
Main difficulty finding work	Time on government support in last two years
Discouraged jobseekers	
Whether would like a job	
All reasons not looking for a job	

Source: Derived from ABS (2005b)

In general, it appears that questions relating to labour market topics are very similar (virtually identical) in the community and non-community questionnaires (see ABS 2005b and the list of variables in Table 10.2). While there may be some effects generated by differences in the data collection method (CAPI versus paper-based questionnaire), we do not anticipate this will have introduced major biases. While those analysing the data will need to carefully consider the extent to which the remote and non-remote data is comparable for their particular application, our reading of the questionnaires suggests that there is no particular reason for expecting there to be comparability issues.

The questions are also, in large part, standard ABS questions. This allows comparative studies of labour market outcomes for Indigenous and non-Indigenous Australians using the 2002 NATSISS and other data sets, such as the GSS.

An illustration of new information available from the 2002 NATSISS

The 2002 NATSISS, for the first time, provides information on how long CDEP participants have been participating in the CDEP scheme. Information of this type is important in assessing whether CDEP employment is a destination or a stepping stone to mainstream employment. While there are some ambiguities in the NATSISS question, 'How long have you been on CDEP', it does provide valuable data. One difficulty with the question is that it is unclear whether participants who had multiple spells of CDEP would give the duration of CDEP participation as from when they first participated in the scheme or whether it would be from when they most recently started on the scheme.

The length of time that participants spend on the CDEP scheme varies across regions. In very remote areas, 40.6 per cent of participants had been on the CDEP scheme for five years or more and 21.8 per cent had been on the CDEP scheme for less than one year (see Table 10.3). Similarly, in remote areas, many participants had been on the scheme for a number of years, although the average duration was shorter. In non-remote areas, only a minority of participants (15.2%) had been on the scheme for five years or more and 38.0 per cent had been on the scheme for less than one year.

When interpreting these figures it should be kept in mind that the length of time that a person can be on the CDEP scheme is constrained by the length of time a CDEP scheme place has been available to them. On average, places have been available for longer in remote and very remote areas. It will also depend on the age of the participant, although this could be taken into account in a more sophisticated analysis of the data.

Table 10.3. Duration on CDEP by region of residence, 2002[a]

Length of time on scheme	Non-remote	Remote	Very remote
	%	%	%
Less than 1 year	38	29.7	21.8
1 to less than 2 years	17.4	10.8	14.7
2 to less than 3 years	14.1	13.5	12.2
3 to less than 4 years	8.7	10.8	7.1
4 to less than 5 years	6.5	10.8	3.6
5 years or more	15.2	24.3	40.6
Population (no.)	9 200	3 900	21 100

a. Table population is CDEP participants.
Source: The 2002 NATSISS, derived from Altman, Gray and Levitus (2005: Table 5)

Another important topic on which the 2002 NATSISS provides new information is participation in vocational education and training (VET). Participation in VET is an important way in which those with low education can increase their skill level and improve their labour market outcomes. Of particular interest is the extent to which CDEP scheme participants receive VET, and hence are improving skill levels and chances of finding mainstream employment. This is an important policy objective of the scheme.

Table 10.4 presents information on participation in the VET sector by labour force status and region in the previous 12 months. There are relatively high rates of participation in VET in the last 12 months in all areas, although rates in very remote areas are half those in major cities. The CDEP employed overall have lower rates of undertaking VET than the mainstream employed. The only exception is in major cities where 54.8 per cent of the CDEP employed undertook VET, compared to 46.9 per cent of the mainstream employed.

Table 10.4. Participation in VET in the last 12 months, by labour force status and region, 2002[a]

	Major cities	Inner regional	Outer regional	Remote	Very remote
	%	%	%	%	%
CDEP	54.8	45.8	45.9	30.8	18.0
Mainstream	46.9	51.7	52.8	57.5	43.2
Unemployed	27.5	21.2	25.5	26.1	13.6
NILF	2.8	2.1	7.1	3.7	1.6
Total	31.0	28.5	30.5	29.0	15.9

a. Table population is Indigenous persons aged 15–64 years.
Source: The 2002 NATSISS, derived from Altman, Gray and Levitus (2005: Table 10)

A limitation of the survey

All surveys have both strengths and weaknesses, and the above discussion has highlighted aspects of the former with respect to the 2002 NATSISS. What now follows considers some deficiencies of these data, in particular the lack of useful information concerning labour market experience. Specifically, the data set has no measures of either the length of time individuals have spent in paid employment (general labour market experience) or how long employed individuals have been in their current place of work (tenure). The discussion now examines the potential significance of the omission from the data of measures of general labour market experience).

An important focus of modern labour economics concerns the role of skills or, to use the accepted parlance, human capital. Human capital is seen to be a major—even *the* major—contributor to individuals' success or otherwise in the labour market. There are two important aspects of human capital: formal education and the skills acquired by individuals from on-the-job training. In both areas, there are significant issues associated with measurement, since the pure human capital aspects of both education and training are not directly observed.

Labour market experience is typically represented in surveys like the 2002 NATSISS by the length of time spent in paid employment. Unfortunately, this variable is unavailable in the survey, and this raises the possibility that labour market statistical analyses of the 2002 NATSISS will provide inadequate, even misleading, results concerning the true determinants of Indigenous labour market success or failure.

Not having information on labour market employment history can be seen to be a major weakness of the 2002 NATSISS. In part, this is because Indigenous Australians have much higher rates of movement between labour force states than non-Indigenous Australians (Gray & Hunter 2005) and have much more interrupted labour market histories. For example, using a longitudinal sample of Indigenous job-seekers, Hunter, Gray & Jones (2000) find that 33.6 per cent of Indigenous males and 37.6 per cent of Indigenous females had been employed for less than 25 per cent of the time since leaving school. Only 16.5 per cent and 18.0 per cent of Indigenous males and females respectively had been employed for more than 75 per cent of the time since leaving school.

In order to illustrate the extent of the potential problem associated with the omission of measures of labour market experience from the 2002 NATSISS, we have examined econometric modelling in two areas: wages and being in employment. Our aim is to demonstrate the likely empirical importance of having to use the wrong variable. Our approach is to use an alternative data set that contains both a poor and a better measure of labour market experience. The poor measure is the length of time individuals could have spent in the labour force after finishing formal education, and the better measure is the number of years an individual has actually spent in paid employment. The models are estimated using both labour market experience measures and the results compared. One such data set can be derived from the HILDA survey.

We have chosen the female sample, since the potential significance of not having the more correct experience measure will be greater for groups with less attachment to the paid labour force, such as women (and Indigenous individuals). The econometric models are now briefly described.

Wage determination exercises take many forms, with the most basic human capital approach being represented by the following equation:

$$Wage = a + b*EXP + c*EXP2 + d*YOS + e$$

Where wage is the log of the hourly wage received by the individual, EXP is the number of years of paid employment, and YOS is the number of years of formal education. EXP2 is the square of the experience term, which is included because it is believed that the wage-experience term is non-linear. Table 10.5 compares the coefficients from the estimation of this wage equation (with the log of wages as the dependent variable) for the 2002 NATSISS specification and the HILDA specification.

Table 10.5. OLS wage regressions[a]

Explanatory variables	2002 NATSISS	HILDA
EXP	.0203	.0287
EXP2	-.000383	-.000758
YOS	.0572	.0525
Constant	1.618	1.689
R^2	0.11	0.12

a. All coefficients are significant at the 1% level.
Source: Author's calculations

While the results are apparently similar for the two specifications (certainly the coefficients on years of schooling are very close), closer inspection suggests that at low levels of measured experience there are significant differences in the wage relationships. This is illustrated in Table 10.6, which shows the percentage change in individuals' hourly wages for additional years of experience at different levels of experience.

Table 10.6. Effect of experience on wage (percentage)

Experience (in years)	NATSIS	HILDA	Percentage difference
1	1.95	2.72	40
5	1.65	2.11	28
10	1.25	1.35	7

Source: Author's calculations

The results of Table 10.6 suggest the following:

• At one year of experience, the effect of an additional year of experience on wages is estimated to be 1.95 per cent using the (poor) measure of experience available from the 2002 NATSISS, compared to about 2.9 per cent using the (better) measure of experience available from HILDA. This difference can be argued to be the very large difference of around 40 per cent of the NATSISS coefficient.

• At moderate levels of experience (e.g., five years), HILDA still results in a higher wage-experience relationship than that found for the 2002 NATSISS, but the difference has been reduced to about 28 per cent.

• At high levels of experience (10 years), there is effectively no difference found between the wage-experience estimates.

We then repeated the above exercise with respect to estimating the determinants of whether or not a person is employed. The typical econometric approach used in this area takes an equation of the following form:

$$EP = a + bEXP + cEXP2 + dEDUC + eDEMOGRAPHY + e$$

Where EP is the probability that an individual is employed, EDUC are measures of education and DEMOGRAPHY reflects demographic factors. In our exercise, DEMOGRAPHY includes measures of marital status, whether or not the person is an immigrant, and the presence and age of children. The major relationship sizes for both specifications are available from the authors, and the experience effects are now shown in Table 10.7.

Table 10.7. Effect of experience on probability of employment (percentage)

Experience (in years)	NATSIS	HILDA	Percentage difference
1	2.08	4.70	226
10	1.03	2.70	262
25	0.80	0.60	75

Source: Author's calculations

The data of Table 10.7 suggest strongly that the poor measure of labour market experience available from the 2002 NATSISS has a significant potential to be misleading with respect to the effects of labour market experience on employment. The following results can be highlighted:

- At one year of experience, the 2002 NATSISS estimation suggests an additional year of measured experience increases the probability of employment by about two percentage points, but the (more accurate) experience measure from HILDA suggests that the relationship is more than double this, at nearly five percentage points.
- At 10 years of experience, the estimated differences between the two data sets in the role played by labour market experience is even higher: about one per cent for NATSSISS 2002, and nearly three percentage points for HILDA, a difference of over 250 per cent.
- At very high levels of labour market experience, 25 years, the apparent problem with using the NATSSIS 2002 experience measure has been reduced considerably, to the extent that the poor experience measure now apparently overstates the effect of experience on employment probabilities (0.8 compared to 0.6 from HILDA).

These comparative exercises make it apparent that the statistical problem associated with the omission in the 2002 NATSISS of a good measure of labour market experience are potentially very important. By comparing the same modelling with results found with a data set which has available a better measure of experience, it is clear that the 2002 NATSISS understates the value of experience for wages, and that this understatement becomes less as the experience measure increases. Similarly, results on the determinants of employment using the 2002 NATSISS seem to importantly get the story wrong with respect to the true role of experience. And, as with wages, the extent of the problem seems to be greater at the lower levels of experience.

It is important to record that the interpretation difficulties associated with the 2002 NATSISS not having an accurate measure for labour market experience seem to be confined to estimation of the true role of experience. In other words, the modelling and data problem has not affected estimates of the role of variables such as education with respect to wages, and education and demography with respect to the determinants of employment. This suggests that even though researchers are unlikely to be able to show with accuracy the effect of experience on labour market success, there are no associated difficulties for determining the true role for Indigenous labour market performance of other critical variables.

Concluding comments

The 2002 NATSISS provides a valuable new source of data on labour market issues for Indigenous Australians. It provides some data on labour market issues that has not previously been available. It also repeats much of the labour market content of the 1994 NATSIS and may allow the estimates made using the 1994

NATSIS to be updated, and the robustness of findings from the earlier survey tested.

This is a very valuable data source, although there is no direct information on critical variables such as labour market experience and tenure in the current job. These are important variables for understanding many labour market relationships, and their absence will likely restrict the value of some types of analysis.

While the 2002 NATSISS survey will certainly advance our understanding of labour market outcomes, the cross-sectional nature of the survey will make the identification of some causal relationships quite difficult and, in some cases, impossible. A longitudinal labour market study for the Indigenous population needs to be considered seriously.

11. Asking the right questions?

Bob Gregory

Employment and the hard policy choices

I am very pleased to be here today. I have been so impressed over the last decade or so by the progress made in collecting and analysing Indigenous economic and labour market data that I wanted to come and say how much the situation has improved and how important CAEPR and its Director, Professor Altman, and the ABS, have been in pushing hard to improve the empirical foundations upon which sound policy can be built.

I, too, have made some small contribution to this development. Dr Boyd Hunter and Dr Anne Daly were my students and they have played a significant role in advancing knowledge in this important area. Perhaps I might share a little reflected glory from their fine publishing records. Apart from this, however, my contribution has been small. So I am very much an outsider with all the advantages of that position; I can speak in a tone of voice with great authority without knowing what I am talking about; I can leave the conference very quickly without being missed; and, lastly, the chance that anything I say will adversely affect my career is very slight. If I was actively working on Indigenous economic policy, and building a career in this important area, then I might choose to be more circumspect and a little more humble and politically sensitive in my remarks.

So, I thought that the best thing I could do as an outsider, and remembering my state of ignorance, is to say the sort of things that are often said behind closed doors, or at dinner parties, but not usually said too explicitly and openly by experts who are working on Indigenous affairs. The public remarks of insider experts seem to me to be too optimistic in assessing past progress, too optimistic as to likely future outcomes, and to place insufficient emphasis on economic outcomes. The more pessimistic, and I believe more realistic comments I am about to make, need to be laid out clearly because I believe that Indigenous economic and employment policy is about to change quite markedly. It is possible already to discern three major initiatives and these policy directions will continue, probably at an accelerating pace.

Firstly, there will be large political shifts in the relationship between government and the Indigenous community. Relationships are becoming more concerned with economic issues and less concerned with developing an Aboriginal identity and with gestures of reconciliation. The Federal government has clearly decided to reduce its support for Indigenous political development and to change the way it seeks policy advice from the Indigenous community. This process has begun with the abolition of ATSIC. My guess is that the golden period of broad-based Indigenous political influence is over, at least at the federal level. It is worth noting that there was no widespread political opposition to recent changes.

Secondly, there will be important shifts in Australian welfare policy, not prompted by Indigenous outcomes but, nevertheless, very important for Indigenous living standards. These changes, in the short run, must lead to lower Indigenous incomes. Much of the income that flows to Indigenous people comes from the Australian Government as income transfers through the Australian welfare system, and policies to reduce welfare expenditure across the board will impact disproportionately on Indigenous people. The hoped-for trade-off from these initiatives is that lower real welfare incomes in the near future will generate individual responses that will lead to higher real employment incomes in the far future. The hoped-for response is that individuals will substitute higher employment income for lower welfare income. To offset these short run losses of welfare income would require disproportionate and wide-ranging employment responses, outcomes which seem unlikely when placed against past trends.

Thirdly, there will be large changes made to the CDEP scheme, which is the major employment growth centre for Indigenous people.

In these comments I focus primarily on the economic issues of employment and welfare reform and put aside political changes.

Employment

I wish to make two major points about Indigenous employment. Firstly, Indigenous mainstream employment is extremely low. Secondly, if CDEP is set aside, there is no evidence that across-the–board employment prospects are improving. These employment outcomes are extremely worrying and disappointing. Despite large amounts of public expenditure to improve Indigenous employment levels, and despite very significant improvements in Indigenous school attendance and education levels, there has been very little employment growth outside the CDEP scheme. In this dimension, our policies seem to have failed spectacularly.

The parlous state of Indigenous employment is illustrated in Table 11.1, which presents various estimates of the employment/population ratios for adult Indigenous Australians taken from a range of publications. The employment ratios listed in the first set of data are taken from the census and presented by Hunter & Taylor (2004). In 1991 the Indigenous employment–population ratio for adult males, aged 15 to 64 years, and including CDEP employment, was 37.6 per cent. Just over one in three Aborigines had a job. By 2001 the employment ratio has lifted to 40.4 per cent. This is an improvement but the employment rate is still very low. Altman, Biddle & Hunter (2005) take the data back to 1971 and the employment–population ratio of 2001 is clearly lower than thirty years earlier.

Table 11.1. Various employment-to-population ratios for adult Indigenous Australians

	1971	1981	1991	2001	2011
	%	%	%	%	%
Total employment–population					
Taylor/Hunter, Hunter/Kinfu/Taylor			37.6	40.4	36.0
Altman/Biddle/Hunter	42.0	35.7	37.1	41.4	
Mainstream employment–population					
Taylor/Hunter, Hunter/Kinfu/Taylor,					
Altman/Biddle/Hunter	42.0	35.7	32.9	29.5	25.8
Non-elite mainstream employment–population					
Gregory			28.4	24.1	19.6
Other employment–population outcomes					
Altman/Biddle/Hunter					
Full-time	32.9	19.5	21.9	21.6	
Private	29.7	17.2	20.5	22.9	

Sources: Taylor & Hunter (1998), Hunter, Kinfu & Taylor (2003), Altman, Biddle & Hunter (2005), Gregory (own calculations)

The second set of data uses a concept I call mainstream employment. Mainstream employment excludes CDEP from the employment data. These employment estimates suggest that in 1991 the employment–population ratio was 32.9 per cent and by 2001 it had fallen to 29.5 per cent. It is worth spending some time emphasising how low this number is. Suppose you were to undertake an Australia-wide survey and the first question posed of the respondent is 'Are you an Indigenous Australian?'. If the answer is yes, you could bet that the respondent was not employed in the mainstream economy and you would be right 7 times out of 10. Perhaps even more importantly, over the last three decades, Indigenous employment (net of CDEP) has moved backwards. The various policy initiatives—primarily additional Indigenous education expenditure, increased Indigenous access to welfare support and extensive Indigenous employment programs—have been associated with no net employment gains relative to population growth.

To construct the third set of data, I subtract from the mainstream employment–population ratio my estimates of the number of Indigenous adults in a sub-group that I call the Indigenous elite. I define the elite as those Indigenous people who report employment income that places them in the top 30 per cent of the Australia-wide employment income distribution. The number of Aboriginals who are in this category has grown over the last few decades and it is one area of great success. There has been good progress here and there is now a significant, but still very small, group of Aboriginals who do well in the mainstream economy, although most of this group are associated with various Indigenous economic and political activities. It is important to subtract these successful individuals from the employment data because I want to emphasise what has been happening to the non-elite. For non-elite Indigenous people, the employment–population ratio was 28.4 per cent in 1991 and it has now fallen to 19.6 per cent. So, after removing the small employment increase of those with high incomes, 80 per cent of the remaining Aboriginals do not have employment in the mainstream economy. This is an appallingly low figure.

Finally, in the last data set presented in Table 11.1, I list estimates of those employed full-time and those employed in the private sector. Both of these categories of employment have fallen.

Demand and supply

Why is the Indigenous employment record so bad? The Australian Government has spent considerable amounts of money and invested in a wide range of Indigenous employment programs and yet we do not seem to be able to significantly increase the number of Aboriginals involved in mainstream employment. Why not?

Most people address this question in terms of two main explanations, each with very different policy implications. One suggested explanation is that, by and large, Indigenous people choose not to be employed in the mainstream economy (the supply side explanation). This is done in various direct and indirect ways. One decision is the place of residence. Perhaps this is for family reasons or the desire to live a life closely based on traditional values—many Indigenous people choose to live in traditional remote areas where there are no jobs and thereby choose not to move to locations where there are employment opportunities.

Perhaps Indigenous people also choose not to search too hard for a job? Perhaps the traditional values that lead to income sharing reduce the incentives to accept well paying jobs on mine sites, for example, as well paid individuals inevitably have to distribute so much of their wages to other members of their community.

If supply side behaviour is an important reason for the low employment outcome then it is important to understand why this choice is being made. Is this choice strongly influenced by financial incentives that favour the non-seeking of jobs and non-involvement in the mainstream economy? In this respect, are unemployment benefits too high, so that Indigenous people are better off financially without employment? If financial incentives are important then employment may respond significantly to supply side policy changes. If supply decisions are based not on financial outcomes of different choices but on different personal considerations, then supply side policy initiatives aimed at financial incentives, such as reduced welfare payments, are unlikely to be effective at increasing employment. They will just reduce Indigenous income, with the associated increase in financial hardship.

The other explanation for low Indigenous employment levels is that Indigenous people want to be employed in the mainstream economy, perhaps at much the same rate as anyone else, but they cannot find a job offer to accept (the demand side explanation). When Indigenous people apply for jobs, they are placed too low on the candidate ranking list and rarely get the chance to accept a job. Their low ranking may be the result of discrimination or their poor skill levels.

This bald distinction between demand and supply explanations for low employment is often made behind closed doors. Different experts often hold one or other of these positions with considerable conviction, but very rarely is the distinction made starkly at conferences like these. The natural response of experts is to blur the distinction between demand and supply, or to emphasise the demand side alone, which they judge to be more politically acceptable. But little is served by this behaviour because these two classes of explanation lead to very different policies. We need to be clearer as to why we are observing poor employment outcomes. Indeed, the relative importance of these two explanations in the Australian community more generally is currently being actively debated among experts and politicians. The consequent judgment to place more emphasis on supply is beginning to impact on policy in non-Indigenous areas of the labour market.

Welfare reforms (the supply side of the labour market)

The important distinction between demand and supply and their impact on recent policy developments can be illustrated by considering proposed changes in welfare policy. Consider, for example, policy changes for lone parents. Lone parents, by and large, receive most of their income from the welfare state in much the same way as Indigenous people. Virtually no lone parent has a full-time mainstream job. Although we do not know the exact proportion, it is probably something of the order of 10 to 15 per cent. About 40 per cent of lone parents—a higher proportion than in the Aboriginal community—work part-time. Recent policy developments for this group might mean future changes for Indigenous people.

In broad terms the policy judgment has been made that the reason why lone parents have low employment rates is that they are choosing *not* to work in the mainstream economy because the welfare system is too generous. The policy judgment has been made that there is a need to change the financial balance between work and welfare.

Mr Kevin Andrews, the Minster for Employment and Workplace Relations, has said recently:

We now live in an economy where we don't so much have a shortage of jobs as a shortage of workers. In fact, it has never been a better time in Australia for those who want to work ... We believe that if you can work you should work. (Andrews 2005)

In response to this view, the government has recently decided to change welfare policy for lone parents so that once the youngest child turns seven, lone parent income support will be reduced from a lone parent pension payment to an unemployment allowance payment. For lone mothers who do not work, and do not qualify for a pension, their weekly income will be reduced by $29 and, in the future, indexed only for price increases and not for the larger increase of wages. Furthermore, if the lone parent combines work and welfare, the loss of income will probably be of the order of $100 per week. At the same time, the lone parent will be subject to all the work-for-the-dole requirements. The object of the new policy is clear. Life should be less financially comfortable for welfare-reliant lone parents with a youngest child over seven years of age.

Consider, as another example, recent policy changes for those who may receive invalid pensions. Here government policy is following a similar path to that followed by the lone parent initiatives. There is a clear presumption that it is the supply side that is responsible for the lack of employment among those with a disability and that a significant proportion of those on invalid pensions are choosing not to work. From June 2006, it will be much harder for individuals to be placed on invalid pensions.

These examples matter for Indigenous people. They matter because they are illustrative of a significant shift in prevailing policy attitudes. There is an increasing conviction that the large number of individuals receiving welfare support is a reflection of individuals being unwilling to work because welfare payments are too high. It seems inevitable that these attitudes will increasingly begin to impact on Indigenous economic and social policy. These examples also matter because Indigenous people, with their heavy reliance on government income support, will be directly and severely affected by the current changes.

It is fairly obvious that we need better data to throw light on the extent to which Indigenous people want to be employed in the mainstream economy but cannot find jobs (the demand side) and the degree to which the low employment level is the result of the impact of traditional values and/or financial incentives (the supply side). This is a crucial issue, in that any change in financial incentives directed towards supply responses will make Indigenous people financially worse off if there is an insufficient positive employment response.

The Fair Pay Commission (the demand side of the labour market)

On the demand side of the labour market, the Australian economy has not been generating sufficient unskilled jobs over the last three decades. Under these circumstances, Indigenous Australians have found it difficult to receive job offers. In other words, there has been a strong demand side influence on Indigenous employment outcomes.

Here the policy stance is clear. Government is of the view that wages are too high for the unskilled and that something should be done, over time, to reduce the relative living standards of the group. To facilitate this process, the government has introduced a Fair Pay Commission to attempt to prevent significant pay increases for the low paid. This role has not been explicitly announced, except to say that employment considerations will play a larger role in wage-setting than in the past. In a situation where there is heavy unemployment and a lack of job opportunities among the unskilled, this must mean that unskilled pay levels will fall, at least in relative terms.

It is important that we have a firm view as to the relative importance of demand and supply side influences. If the main problem is the demand side, then reducing welfare incomes to change supply side incentives will have weak employment effects and mainly reduce incomes. Similarly, if the supply side is the problem, attempting to create additional jobs for Indigenous people by lower wages will also be relatively ineffective. Lower wages will just reduce the income of those employed without the beneficial effect of extra jobs. So, we should ask, 'What can we learn about the relative importance of demand and supply influences from the NATSSIS data?'. I think the answer is that we can learn very little. We need more direct information on the rate of job offers to Indigenous people and the rate at which Indigenous people fail to search for jobs and the reasons why.

Further data issues

CDEP

The CDEP scheme provides a permanent source of income at the level of unemployment benefits, plus an infrastructure loading that reduces poverty in remote areas. In my view, the CDEP scheme is currently the most important policy instrument available for affecting employment outcomes in remote communities in the short run. CDEP has been the major employment growth industry for Indigenous people, and it is now extremely large, accounting for most of the employment growth of the last two decades (the difference between row 2 and row 3 in Table 11.1). We all need a clear view of what we think about CDEP and the incentive structure it embodies.

In financial terms, and in the short run at least, the CDEP scheme puts in place incentives that encourage people not to be employed in the mainstream economy and not to move from traditional locations. But the existence of incentives to remain in remote communities does not necessarily mean that large numbers of Indigenous people respond. We do not know how many Indigenous people are making the financial judgment that life in a remote area on CDEP is much better than moving and trying to find a job somewhere else. If there is a significant long-run rejection rate of mainstream jobs because of CDEP, this must be against the long-run economic interests of the Indigenous community. On the other hand, if CDEP is a scheme where community life is made better for people who really have *no* employment alternatives, then we should be less willing to restrict its further growth.

To adequately discriminate between these two views of the impact of CDEP on long-run employment opportunities and income growth, we need more data that relates to individual CDEP projects, their participants and the linkages between CDEP growth and mainstream jobs.

Program evaluation

One weakness of NATSSIS-type of data collected by the ABS is that it is not well suited for program evaluation. To do program evaluation well requires combined administrative and survey data, and the responsibility for putting this type of data together must rest with the relevant program departments. Let me give you some indication of the type of data needed.

The Department of Employment and Workplace Relations (DEWR) has recently provided important administrative evaluation data as part of the Indigenous Employment Policy Evaluation Stage 2. Administrative data like this provides us with a number of important lessons that we cannot learn from ABS data alone. Although the DEWR administrative data in the public arena is only a small fraction of that which is needed, DEWR is, nevertheless, to be congratulated on making some data available.

The first lesson we learn from a consideration of the DEWR data is that administrative data re-emphasises how difficult it will be to develop effective employment programs. The data makes clear that Indigenous job-seekers are extremely poorly qualified. They have very low education levels and very low levels of English competence (see Table 11.2).

Table 11.2. Selected characteristics of the Job Network eligible population, 2002–03

	Indigenous	Non-Indigenous
	Per cent of participant in categories	
Male	63.3	62.6
Less than Year 10 education	47.7	22.3
English speaking ability less than good	10.6	6.2
English writing ability less than good	28.3	12.7
From a remote or very remote location	33.7	1.7

Source:DEWR (2003: 28)

The second lesson we learn from administrative data is that the juxtaposition of administrative data with the NATSISS and ABS type data helps prevent over-optimistic interpretations of program evaluation results. Consider the following. Over the last five years, the annual growth of mainstream non-CDEP Indigenous employment revealed by ABS census data is of the order of 1000 people per year (row 4, Table 11.3 and Gregory 2005). DEWR, however, states that during a typical year, 38 000 Aborigines are referred to Intensive Assistance (see Table 11.4). This referral rate, as a ratio of the increase in mainstream employment, is of the order of 38 to one, which suggests that any job finding that can be linked to participation in the job network must be largely the replacement of one employed Indigenous person for another. Intensive Assistance seems to add very little to the stock of Indigenous employment.

Table 11.3. Employment and population changes for Indigenous Australians (in 000's)

	1991–1996	1996–2001	2001–2006	2006–2011
Population increase	23.5	33.6	40.6	43.0
Employed	11.9	16.9	9.2	8.6
CDEP	7.5	11.8	3.8	2.7
Mainstream non-CDEP	4.4	5.0	5.5	5.8
Mainstream non-CDEP, non-elite	1.0	1.4	N/A	N/A

Sources:1991–1996:Taylor & Hunter (1998), Gregory (own calculation),1996–2011:Hunter, Kinfu & Taylor (2003), Gregory (own calculation)

Table 11.4. Hypothetical estimates of labour market circumstances 16 months after being referred to Intensive Assistance

Commenced			20 900
Employed due to participation	1360		
Employed anyway	3200		
Total employed	4560		
Unemployed		16 340	
Did not commence			17 100
Employed due to compliance	1910		
Employed anyway	2460		
Total employed	4370		
Unemployed		12 730	
Totals			
Employed	8930		
Unemployed		29 070	
Total Referred to Intensive Assistance			38 000

Source: DEWR (2003: 76)

The third lesson we learn from administrative data is that when individuals are directed towards Intensive Assistance, which is very expensive and designed to help them find employment, many individuals choose not to enter the scheme. For example, of the 38 000 Indigenous people referred to Intensive Assistance each year, just over 50 per cent (20 900) commence the program of Intensive Assistance. Around 17 100 say no thanks and refuse to enter the scheme. Does this mean that Indigenous people are judging that Intensive Assistance is not useful or does it mean that they do not want jobs?

The fourth lesson we learn is that of those who were referred to the scheme, 4560 became employed. Of those who were referred but did not commence the scheme, the increase in employment was approximately the same, 4370. This suggests the Intensive Assistance scheme has a very small impact, and perhaps no impact at all. DEWR, however, went further and suggested that among those who entered the scheme, only 1360 were employed as a result of participation. The other participants who became employed would have found employment without participating in the scheme. This estimate implies that 4 per cent of these who enter Intensive Assistance receive a job because of the scheme. This is a very low rate of success.

Finally, Table 11.4 indicates a nice point to think about. DEWR suggests that the threat of entering Intensive Assistance leads to individuals accepting jobs that they would not otherwise have done. The employment response to this threat suggests an interesting result; namely, that the success rate from the threat of having to participate in Intensive Assistance is greater than the success rate from participating in the scheme. This prompts the thought that from a financial point of view, more emphasis should be placed on threats which are inevitably cheaper than providing training.

To sum up, the administrative data indicate that employment training schemes are relatively ineffective. This pessimistic result is not unique to programs directed toward Indigenous people. If non-Indigenous program evaluations are conducted, they produce similar results. It appears, therefore, that the lack of employment growth among Indigenous people is not easily fixed by policy initiatives that focus on skill development and job-finding skills for adults. Intensive Assistance is the Rolls Royce of job training policy. Government puts a lot of money into the program and yet it seems to pay very low returns. If, as a community, we are to do more in this policy area of developing skills and job-finding ability among the unemployed, we will need to substantially improve these schemes—a task that is, perhaps, too difficult.

Concluding comments

To conclude, I think we are well placed in terms of data that describes the difficulties that Indigenous people face, and in terms of data that documents their deep economic and social problems. My own feeling is that we need to know more about the interplay of culture and Indigenous decision making and how to create an environment within which Indigenous people can make better choices for themselves. To be thinking of poverty and employment problems in economic terms alone is clearly inadequate. We need to know more about motivation and the mechanisms that can create communities with more productive outcomes.

Policies for Indigenous young people of exceptional talent seem to be working well. For this group, there are efficient and effective pathways into professional sport, into universities, into business activities and into the public policy arena. And things are getting better each year.

What is not working at all are policies for the majority of Indigenous people and for the young who do not relate well either to a school environment or to the discipline of a work environment in the mainstream economy. How to design effective policies for these groups and how to collect the right sort of data to help develop good policies is not straightforward.

Finally, as indicated earlier, we need more administrative data that relate directly to policy interventions and directly to policy outcomes.

12. The real 'real' economy in remote Australia[1]

Jon Altman, Geoff Buchanan and Nicholas Biddle

The Productivity Commission's recent report Overcoming Indigenous Disadvantage, notes that the vision behind the report is that Indigenous people will one day enjoy the same overall standard of living as other Australians (SCRGSP 2005: 1.2). This admirable goal is a reflection of the Howard Government's commitment to practical reconciliation; that is, to equality in health, housing, employment and education outcomes between Indigenous and non-Indigenous Australians. The futuristic reference to 'one day' suggests that the goal may indeed be more visionary than policy realistic.

Such a commitment by government is not new and was first articulated by the Aboriginal Employment Development Policy of 1987 (Australian Government 1987). Altman and Allen (1992: 147–8) noted before the 1994 NATSIS that while government policy aims to provide employment and income equality for all Indigenous Australians, it ignores the contributions made by 'the informal economy', or what we refer to here as the customary sector.

This chapter focuses on a different livelihood vision—one that seeks a sustainable future for Indigenous people residing in remote and very remote Australia, and one that accords with the culturally-informed aspirations of many. Our view is that measured equality, based on mainstream notions of development, will be impossible to deliver in many remote and very remote parts of Australia where Indigenous Australians reside, often on Indigenous-owned and managed lands. Rather than envisioning Indigenous futures that are limited to mainstream notions of development in either of the private (market) or public (state employment and welfare) sectors, we focus here on an economy that includes a third—customary or non-market or Indigenous—sector.

This framework has been termed the hybrid (three-sector) economy (Altman 2005b), but it has other nomenclatures. For example, Gibson-Graham (2005) refers to the diverse or community economy. And the feminist literature identifies a somewhat different three-sector economy that has crucial resonance with the model used here (Cameron & Gibson-Graham 2003; Ironmonger 1996). The

---

[1] It should be noted that the results presented in this chapter differ from those presented at the Indigenous Socioeconomic Outcomes conference held on 11–12 August 2005. There was an error on the 2002 NATSISS CURF regarding the fishing or hunting in a group results that did not enable us to accurately calculate the denominator for our proportion estimates. Our conclusions, however, remain unchanged.

Canadian Royal Commission on Aboriginal Peoples (1996) used the term 'traditional-mixed economy'.

Altman (2003b) contends that the real economy in many remote Indigenous contexts is the hybrid economy. The term 'real' economy has come to some prominence in the influential writings of Noel Pearson (2000, 2005), although he uses the term to mainly refer to the market economy. Thus, the term can have very different meanings and has been the subject of considerable debate within feminist discourse (e.g. Gibson-Graham 1996), as well as in policy debates on achievable Indigenous economic development (Altman 2003b, 2005b; Canadian Royal Commission on Aboriginal Peoples 1996). Both approaches seek to embed economic representation within a broader and more inclusive frame. For example, within feminist discourse it has been argued that the real economy has to include the contributions made by unpaid productive activities, predominantly performed by females. From this feminist perspective, economist Duncan Ironmonger (1996) looks to quantify the market and household sectors of the economy and thereby 'feminise the economy'. In a similar way, we seek to include and quantify a predominantly unpaid and unrecognised Indigenous component in the economy—we 'Indigenise' the economy, at least in remote and very remote Australia, regions for which some data are available.

One important reason for conducting the 2002 NATSISS was to gather information that is distinctly Indigenous. In the search for the real Indigenous economy in remote and very remote Australia, we are highly reliant on official statistics to move beyond community or small group case studies. In this chapter we explore what can be documented about customary activity in Australia using the 2002 NATSISS. Given that Indigenous Australians now own over 20 per cent of the Australian continent (Pollack 2001), we explore what can be ascertained about their non-market activities on this land in order to inform debates over whether there is development benefit from land rights and native title. This debate, which has some currency, tends to ignore the reality that the market economy is very limited in remote regions and fails to quantify the non-market (see Hughes 2005b; Hughes & Warin 2005).

The paucity of data available on the customary sector marginalises Indigenous productive participation (Altman 2003b, 2004, 2005b), positioning it as 'other' to 'real' economic activity (see ABS 2005a, where it is reported as 'social and sporting activity'). Such marginalisation facilitates poorly informed criticism of remote Indigenous communities by those who are opposed to the contemporary reality of a customary sector that is often underwritten by some state income support (e.g. see Hughes 2005b; Sandall 2000).[2] At the most fundamental level,

---

[2] The ideological opposition to the legitimacy of the customary sector referred to here, which is described by Gibson-Graham (n.d.) as 'capitalocentric', is outlined well by Gibson (1999: 3) with reference to what she terms 'community economies'.

we interrogate the 2002 NATSISS to see if it provides information about the magnitude and significance of the customary sector that can inform academic and policy debate. Gibson-Graham (n.d.: 2) note that what is at stake in conversations about rethinking 'economy' is 'who and what is seen to:

1) constitute the economy, and

2) contribute to economic development'.

How can 2002 NATSISS data help answer such questions?

A brief overview of the customary sector and the hybrid economy model

The term 'customary sector' is similar to other widely used terms such as subsistence, non-market, non-monetary, informal, non-mainstream, cultural, and traditional-economy (see, inter alia, Altman & Allen 1992; Canadian Royal Commission on Aboriginal Peoples 1996; Fairbairn 1985; Fisk 1985; Morgan, Strelein & Weir 2004; Smith & Roach 1996). As outlined by Altman (2005b), the customary sector is made up of a range of productive activities that occur outside the market that are based on cultural continuities and specialities. These include activities such as hunting, fishing and gathering, production of art and crafts, and land, habitat and species management.[3] Although the customary economy is not monetised (or marketed) by definition, a proportion can attract a dollar value, as with the sale of Indigenous art (Altman 2005b). At the core of the hybrid economy model is recognition of the linkages and interdependencies between the market, state and customary sectors, as shown in diagram 12.1 below. Our focus here is on segment 2, but also on segments 4, 6 and 7.

---

[3] Fairbairn (1985: 327–29) provides an inventory of subsistence output in a Samoan context, including foodstuffs, buildings, capital works (canoes, road building, land development), craft industries and miscellaneous products, law and order functions of village councils and chiefs, certain kinds of intra-household works (water carrying, house cleaning, religious services), social security aspects of village life, the home processing of foods, Indigenous medicine and funerals). Fairbairn defines subsistence as embracing all non-monetary economic activities like the definition of the customary economy used here.

Figure 12.1. Conceptual representation of the hybrid economy

The Hybrid Economy

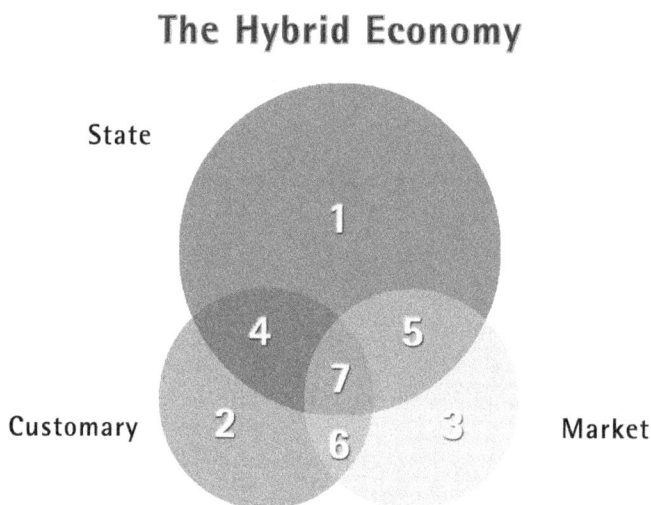

These linkages and interdependencies are often overlooked in Indigenous policy. A recent example of such an oversight can be found in the Productivity Commission's *Overcoming Indigenous Disadvantage* report mentioned above. The chapter 'Economic participation and development' recognises fishing, hunting and gathering as 'important economic activities for some Indigenous people living on Indigenous-owned land' and that the 'customary economy could have significant value for Indigenous people' (SCRGSP 2005: 11.20). However, by limiting its perception of such activity to being 'an affordable source of fresh and healthy food' the Productivity Commission focuses only on the direct subsistence benefits of such activity—it ignores culturally-informed interpretations of its significance and wider spin-off benefits that might be generated. Opportunities to recognise and promote economic development on the Indigenous estate through commercialisation of the customary are also overlooked (Altman 2005b).

In much of its report, the Productivity Commission relies heavily on data from the 2002 NATSISS. However, in its discussion of the customary sector, no reference is made to survey results. Similarly, in its main publication about the 2002 NATSISS, the ABS (2004c) similarly fails to report data it collected on hunting and fishing activities. More recently, in *Australian Social Trends 2005*, the ABS (2005a: 52–57) reports on fishing or hunting, but as a selected social activity rather than economic activity. It is worth asking whether this was because the 2002 NATSISS did not go far enough in assessing customary activity as economic activity. As a first step in exploring this issue, we examine the questions that the 2002 NATSISS asked in relation to the customary sector.

2002 NATSISS questions and results

We focus here on three key issues in the 2002 NATSISS that relate to the customary sector and its interactions with the market and state. These are included in the 2002 NATSISS through variables on:

- fishing or hunting as a group activity
- participation in and payment for cultural activities, and
- the ability to meet cultural responsibilities while in employment.

Fishing or hunting in a group

Under 'Culture', the 2002 NATSISS asked those in the CA sample, 'in the last three months, have you done anything else with other people?' There were eleven possible responses, of which one was 'going fishing or hunting in a group'. While a similar question was asked in the NCA sample in terms of involvement in social activities, 'going fishing or hunting in a group' was not a listed response.

The two components of the 2002 NATSISS sample selection and survey design—CAs and NCAs—have been described and discussed in detail by Biddle and Hunter (in this volume). In terms of sample size and geographic coverage, the 2002 NATSISS survey methodology has meant that information on fishing or hunting in a group was only gathered from the 2120 CA respondents (i.e. individuals residing in discrete communities and outstations in parts of Queensland, SA, WA, and the NT). No information was gathered on such activities from the vast majority of respondents to the 2002 NATSISS—a total of 7362 individuals in non-community remote areas (1997) and in regional areas and cities (5242). As a result, any discussion of Indigenous involvement in fishing or hunting activities based on 2002 NATSISS data is limited to the CA sample, the majority of whom were from very remote, as opposed to remote, Australia (see Biddle & Hunter in this volume).

According to the 2002 NATSISS, 82.4 per cent of the CA sample aged 15 years and over answered that they fished or hunted in a group in the past three months. This represented 39 400 Indigenous Australians. A slightly higher, though not significantly different, proportion of males reported that they fished or hunted in a group than females (84.0% compared to 80.8% respectively). There was no difference between those aged 15 and 34 years (84.2%) and those aged 35 to 54 (84.9%). There was, however, a significant difference between the proportion of those aged 55 years and older who said they participated (66.7%) compared to the rest of the population.

Table 12.1. Percentage of Indigenous population in CAs who did and did not fish or hunt in a group in the last three months, by recognising and living on homeland, 2002[a]

	Does not recognise homeland	Lives on homeland	Does not live on homeland	Total
Did not fish or hunt in a group	25.8%	15.3%	18.9%	17.6%
	(800)	(3400)	(4200)	(8400)
Fished or hunted in a group	74.2%	84.7%	81.1%	82.4%
	(2300)	(18 900)	(18 100)	(39 400)
Total	6.7%	46.7%	46.7%	100.0%
	(3200)	(22 300)	(22 300)	(47 800)

a. The relevant population numbers are provided in parentheses.
Source: Customised cross-tabulations from the 2002 NATSISS CURF

The data from 2002 NATSISS suggests that recognising and living on one's homeland may influence a person's participation in the customary sector. However, the sample sizes are too small to make any definitive statements (see Table 12.1). Of the 3 200 individuals who did not recognise their homelands, 74.2 per cent said that they fished or hunted in a group. Of those who did recognise their homeland, those who lived there (22 300 individuals) participated most in fishing or hunting in a group (84.7%). Those who recognised, but did not live on, their homeland (22 300 individuals) had a lower participation rate (81.1%). None of these differences were significant at the 5 per cent level of significance. Given the small sample sizes, this is not surprising. So, although the figures suggest a positive economic benefit from land rights and native title (and customary harvesting rights under s.211 of the *Native Title Act 1993*), that is ignored in current public discourse on land rights and development, the survey methodology makes it difficult to properly examine such an important issue.

A person's participation in paid work may be associated with their participation in fishing or hunting. On the one hand, employment that is not directly related to fishing or hunting may take time away from such activities. On the other hand, employment may give people economic resources that facilitate such activities, or alternatively some people's employment may be directly related to fishing and/or hunting. Of those who were employed (24 800), 86.8 per cent reported that they participated in fishing or hunting in a group. Of those who were not employed (1600 unemployed, 21 400 not in the labour force), 77.8 per cent said that they did so. This difference was statistically significant.

Table 12.2. Percentage of Indigenous population in CAs who fished or hunted in a group in the last three months, by industry and hours worked, 2002

	Public sector	Private sector	CDEP	Total
	%	%	%	%
Full-time	90.0	71.4	85.7	83.9
Part-time (< 35 hours)	91.7	80.0	87.3	88.0
Total	90.6	75.0	87.1	86.7

Source: Customised cross-tabulations from the 2002 NATSISS CURF

Of those who were employed, there were inconclusive differences by sector of employment and the hours worked. Table 12.2 shows that those employed full-time are slightly less likely to participate in fishing or hunting than those who were employed part-time. However, this difference was not significant. The type of work also made a difference. Of those in non-CDEP employment, those employed in the public sector were more likely to participate than those who were employed in the private sector, with those employed in CDEP somewhere in between. None of these differences were significant

The role that CDEP employment plays in facilitating such activities as compared to not being employed is a consistent finding between the 1994 NATSIS and the 2002 NATSISS (despite different survey questions). Hunter (1996: 60–61) showed that those on CDEP were most likely to engage in hunting, fishing and gathering bush food in 1994. However, given the smaller sample size and change in survey methodology in the 2002 NATSISS, much less can be said about differences between employment sectors and geographical categories than was possible from the 1994 NATSIS.

Participation in, and payment for, cultural activities

Another aspect of the hybrid economy, which the 2002 NATSISS gives us some information on, is participation in, and payment for, cultural activities. Unlike fishing or hunting in a group, this question was asked in CAs and NCAs, making it possible to differentiate between remote Australia (incorporating remote and very remote Australia) and non-remote Australia (incorporating outer and inner regional Australia and major cities). However, it is not possible to differentiate between remote and very remote areas using the standard output from 2002 NATSISS.

Under 'Culture', the 2002 NATSISS asked respondents—with only a minor difference in wording between the CA and NCA questionnaires—whether, 'including activities done as part of your job, in the last 12 months did you: make any Aboriginal or Torres Strait Islander arts or crafts; perform any Aboriginal or Torres Strait Islander music, dance or theatre; and/or write or tell any Aboriginal or Torres Strait Islander stories?' Those who answered 'yes' to one or more of the above were then asked, 'Were you paid, or will you be paid, for your involvement in this activity/any of these activities?' and if so, for 'Which activities?', with more than one response being allowed.

Table 12.3 shows the three types of cultural activities that are grouped in the 2002 NATSISS and the participation rates for remote and non-remote populations. Of those who did participate, the proportion paid for doing so is also presented.

Table 12.3. Percentage of Indigenous population who participated in, and were paid for, various cultural activities, by remoteness, 2002

	Arts or crafts	Music, dance or theatre	Writing or telling stories
	%	%	%
Non-remote			
Participated	15.1	7.5	12.4
Paid (of those who participated)	20.5	34.9	20.8
Remote/very remote			
Participated	19.1[a]	10.4[a]	12.9
Paid (of those who participated)	51.9[a]	26.3	23.7

a. The remote/very remote estimates that are significantly different from the non-remote estimates at the 5% level of significance.
Source: Customised cross-tabulations from the 2002 NATSISS CURF

Table 12.3 shows that the type of cultural activity that people participated in is somewhat different by region. Those in remote or very remote regions were more likely to participate in 'arts and crafts' as well as 'music, dance or theatre' compared to those in the non-remote population. The proportion that participated in writing or telling stories was not significantly different. Within those who participated, over half of the population in remote or very remote areas were paid for the activity. This was much higher than the proportion for the non-remote population and significantly higher than the proportion that was paid for the other two activities.

Ability to meet cultural responsibilities while in employment

The final variable examined is the ability to meet cultural responsibilities while in employment. Under 'Employment', 2002 NATSISS asked all employed respondents (including those employed in CDEP), 'Because you work, is it possible to meet all your cultural responsibilities?'. As with participation in, and payment for, cultural activities, this question was asked in both CAs and NCAs, thus allowing for differentiation between remote and non-remote Australia. In NCAs, at least, cultural responsibilities were described to respondents as including 'such things as telling traditional stories, being involved in ceremonies and attending events such as funerals or festivals'.

Table 12.4 gives the number of people (who were employed) who: did not have cultural responsibilities; had cultural responsibilities and were able to meet them; and had cultural responsibilities but were unable to meet them.

Table 12.4. Presence of, and ability to meet, cultural responsibilities while in employment, by remoteness (number of persons), 2002

	No cultural responsibilities	Able to meet responsibilities	Unable to meet responsibilities
	%	%	%
Remote/very remote	7732	27 600	4402
Non-remote	31 578	35 360	23 592

Source: Customised cross-tabulations from the 2002 NATSISS CURF

Table 12.4 shows that, for the total surveyed population who had cultural responsibilities while in employment, 69.2 per cent were able to meet them. There were significant differences by remoteness with 60.0 per cent of the non-remote population able to meet their responsibilities, compared to 86.2 per cent in the remote population.

Table 12.5. Percentage of population who were able to meet cultural responsibilities, by industry and remoteness, 2002

	Public sector	Private sector	CDEP	Total
		%	%	%
Remote/very remote	79.2	62.4	93.7	86.3
Non-remote	61.2	54.3	78.5	60.0
Total	65.9	55.3	89.6	69.2

Source: Customised cross-tabulations from the 2002 NATSISS CURF

As Table 12.5 shows, industry sector can influence these differences.

CDEP employment appears to be the most conducive form of employment to allow Indigenous people to meet their cultural responsibilities, followed by employment in the public sector. Apart from the difference between public sector and private sector employees in non-remote Australia, all these differences were significant at the 5 per cent level of significance.

Table 12.6. Percentage of each State/Territory engaged: in fishing or hunting in a group; paid; and unpaid arts and crafts activity, 2002

State	Remote and very remote area who fished or hunted in a group	Total State who participated in arts/crafts	Those who participated in arts/crafts who were paid for doing so [a]
	%	%	%
NSW	N/A	16.4	15.4
VIC	N/A	14.4	17.0
QLD	82.4	14.4	25.9
SA	76.0	20.9	43.0
WA	80.1	16.4	33.3
NT	83.5	19.5	67.6
ACT/TAS	N/A	11.4	15.6
Australia	51.0	16.2	30.6

a. The percentages in this column represent those who reported that they participated in arts/crafts.
N/A Refers to those States or Territories without a remote or very remote population in the sample.
Source: Customised cross-tabulations from the 2002 NATSISS CURF

In Table 12.6 we provide some slightly different summary data by State/Territory on fishing or hunting, participation in arts and craft, and in being paid for arts and crafts. Bearing in mind that the question on fishing or hunting was limited to CAs, it is instructive that such activities are prominent in jurisdictions with land rights and sea access. It is also interesting that those States that have most land rights and that have benefited most from arts marketing support also have the greatest proportion paid for arts and crafts participation. The Northern Territory stands out in both.

It is most disappointing, in terms of the data on fishing or hunting, that a more comprehensive and detailed geographical analysis of such activity is not possible under the 2002 NATSISS, especially when compared to what is possible using data from the 1994 NATSIS (Smith & Roach 1996: Fig. 6.1).

Shortcomings in the 2002 NATSISS

There are shortcomings in the 2002 NATSISS's capacity to generate useful data on the customary sector. These are explored here with a focus on the data pertaining to fishing and hunting.

Coverage of Indigenous population and key customary sector activities

We identify three key shortcomings in the 2002 NATSISS's coverage of economic activities with customary links. Firstly, and most importantly, it is unclear why 'fishing or hunting in a group' was not included as an option under the involvement in social activities question in NCAs, especially when a similar question on 'hunting, fishing and gathering bush foods' was included in all areas in the 1994 NATSIS (Hunter 1996; Smith & Roach 1996). Subsequent research shows that the customary sector is of economic importance to non-remote Indigenous people, especially in inner and outer regional areas. For example, Gray, Altman and Halasz (2005) estimated that the value of wild resources harvested by members of the Indigenous community of the Wallis Lake catchment in coastal NSW was between $468 and $1200 per annum, accounting for between 3 and 8 per cent of the gross incomes of Indigenous adults. The 2002 NATSISS clearly ignores customary activity known to occur in these areas by not allowing respondents in NCAs to indicate their participation in fishing or hunting activities.

Secondly, it is unclear why 'gathering bush food' was not either included in the above question as it was in the 1994 NATSIS, or asked as a separate option. As women are the predominant gatherers, there is a clear possibility of gender bias here. Third, a very significant aspect of the customary sector—the role of Indigenous people in land and sea management and biodiversity conservation—is ignored in both the 1994 NATSIS and the 2002 NATSISS (Altman 2005b; Smith & Roach 1996).

Economic versus cultural activity

The inclusion of hunting, fishing and gathering activities in the category 'Employment and income' in the 1994 NATSIS was positive. However, its inclusion under the sub-category of 'voluntary work' was not (Smith & Roach 1996), especially as it was included in a list of other voluntary work activities that could not be considered economic activities with customary links.[4] The 2002 NATSISS potentially ignores the economic significance of such activities by addressing them only as cultural activities. Similarly, while recognising that monetary benefit may be gained from activities such as the production of art and crafts, the 2002 NATSISS categorises such activities exclusively as cultural rather than economic.

As shown in the recommendations section below, with reference to the Canadian Aboriginal Peoples Survey 2001, such activity can be recorded under more than one category.

Group versus individual activity

It is unclear why 2002 NATSISS focused on fishing and hunting as a group activity to the exclusion of individual activity, especially as the question was not asked in this way in the 1994 NATSIS. Indicative fieldwork undertaken by Altman (2003a) in January 2003 indicates that 20 of 40 harvesting events recorded at Mumeka outstation were by individual hunters or fishers who were also, coincidentally, the most productive. While we are not suggesting that these individuals would not have participated in a group harvesting activity within a three-month period, it is unclear why this distinction is made in circumstances where people clearly hunt and fish alone. The question appears to be primarily concerned with Indigenous people's involvement in group activities at the possible risk of excluding accurate information on the labour individuals had invested in these activities. [5]

Seasonality

It was an improvement in the 2002 NATSISS CAs interview to specify a set time period of three months in terms of fishing or hunting activities, compared to the unspecified time period in the 1994 NATSIS question. However, it is still

---

[4] Smith & Roach (1996: 74) noted that a number of definitional issues needed to be reconsidered with regard to the voluntary work question in the 1994 NATSIS arguing, for example, that 'subsistence activities are clearly significant but should be treated separately to voluntary work in a questionnaire schedule, being more appropriately classified as unpaid own account production in the informal economy'.

[5] It is worth noting with regard to the issue of labour effort expended, that the question on hunting, fishing and gathering bush food as voluntary work in the 1994 NATSIS also delivered information on the amount of time spent on these activities. This information would prove extremely useful in determining labour input into such activities. Again, it is unclear why the 2002 NATSISS did not ask for this information.

inadequate. The primary reason for this is the influence seasonal variability may have on hunting, fishing and gathering activities and the consequent potential for seasonal bias in recording for a period under 12 months or a full seasonal cycle (Altman & Allen 1992: 145; Altman, Gray & Halasz 2005: 15). [6]

Lack of comparability between the 1994 NATSIS and the 2002 NATSISS

A significant shortcoming of the 2002 NATSISS is the inability to confidently compare data about customary activity with the earlier 1994 NATSIS data. Biddle and Hunter (in this volume) have alluded to this shortcoming.

Restrictions on geographic analysis

A major problem with the standard output from the 2002 NATSISS is the inability to differentiate remote areas from very remote areas. Furthermore, the publicly available unit record data confuses the concept of remote and non-remote with CAs and NCAs. Especially with regard to the social activities question, the ABS must allow users to identify those who were and those who were not given 'fishing or hunting in a group' as a response option.

Recommendations for NATSISS 2008

We limit our recommendations here to the improvement of data on hunting, fishing and gathering activities in the next NATSISS. Biddle and Hunter (in this volume) point out that the change in the way information was gathered on these activities makes comparison between the 1994 and 2002 surveys 'uninformative, if not meaningless'. While we recommend some significant changes to survey questions on customary activities, we believe it is important that some comparability is maintained between the 2002 NATSISS and the NATSISS scheduled for 2008. This can be done by repeating the 2002 NATSISS questions under cultural activities, but by adding one or two more specific questions under Employment and/or Income. This of course needs to be done while taking into consideration the time burden on respondents of such surveys. However, the importance of such activities to the access to resources of many Indigenous Australians necessitates asking the right questions.

There are some straightforward recommendations on how information on this range of activities should be collected in NATSISS 2008 which will not substantially increase respondent burden:

- Information needs to be collected for all Indigenous Australians and not be limited to just those in CAs.
- Information needs to be gathered as economic activity, not just as cultural activity. One aspect of the question used in the 1994 NATSIS but not in the

---

[6] This is not to deny the fact that the longer the time period, the greater the potential for recall error.

2002 NATSISS that is worthy of re-inclusion is information on labour effort expended on customary sector tasks in hours per week.

- In an economic context, the question should not be confined to conducting these activities in a group.
- The question needs to include gathering of bush food as an activity, as was done in the 1994 NATSIS.
- Information needs to be gathered on the basis of activity over the last 12 months in order to overcome problems of seasonal bias.

The Canadian Aboriginal Peoples Survey 2001 provides guidance on how questions on hunting, fishing and gathering could be collected in order to better capture their significance as economic, as well as cultural, activities. The 2001 survey comprised a core questionnaire as well as distinct supplementary questionnaires for the Metis and Arctic communities (Statistics Canada 2003). Under 'Section C – labour activity' the core questionnaire asks whether respondents have hunted, fished, gathered wild plants and/or trapped for either food, pleasure, commercial use or other use (medicinal, ceremonial) over the past 12 months. Unlike the 1994 NATSIS and the 2002 NATSISS questions, hunting, fishing, gathering and trapping are each listed separately. Collecting information on the purpose of activity seems advisable, given uncertainty as to whether such activity is being undertaken for subsistence, commercial or recreational purposes. [7]

The Arctic supplement to the Canadian survey devoted an entire section to 'Household and harvesting activities' in which much more detailed information was gathered on commercial and unpaid household activities related to the harvest, preparation and consumption of 'country food'. Information was gathered on inputs, including food preparation and hunting equipment repairs carried out in the home and household investments in harvesting equipment. Data collected on outputs included whether 'country food' was eaten in the household, shared outside the household, exchanged or sold; the proportion of household food that was made up of 'country food'; and household income from the sale of 'country food' and other harvested products.

The Arctic supplement also asked for more information than may be feasible in NATSISS. And asking respondents to estimate actual amounts of income derived from harvesting activities may not produce reliable data. The Metis supplement

---

[7] Altman & Allen (1992: 138) note that while participation in the informal economy is not limited to Indigenous Australians, the major distinction is 'that for those Aboriginal and Torres Strait Islander people who continue to live off the land, subsistence is their 'primary' work and income-generating activity'. Hunter (1996) suggests that caution should be exercised in interpreting hunting, fishing and gathering activity automatically as customary activity. Hunter (1996: 60) warns that hunting, fishing and gathering in urban areas requires a certain level of income to allow engagement and may be more accurately interpreted as a consumption activity as opposed to a productive economic activity providing an alternative to market sector employment..

provides an alternative approach by asking respondents about the importance of income from hunting, fishing, gathering, guiding, trapping and/or art and craftwork production in making ends meet in the household, using a four-point scale from 'not important at all' to 'very important'. In the Australian context, land, habitat and species management activities could be included in such a question.

We would gain a much more comprehensive picture of the economic importance of customary activity within Indigenous households in Australia if we collected information on participation in, and the purpose of, such activities (as in the Canadian core questionnaire); labour effort expended on them (as in the 1994 NATSIS); and the contribution of such activities to household income (as in the Metis supplement).

Conclusion

In this chapter we have examined the 'real' economy in remote and very remote Australia, where 27 per cent of the Indigenous population resides. Of this population, 17 per cent resides in discrete Indigenous communities, frequently on Indigenous-owned land and very remote from markets and mainstream employment and business opportunity. The customary, Indigenous or non-market sector of the economy makes contributions to Indigenous people's livelihoods that are not reflected in standard statistical collections like the five-yearly Census of Population and Housing.

The 2002 NATSISS provides strong statistical support for the view that the real economy in remote Indigenous Australia is made up of three, rather than two, sectors. The 2002 NATSISS information reinforces a view that other ABS statistics that ignore the non-market sector understate the extent of Indigenous economic participation and wellbeing. The policy ramifications of this finding is that the customary sector might provide economic opportunity, and that major programs like the CDEP scheme, as well as land rights and native title rights, might be useful instruments to facilitate enhanced customary participation with positive livelihood outcomes.

The 2002 NATSISS has generated some important information on hunting and fishing in some contexts; on paid and unpaid cultural work; and on the impact of employment on the ability to meet cultural obligations. As social scientists, we welcome new data sets that allow testing of new ways of looking at the Indigenous economy. We are naturally very disappointed that these data do not go far enough, and strongly encourage the ABS to enhance efforts to better capture customary activity throughout Australia in 2008.

13. Panel Discussion: Diverse perspectives on the evidence

Larissa Behrendt, Tom Calma, Geoff Scott, (with introductory remarks by Jon Altman)

Jon Altman

The aim of this panel discussion is to get some diverse perspectives on the evidence. To do that we have arranged a panel of three people who will evaluate the evidence on socioeconomic outcomes for Indigenous Australians as articulated in the first day of this conference and then look at those with respect to their diverse professional and academic experiences. We thought it would be very useful at the end of Day 1 of this conference to have some discussions that would tease out how and, indeed, if statistical collections can add substantially to the debate on Indigenous policy. We are sure these statistics can do that, and I am sure the panel would agree with that. One of the main questions to be addressed is how large scale statistical collections can be improved to collect more accurate information about Indigenous circumstances so that policy can be better informed to address Indigenous disadvantage and need. What I will do is introduce my distinguished panel in the order that they will speak, and invite them to speak for up to 15 minutes on what they have heard today.

The three panelists are: Larissa Behrendt, Professor of Law and Indigenous Studies and Director of the Jumbunna Indigenous House of Learning, University of Technology, Sydney; Geoff Scott, who has had a long and distinguished career, and is currently holding two positions: Chief Operating Officer of the New South Wales Aboriginal Land Council in Sydney, and Distinguished Professor in Public Policy Research at University of Technology in Sydney; and Tom Calma, the Aboriginal and Torres Strait Islander Social Justice Commissioner, and Acting Race Discrimination Commissioner with HREOC.

Larissa Behrendt

I would like to begin by acknowledging country and Matilda House's welcome this morning and paying my respects to the Ngunnawal people, and to the many Aboriginal nations who are represented at this conference among the audience and participants.

I've been asked to make a few reflective comments, and I do that as someone who has a legal background, rather than being a statistician or a policy maker.

In the law reform work we do at Ngiya, where we are primarily trained as lawyers, we are very reliant on statistical analysis as a part of the methodology that we sometimes engage in. But more often we rely on people like Don Weatherburn and the ABS, the Productivity Commission and the NATSISS to be able to have a stronger base from which to articulate what we hear anecdotally. This base data allows us to counterpoint the rhetoric and ideologies that we see influencing the legal and policy agendas we are working on within our research unit.

One of the things Jon's speech reminded me of this morning is that there are many key roles that are being lost with the abolition of ATSIC and he touched on one when he mentioned its very vital role in collecting data that was independent of government. It seems to me that the ability to provide a form of independent advice from government has been lost in many of the roles that ATSIC used to play. So, for example, we saw it play a similar role in relation to the issue of native title. We saw it play a similar role in relation to the way it analysed Australia's performance under our key human rights organisations. And we saw it in the way it focused in this era of reconciliation on a more developmental rights framework, including agendas such as a treaty. These viewpoints were in many instances much more reflective of Indigenous people's perspectives on policy issues than of the Australian Government's, and often stood in stark contrast or opposition to federal government policy. So, with the eradication of that national representative structure, there has been an increasing feeling amongst Indigenous communities and families of incredible disempowerment and I guess also too we are often hearing comments about how quickly these changes have come along and they are saying 'I didn't realise it was a new era because the changes have happened so rapidly'. With the fast pace that ATSIC was abolished and a new system was put in place, many Indigenous people felt left out of that change because indeed they were not consulted about it or included and they felt left behind by these enormous changes happening within the sector.

Despite the rhetoric of government that the abolition of ATSIC structures has meant government can now directly engage with Aboriginal and Torres Strait Islander communities, the questions that we've heard asked most often is one that Jon Altman raised this morning, and that is: who is it that government is actually engaging with? Many commentators have noted there is a real inconsistency with policy on the one hand that it is interested in greater interaction and negotiation with communities through things like SRAs but on the other hand seeks to destroy a regional representative model that would have facilitated that negotiation.

It is interesting to note that the areas that seem to have done the best in navigating SRAs are areas like Murdi Paaki in New South Wales where there was a very strong ATSIC regional council and their ability to navigate the changes has very much been led by the very strong regional council work that was being done there. We need to remember that there is evidence, both here (much of it done by CAEPR) and overseas (particularly in the mainland of the United States and Canada) that shows that socioeconomic outcomes are best improved when Indigenous people are included in priority setting, policy development and program delivery. It is not simply an ideology of self-determination that shows that: it has been actually proven by research. This evidence of the importance of having Indigenous people involved in the administration, development and planning of policy has been overlooked in the rush to enter SRAs.

The underlying issue that can be gleaned from Jon's observations this morning is that much of the data that would assist communities in making informed choices about entering into those sorts of agreements—what areas they might want to cover, what additional services they might need and seek, what sorts of priorities they might want to focus on—has not been provided to allow Aboriginal people to enter into these agreements in an informed way. Nor is there a framework for assessing those agreements to monitor their effectiveness or their fairness.

I would also note that one of the things we have noticed from our research work is that, in addition to the sorts of enquiries we get from community people about what SRAs are, how do we figure out if we want to be involved in them, and what sorts of things should we be looking at? We are particularly interested in the number of people who work within various State and Territory government departments who seem to be asking the same sort of question about the process, the substance and the evaluation of SRAs.

One of the key problems is that the SRAs have been led by the ideology of mutual obligation and other aspects of Indigenous policy have been driven by an ideology of mainstreaming. Ideologically-driven policy is not always in line with research-based policy—in fact, it rarely is—but the latter has a better chance of addressing socioeconomic disadvantage than the former.

I want to conclude by flagging three ways in which we can improve these outcomes and use the sort of statistical analysis that we have been talking about today.

Firstly, the ATSIC legislation provided for a regional planning process. With the abolition of ATSIC and the regional councils, we have lost a mechanism by which we can engage in that sort of analysis and planning within each of the regions. I think it's really important that we re-establish that as a framework for working through any negotiation process that is intended by the SRAs. It is true to say that the promise of the regional planning processes that were articulated in the ATSIC legislation were never actually met. But I think it provides a framework for the groundwork you would need to do to actually start effectively engaging communities and targeting priorities in different areas. The example I would use of a really good regional mapping process is the

Fitzgerald Report in Cape York—the Cape York Violence Study—which, even though it was initiated by concerns about levels of violence and substance abuse, did a mapping exercise that was really thorough in terms of looking at issues of governance and the involvement of State and federal agencies in the area. It's an excellent example of how you could map out a regional area. So I think you need to go back and do that planning process that was foreshadowed but seems to have been lost in the impetus to try and come up with large numbers of SRAs without doing that sort of groundwork beforehand.

The second thing I would flag is the need to rethink the way in which SRAs are approached. There is no doubt that the idea of negotiating with Indigenous communities is an important one. I have already mentioned that there is a great deal of evidence to show that you need that sort of involvement to ensure programs are going to be successful. In theory, the notion of SRAs should actually be able to provide some way of engaging meaningfully with Aboriginal communities. I think one of the real concerns is that that promise has been lost in the way that SRAs have been approached in practice. I think we need to rethink the way the process for engaging in SRAs is thought out and make it much more transparent. This process needs to include issues like who actually has the mandate and who is making the agreement. We need to ensure there are mechanisms in place to monitor the content, to ensure they are not breaching basic human rights or that there is no bargaining for essential services. Also, there should only be a commitment made to provide something as part of the exchange, particularly by governments, if there is a capacity for them to do so. You will all be familiar with the stories of kids turning up to school as part of their agreement, only to find out there were not enough teachers or classrooms. There also needs to be more attention given to the issue that was raised in this morning's presentation about that real need to come up with a framework for monitoring and evaluating those agreements to ensure that the outcomes are actually beneficial to Indigenous people, and they are making some kind of impact.

The third thing I would raise is the way in which we need to empower communities to work alongside the statistical research that we do. There is a need to increase the number of Indigenous people in the public service. One of the really unfortunate side effects of the abolition of ATSIC and the move from ATSIS into the Office of Indigenous Policy Coordination (OIPC) was the loss of numbers of Indigenous people working within the public service, particularly across the Senior Executive Service. We have actually seen a large drop in the number of Indigenous people who were transferred from OIPC into other government departments but found that this was a very different environment than the one they signed on for, so they have quietly left. I think that has been a real step back for us, because it has been an important development that more and more of our people have been working within the policy area. It's never an easy choice to make as an Indigenous person, but I really respect people who have done that. It is a real shame that we have lost so much of that capacity under the new arrangements. Ensuring those numbers increase in the future will be another big challenge.

Geoff Scott

I would like to start by acknowledging country and the Ngunnawal people on whose land we gather today. I would like to say this forum has been quite useful. I found 70 per cent was quite useful, and 30 per cent was completely over my head. But it also raised issues where I should take a lot more interest and also some concerns. But the honesty about the data and its usefulness is very enlightening.

I would like to pick up the first point in Jon Altman's paper this morning: that the central tenants of policy have moved over the last few months. The principles have moved. I'm sorry, there is no principle, it's about mainstreaming, mutual responsibility and whole of government. That's a positive move that was pursued by ATSIC for a number of years but was not pursued by government. What I think it does acknowledge with government is that they have continued to fail in service delivery. For the amount of money going in and the effort involved, the results have not been what were intended and there are reasons for that. The national commitment to improve services to Aboriginal and Torres Strait Islander people was put in place in 1992 and confirmed again by the Council of Australian Governments (COAG) in 1994, and then they came up with the COAG Agreement in 2000.

What was missing in all of those issues was the authority of central government: it was not there. It was an impediment to service delivery, an impediment to research, it was an impediment to access to data. That is one thing I think you will be facing in your studies as well. That is there now, but it only came about when ATSIC was removed from the environment. The accountability provisions of government have changed as well. The accountability provisions now are not measuring performance as disbursement, but accountability as being accounting. They were the measures imposed on ATSIC.

They've moved now to accounting for activity, and for some outcome performance, which is a positive development. It's a shame it has taken so long to come out. The point that should be made here, about the way ATSIC went about its business, is that the critical insider is not tolerated in this government. It is not tolerated in today's society. If you criticise government, you will suffer. I think most people here today are aware of that. Today it's about controlling the media. There are positives in the environment. If we make our comments about the history of what has happened, then greater minds will look at that and make their assessments in the future when the venom goes out of the debate.

The shift to a greater focus on communities and regions is not a negative development. But it needs a fundamental policy framework to underpin it. It needs a policy framework underpinned by research and analysis. I do not know how you identify what a community is. In a lot of Australia, it is a very nebulous concept. Who are you talking to, who do you have an agreement with, who makes up your community, and who's got authority? And I'm sure the people in OIPC and the Indigenous Coordination Centres (ICCs) are grappling with that now.

It's also about identifying the issues, and needs and concerns, but also both the symptoms and causes of those. Identifying the existing service mechanisms, identifying the baseline data, and then what milestones you are looking for. What are the impacts, and how we measure the outcomes? That's why, in terms of today, the 2002 NATSISS survey (and its outcomes) is just one aspect of the whole package that could be backed up by the case studies, by the longitudinal studies. But it is a very useful mechanism. The challenge we face is to meet this emerging requirement. The Productivity Commission work was mentioned a couple of times this morning. What I think that report and the Commonwealth Grants Commission confirms is the dearth of data availability, of access to data and of analysis.

One massive problem I see in the current policy framework is that we develop all these frameworks of what the indicators were. They are sitting there up at the national level.

On the ground we've got SRAs. There's no link, and that should be a worry to us all. We have lots of activity. That's not to say SRAs are bad. They are potentially a useful mechanism, but by themselves they are quite dangerous. In saying that, I would like to make the point that outcomes and impacts is a function of a number of policy initiatives, and the resources applied to those initiatives. It's about incentives and penalties, and service accessibility and strategies. It's not about one program. Some of the data presented this morning actually brought that out. I was in the States recently. Listening to the TV, advertising for a drug, they say take drug A and it will fix B (your condition), but by the way, here are the side effects, and they read off a long list. An interesting analogy, I would suggest.

But on those issues this is the problem of the way in which policy frameworks are implemented. We have no way of measuring the intended or unintended consequences of a policy initiative. Some of the work this conference is doing in trying to identify some of those relationships is extremely important. It is a body of work that has not been done. This is exacerbated by the silent mentality of government agencies, at both Commonwealth and State levels. It is a challenge for the new environment. They are trying to work in practice but on the ground it is not working very well. I do not think you can ask a Minister to give away his power over money that he is responsible to government for. Nor will a Secretary give away decision making power over something that his Minister will hold him accountable for. So you have to change the very tenets of government policy to give this process a chance to work. The other point I would like to make here is that we have to try to understand in this policy context the difference between development and intervention. In this country, most of the policy initiatives you see are interventionist. The SRAs are interventionist. The work in the Cape York report focused on interventions, and mentioned the difference between development and intervention, but did not take it further.

Another important point that was raised this morning was about social capital, about human capital. Intervention to date has focused on services and infrastructure. You need to build a sustainable social capital framework to underpin that. In Australia when you mention intervention it is a dirty word, but overseas you can do it. We need to build the necessary framework for that.

Boyd Hunter raised the issue this morning about poverty. The whole poverty debate and how narrow and immature it is in this country. It's not just income. It's about powerlessness and exclusion. It's about capacity. Without taking on poverty and all those related aspects of exclusion, one will not succeed.

One point that was raised this morning was mobility. It's very important in today's emerging policy environment, especially the policy arising from the reports issued by the CGC (2001). When the Commission was doing their work, they tried to do an absolute needs-based measurement and they were told no, you're not allowed, we want a relative one. It's all about an ideology of moving specific funding for Indigenous affairs to the north, to the more remote areas, and then requiring government line agencies to fill the gap left behind by the other programs. So it's very important from that point of view. Coming out of that presentation this morning is a number of factors looking at what mobility was. It's a function of security and social capital. If someone's not scared to move, they will do it. If they are, they will stay at home. And that's no different to anyone else. That's human nature. Everyone likes the status quo. They will not move unless they see a positive in moving. It's those issues that are brought out of that point for me.

On what the drivers and levers are, have you seen recently there is a move on having home ownership on community lands? There is a move on promoting scholarships to help people to leave their communities for education. This concept of orbits has been put up now and bandied around. We have transient employment issues where the great new initiative for Aboriginal people is fruit picking, as suggested in a CIS study. Send people down south at a cost of about $12 000 to earn $4000. I do not know where the economies come in there, but it made a good story.

Another issue is child mortality which was very interesting this morning as well. I have put it that this study is going to be very useful for someone who is a rational policy maker. There is data being presented here that is quite useful in explaining the relationship between the different factors. Coming out of the child mortality debate this morning was the relationship between a stable and safe and secure home environment, both from an economic and a health point of view. The discussion about social stresses and the primacy of the home environment were equally important. We need to identify what those positives were.

Another point there is that policy makers drive demand, mobility being one example. We all must be aware of what use the information produced will be put to. There must be an implicit acknowledgment there that the economic growth and development required to sustain people in remote areas is just not there. It's fanciful to keep saying that it is. Government put the people on the excisions and the reserves and now it's come to the reality that it's going to cost a lot of resources to maintain that, and the economies are not there.

There was a point this morning about the CDEP scheme. Be very careful about confusing the vagaries of CDEP. The ABS classifies a CDEP participant as being employed. If CDEP representatives were here they could tell you that from their perspective, CDEP participants are unemployed.

The dual focus on development versus intervention is critical. I keep making that point because I think it is very important. One issue I am very critical of, and have been for some time, has been government's service delivery—it's all from a project mantra. Projects are great because the financier controls the project, the timing, the outcome, and also controls the credibility and the sustainability of the organisations that get the money. I work with an organisation at the moment and they spend most of their time chasing next year's money, not doing what they are supposed to do because of the uncertainty that is derived from that. In terms of capacity, they're the sorts of things to look at. We need to look at what these initiatives are doing. We are pitting remote versus non-remote and we must be careful about those issues. It's a real issue emerging between Aboriginal communities at the moment, between north and south. It's going on behind the scenes, but the debate is going on. And no-one wants to do that, no-one in the south wants to deny someone in the north for the relative needs they have, because they do have those needs, but with a finite amount of resources that debate will happen. And it causes dissension.

If we are going to have sustainable development, we need to focus on people-centred money, or 'hot money'. International development discussions talk about hot and cold money. Cold money is focused on infrastructure and services. Hot money is focused on people. It's about increasing the capacity of people to sustain those services and carry on when the government leaves, or when the project leaves. That debate is not happening here at the moment. It takes a lot of time to heat that money up and make it hot. Just moving on to the large scale collections we are focused on here, and the aggregate data. You have to be careful with those. In terms of methodology we do not lose sight of the indicator of predictors of change, both positive and negative. A bad experience can inform as much as a good experience can at times. You cannot lose sight of that. You cannot lose sight of the detail itself. Usefulness can be the causality of aggregation.

From a policy point of view, the first question I ask when I receive a statistical report or a research report from a research forum is, does it confirm my gut feeling? Does it confirm the observations? If it does, great. If it does not, it is important go through the data and try to talk to someone about it. Try to get a feel for policy makers who have 101 things to focus on, most of which in today's environment is what the Minister wants, not what the public servants want. They work for the government, not for the client.

The 2002 NATSISS survey cannot be used in isolation. It needs to be coupled with case studies and with longitudinal studies, and that is what we are missing at the moment. They should inform each other. The real problem here is issue identification within the extant constraints, be they financial, political or other constraints between governments.

I would like to see a study on the transaction costs between governments in Indigenous affairs. You would find it's enormous. Something I'm sure no Senate would want to embark on, but it is important.

The role of a research forum like this, noting that putting data up that embarrasses people or puts you on the spot will cause you pain, is to keep the bastards honest. It's also about remaining objective. Another measure is utility: What policy has changed or been affected by these studies? I think in terms of looking at the usefulness for what you are doing in the future. If policy-makers are ignoring it, then we can have good press releases and we can have good conferences, but not a lot changes. But it's still important to keep doing the work, keep the flame alive.

The issue for Aboriginal people today is a day-to-day survival issue. Some of the studies will show that in reality when working at a community level and working with people, it is a day-to-day survival issue. We cannot lose sight of that, and the impact of government policy on that.

Tom Calma

Thank you, Geoff. Let me also acknowledge all the Ngunnawal peoples of Canberra today and thank them for allowing us to hold this conference on their lands.

I have found the conference so far very, very useful. A lot of information has been given. I would like to commend CAEPR for being able to pull it together. Firstly, it allows all presenters to provide scrutiny of their peers, and to be challenged about their analysis of the statistics. The objective of influencing the ABS about how they might construct the next NATSISS is critically important. It is good to note that there are a number of bureaucrats here, although some more senior bureaucrats would have been welcome. Many of the people here are in the chain but they are not the ones who are making the decisions, and are not the ones able to influence the decision makers. So we need to look at ways to be able to engage the most senior of bureaucrats, and the most senior of politicians who make the policy or direct the way policy goes, to participate in forums, or at least be informed by forums of this nature.

I would like to give a plug for a seminar I will be conducting in collaboration with the Productivity Commission and Reconciliation Australia next month. It's looking at the Productivity Commission's report *Overcoming Indigenous Disadvantage*. There will be a number of Indigenous speakers in a similar forum to this, looking at it a bit more from a policy perspective and also the engagement of many Indigenous people who are able to give their perspective on various elements of the report. I would welcome you there if you have the opportunity.

From my perspective, both as the Aboriginal and Torres Strait Islander Social Justice Commissioner and previously in bureaucracy, I have found NATSISS to be a useful tool, but not as useful as other reports that are produced, such as the Health and Welfare Survey, CHINS, and the various reports the Productivity Commission puts out. What is most important, and what I find of most value to many of us who are not statisticians or involved in high level research, are the reports from academics from all institutions who are able to do that analysis and are able to provide us with data that we can understand and get a handle on.

While saying that, while the data is useful, it is only one element of trying to address the situation of Indigenous peoples. It was interesting to hear one of the speakers this morning talk about housing. It is often through practical reconciliation trying to look at equity between Indigenous and non-Indigenous people that issues can be addressed. We look at housing as a typical example, where Indigenous people are so far behind in home ownership, either outright ownership or in the purchase of homes. But let's take it back a step—it's only 40 years since we were formally recognised and able to get wages, so we have a lot of catch-up to do. It was only 40 years ago that we got a whole lot of rights that we previously did not have. So, to be able to compare the Indigenous to the non-Indigenous population, and where we are at, is sometimes misleading.

It is also equally a problem when you look at statistics. For the policy developers here, the analysis of the statistics is an analysis of people and their lives and most often they are in fact, never acceptable or show positive improvement. But to be able to develop policies to address situations and to expect overnight outcomes is unrealistic. So we have to look at longitudinal studies. Geoff mentioned, and I fully support the idea of longitudinal studies and specific case studies, and I will talk about that a bit more. But just listening to some of the speakers this morning, it became evident that we need to do more to coordinate some of the numerous surveys that are undertaken nationally by different institutions, because from the user's end of it, it is difficult to know which is the best or most accurate interpretation of the situation of Indigenous Australians.

The danger is people can pick up just one report and focus on one element of information and expect to develop a policy. But while that is not good practice, it is nowhere near as dangerous as a Minister or Prime Minister visiting a community and somebody putting up their hand and saying, 'I would like to own a house', so we get a new policy on home ownership. It's these knee jerk reactions that I think are dangerous for Indigenous people and our advancement, because what is really required from our perspective is to be fully and meaningfully engaged in any process that involves us, and any policies that are being developed that affect us. And that engagement has to be done from the perspective of knowing precisely what the engagement is going to mean and how it is being influenced. Because, to have any sustainable outcome we need to have full engagement, and Larissa and Geoff both talked about this.

I mentioned the longitudinal surveys and case studies which are important. For example, the report on Wadeye recently done by John Taylor was very good, and one would think that should have been embraced by the government as saying that these are the facts, let's get on and deal with them. But it was not necessarily embraced at all—it actually had a negative effect.

We are still waiting on the lessons learnt from the COAG trials. When we consider that the whole of the new arrangements for Indigenous affairs are predicated on the lessons we have learnt from the COAG trials, one would suspect we should have had some document outlining what those good lessons were. But we have not. Unless we get data that is useful and able to be translated into policy that becomes useful, we do not achieve very much. We have sad situations. Palm Island, for example: we all know what's happened on Palm Island, and the investment that has gone into Palm Island, Mutitjulu, the AP Lands, you can go all over the country. Who gets the blame at the end of the day for the lack of advancement? Indigenous people, it's all our fault. When really it's government policy and we are only reacting and surviving through that.

I thought I would ask my Indigenous colleagues here to raise their hand if you have ever participated in a NATSISS survey. We've got two, because I did too. That was the one last year. It was interesting, and this is a bit of a reflection on how accurate are the statistics when we have two out of 30 Indigenous people here who have participated. I heard one in 30 nationally earlier, so maybe that's a good stat.

When we look at Canberra, for example, my situation was interesting. At the time of completing the survey, I had just returned from overseas and this is interesting because mobility statistics is one of the lines of questioning and it indicates that Indigenous peoples are mobile. I must say the NATSISS survey is complicated. The interviewer said it will only take you 20 or 30 minutes, but $1\frac{1}{2}$ hours later I was still going, and still trying to work it out. The questions are complicated and they're multi-dimensional. You think you are going down one track, and then it changes.

It was interesting because for statistical purposes I had just come back from Vietnam, I'd been in temporary accommodation, and by the time I was surveyed I had moved back into my house. I had lived in three locations in the past 12 months. I was fully mobile. So those statistics are sometimes questionable. I think just listening to what was indicated earlier today, there needs to be some consistency in questions asked of people living in urban areas versus those in rural remote areas. But because the form is complicated, the real challenge is making sure that the surveyors are skilled to be able to undertake the surveys in a way that is going to be able to elicit the right kinds of answers. What is also important is to make sure there is consistency across all the surveyors because if there is not, the outcomes will be different because the ways you present those questions will determine how the survey may be influenced. One question it would be interesting to know the answer to, and I don't know whether it's asked, is whether anyone has been surveyed previously. The question is whether the survey is hitting the same people each time, or different people.

In relation to the training of the surveyors, I think its going to be important. I had a question in relation to what I think is a good initiative: the six Indigenous ABS officers who are placed around the country. From my perspective, they would have a good role as an ongoing promotion tool out into the communities and in bureaucracies to be able to get people to understand what the surveys are about. Why that is important, and some may remember this, is that in the late 70s/early 80s we had the Electoral Education Officers, who were itinerant officers, who floated around the country and the Australian Electoral Commission engaged them to promote the electoral system. These people could take on a similar role. There is also a role to start educating the community, and I will touch on this more in my session tomorrow on rights. But it is in relation to prior and informed consent. Do not wait until a month or so before the survey to tell people that you are going to do it. Use it as a process to educate the community well and truly before the event.

Where to in the future? I would like to propose something radical like a triennial Indigenous survey. You might say every six years is bad enough, what are you going to do every three years? I think there is value in it because it is very important to have information that will influence policy makers. I think OIPC has a phenomenal challenge because their job is now to be able to coordinate and inform Indigenous policy. From what basis, I might ask? From the 2002 NATSISS statistics? We do not have any other benchmarks, as they are the benchmarks. A lot has happened in the last couple of years. The economy has turned around. It has gone up, it has gone down. For example, we are seeing many more mining interests engaging people out in communities. If we are going to get some benchmarks, if we are going to determine if there has been an improvement in the life and circumstances of Aboriginal people, we need to have those benchmarks established.

I am sure the ABS is saying it is too hard. One of the solutions may very well be to have a survey that targets remote communities. We have between 1200 and 1300 discrete Indigenous communities. A majority of them are serviced through CDEP. It could very easily be arranged, through the whole-of-government arrangement with DEWR, to engage some of the CEDP workers, train them up, and have them conduct the survey. By the time the second survey comes around in six years time, firstly we will have a benchmark. Larissa and Geoff both talked about governments now saying that a lot of the effort needs to go out into remote communities where the need is greatest—though I'd argue against that. That's one way of doing it, through CDEP, to push it through.

I wanted to mention briefly the issue of needs: needs-based priority or allocation of funding. It is often a misnomer to say 'where the need is greatest'. Statistically it may very well be greatest for those who are most impoverished, those without employment. For policy makers, they also need to consider Aboriginal people, be they in the city or remote areas, who have some capacity to be able to advance. Education is probably the best area to look at. Instead of directing all the programs down to those with the lowest educational capacity, some of our programs need to be directed towards Indigenous people who have some capacity, who are coping. They may be in a gifted program. Those people who are coping, who are in gifted programs, in my view suffer because there is no support for them, it all goes down the other way. Indigenous people have a double whammy. We are expected to be able to find full employment, buy a house—and educate our kids. The moment we start to get anywhere near a decent salary we are penalised because we suddenly lose access to tuition support, study allowance for students and so forth. Often the statistics are distorted because there are many Indigenous kids now who are leaving specific programs like ABSTUDY and going on to Youth Allowance because it is easier to get and you do not have to go through half the hassle you have to do to get Indigenous programs.

QUESTIONS

Boyd Hunter

I have a comment for Larissa, supporting your last comment about declining public sector employment. I have been working on a conference presentation for the Conference of Economists next month, where I will have the pleasure of sitting opposite Helen Hughes. I was looking at the statistics for the change in public sector employment without CDEP over time. There has been a very strong decrease for Indigenous employment for the whole of Australia. It has actually fallen as CDEP employment has increased. There seems to be a greater fall for Indigenous public sector employment than for the rest of Australia. The Public Service Commission's State of the Service report more or less confirms this, with Indigenous employment in the Australian public sector declining consistently since 1999.

Murray Geddes

You indicated there is a major mismatch between the scale of data and information collections and the current focus of policy on the local level—local level variable area agreements and so on. Would you like to explore how we might build that in? For example, a capacity building, evaluation and data component. Though I guess you would expect that those components, if that what was built in, would need to be managed other than by the program managers, who have a vested interest in not getting uncomfortable information. Would you like to comment on some of the explorations, even if you are really not trying to reconstruct ATSIC regional structures in the process.

Geoff Scott

I do not have a direct answer for you. The process employed when ATSIC was in place, was the Office of Evaluation of Audit. That was a useful process because the people had a very positive focus. I have not seen any sort of measure of evaluation of the COAG trials or SRAs today, and that is a worry. It is one of the things that we are trying to give advice to communities on. At this stage, the dearth of information about SRAs is one of the major inhibitors. I do not think even the Human Rights Commission can get access to the data

Jon Altman

I might just make a comment on behalf of John Taylor. The work he is doing in Wadeye involves working with local data collectors. I think there really is opportunity and obligation on researchers who work in Indigenous communities to facilitate the enhancement of the capacity of local people through local organisations to collect data. It is a cliché to say information is power, but if you are going to start getting into SRAs you certainly want three sources of information. You want the community information, the government information, and the independent information, I think that in reviewing any SRAs or COAG trials you need to have that tripartite approach. The issue I think we have all been negligent about, and Geoff has mentioned this frequently in the past and again today, is that we have not really built community capacity for data collection for development purposes. But some places have done it, as Larissa mentioned. Some regions have that capacity and they succeed, so we have the evidence that it works—what we need to do is invest in that capacity. That can sit alongside national surveys, regional surveys, and other data collection exercises but certainly that capacity is needed at the community level. Having said all that, defining communities and regions is difficult. I was quite comfortable with 60 regions once upon a time, then it became 36 regions. Now regions are going to become increasing flexible, and that is going to very hard in relation to historical data. How do these regions or communities fit in with census or ABS geography?

Paul Howarth

A couple of observations, and then a question. The first observation is picking up on something Tom said. The more intensive the condition, the more frequent the observation. If you think about someone who is in intensive care, a lot of time and effort and resources go into keeping track, as regularly as possible, of the situation they are in. There is no more intense policy than the Indigenous situation at the moment and I think it justifies the resources.

Secondly, I am thinking about the two data sets that have been the focus today: the 2002 NATSISS and census information. One of the issues that we come across is the lack of a nationally consistent approach to the way in which national data is used and interpreted. What is a household from an Indigenous perspective? What is a community? Making sure that these definitions are less nebulous would go a long way to help passing on the information to government departments, particularly making sure the data is reasonably objective, which is one of the things we struggle with often.

My third point relates to the methodology of the two main surveys discussed today. One is around the concept that we apply those surveys in the individual household levels. One of the things that we have thought about with some of our policy work in the past is the idea of a survey that can be administered to understand communities from the point of view of isolating and accounting factors of capacity, governance and service provision in a nationally consistent way. I suppose that CHINS does this to a certain degree, and there are aspects of data collection that do this too. We often find it's an important aspect of trying to work out what are the variables, what are the influences.

While geography is one powerful variable, we find that community capacity is also a very powerful variable. I am wondering whether or not there is an idea of consensus around the need for that type of collection to happen as well.

Geoff Scott

Through the Australian Collaboration Foundation, New South Wales Aboriginal Land Council plus other groups around Australia, including Reconciliation Australia, we are looking at that very question.

There has been a predominant assumption that governance is the problem. If you take this view, I think you have jumped two steps down the track instead of trying to identify the issue. Very often it depends what you define as governance, it depends who you spoke to last. At the moment, the capacity of communities is a major determining factor.

Stability of community structures and their organisation is an important issue, particularly whether their structures are community-based in reality. I am currently trying to investigate these issues in terms of identifying organisations around the country that have been succeeding, explore the reasons why they are successful, and look at what governance actually means. Governance is a much bandied cliché at the moment which just serves to confuse people and does not contribute to the debate at all.

Part of the methodology there is how you define success, and from whose perspective—whether it is from the perspective of the financier of the organisation, the client base, and of the peers, and all those issues as well. We are trying to get some definitions that can be useful and comparable. Comparability of data is one of the major problems. The Productivity Commission reports in previous years had a table which identified the comparative data across the country in the different program areas but I notice that it was not in this report.

Peter Radoll

I have a question for Tom in regard to education. You hinted that education may be one key to advancing our communities and you touched on a great subject of what do you do with gifted Aboriginal children. That gave me time to reflect on my time at the Indigenous Education Consultative Body here in the ACT, where we did have some very gifted children. These children are considered gifted from the mainstream perspective. Not only do these kids have to outperform their own cohort or the other Aboriginal and Torres Strait Islander kids in the ACT, but they also have to outperform all the kids in the ACT. To get funding for those children who are probably the most potential leaders at the national level at least, to get funding to assist those, well...you cannot. There is no way to actually get those children any sort of a leg-up, or for the families as well. For a child who finds it difficult to read in kindergarten, you can get a whole lot of resources. How do we address that?

Tom Calma

That is the point of my comment. The policy makers need to influence the Minister to consider investing in that level because we will never get the number of Indigenous people getting through to tertiary studies, getting into the professional level, if we do not start helping those who are capable to go through as well. There still needs to be effort put into those who have the greatest need. But there also needs to be some consideration given to those who have the capacity if we are going to facilitate the process of advancement of Indigenous Australians. Part of the problem is that, even though there are mechanisms in place to coordinate activities, agencies are still operating very independently. There is still a lot of room to be able to work a lot more collegially in approaching these issues.

Larissa Behrendt

I would like to add one thing to that. I think that question goes to the heart of what is really difficult about policy in relation to education, what is really difficult about policy in relation to improving socioeconomic statistics in Indigenous communities, and what is difficult about Indigenous governance.

When we do find instances where community organisations succeed, one of the key factors that seems to be replicated in each of the case studies is the fact that those community organisations have an individual in them who drives that organisation. Where we see regional councils being successful compared to other regional councils, it is usually because of the calibre of the regional council chair, so this success is driven by an individual. It is one of the real difficulties in terms of a holistic approach to improving the socioeconomic conditions of Indigenous communities.

We tear our hair out trying to come up with a formula as to how to improve the socioeconomic conditions of Indigenous families. What we see, particularly through our connections with the university sector, is that when you have an individual person graduate with a tertiary degree, there's an Indigenous family that is never going to live in poverty again. They've got the capacity to earn an income and they introduce a culture of learning into their family which will be disseminated through the generations.

It's really hard to make policies to say that every time we set up a community organisation or structure or a governance system we need to have an individual of exceptional capacity who can drive it. It does go to show that one of the really big challenges is actually focusing on the ability to be able to develop that capacity in the individuals where they arise. That is why the leadership programs that we see building up across the country are really a key mechanism in doing that, particularly when they have a large outreach into a wide variety of communities. These programs need to bring people in to get the leadership skills they need, as well as the intellectual and emotional support they need, to carry those enormous burdens when they go back into their communities.

14. Education and training: the 2002 NATSISS

R.G. (Jerry) Schwab

The impetus for the first national survey of the Indigenous population, the National Aboriginal and Torres Strait Islander Survey (1994), was the 1991 Royal Commission into Aboriginal Deaths in Custody. This Royal Commission highlighted the lack of a reliable statistical baseline from which to assess the experience of Indigenous Australians (Commonwealth of Australia 1991). The survey followed on the heels of two major policy documents, the Commonwealth Government's Aboriginal Employment Development Policy (AEDP) and the Aboriginal Education Policy (AEP), both of which called for coordinated baseline data and improved education and training outcomes. The results of the 1994 NATSIS were dissected and analysed in a special workshop held at the Australian National University in 1996 (Altman & Taylor 1996b). Included in the volume of papers from that workshop were specific analyses of the education (Schwab 1996) and training (Daly 1996) components of the survey and findings. This paper picks up the thread of those two earlier critiques in a review of the education and training questions and results of the 2002 NATSISS.

This paper has three principal aims. Firstly, it sets out to contrast the 1994 and 2002 surveys with regard to focus and content related to Indigenous education and training. Specifically, the paper explores what sorts of questions where asked and data collected in 1994 that were not collected in 2002 and, conversely, what new questions and data emerge in the more recent survey. Secondly, the paper summarises the key education and training findings of the 2002 NATSISS. Finally, the paper attempts to assess the value of the education and training components of any future NATSISS and ask, 'Is it really worth all the trouble'?

1994 NATSIS and 2002 NATSISS: what has been lost and gained?

The 2002 survey is far less ambitious than its predecessor in the areas of education and training. The questions asked of respondents are fewer and consequently the volume of data collected has been reduced. The education and training components of the two surveys are compared in Table 14.1. Specifically, the most recent survey no longer collects data on school participation, where the 1994 survey asked questions about the participation of both preschool and school age children. The earlier survey collected data on parental attitudes and experiences: asking whether parents were happy with the education their children received (and if not, why not); did they feel welcome in the school;

were they involved in educational decision making at the school; and would they have preferred their children attend an Indigenous community-controlled school. Those questions were eliminated from the 2002 NATSISS. The earlier survey included a series of questions, the answers to which could provide a glimpse of the classroom experience of Indigenous students. It asked whether or not Aboriginal and Torres Strait Islander culture and language were taught in school and whether or not children had Indigenous teachers, education workers or community members in their classrooms. Those questions have also been cut from the most recent survey. While the 1994 survey explored the difficulties and barriers to further study, no such questions appeared in the more recent survey and no data were collected on the distance individuals must travel to access educational institutions.

Table 14.1. Comparison of education and training data collected in NATSIS (1994) and NATSISS (2002)

	Data collected in survey (denoted by*)	
	NATSIS (1994)	NATSISS (2002)
School participation (preschool and primary/secondary)	*	
Parental attitudes to schooling	*	
Characteristics of schooling	*	
Highest level of schooling	*	*
Main reason for leaving school		*
Post-school qualifications	*	*
Level and type of post-school study	*	*
Type of educational institution currently attending	*	*
Details of recent training	*	*
Difficulty/barriers in further study or training	*	
Distance to educational institution	*	

The 2002 survey is consistent with the 1994 version in asking about the highest level of schooling individuals achieved, post-school qualifications, the level and type of post-school study and the type of education institution currently attending. There are, however, some entirely new questions in the 2002 NATSISS that provide additional insight into the experience of education and training for Indigenous people. A particularly useful question asked respondents to indicate the main reason for leaving school (see Table 14.3 below). In addition, individuals are asked to provide more detail in 2002 about the training they received. For example, individuals were asked if their training was provided as part of their participation in the CDEP scheme. Overall, the quantity of information gathered in 2002 is less than in 1994. More has been lost than has been gained, a point I will return to later.

Table 14.2. Indigenous persons aged 15 years or over, selected education characteristics, Australia, 1994 and 2002

	1994	2002
	%	%
Has a non-school qualification		
Bachelor degree or above [a]	1.2	3.3
Certificate or diploma [a]	10.6	21.6
Total with non-school qualification [a]	11.8	24.9
Does not have a non-school qualification		
Completed Year 12 [a]	6.8	9.9
Completed Year 10 or Year 11	26.9	28.8
Completed Year 9 or below [a]	43.0	35.2
Total with no non-school qualification	76.7	73.9

a. Statistically significant at the 5 % level
Source: ABS (2004c: Table 6)

2002 NATSISS: an overview of findings about education and training

Table 14.2 compares qualification and highest year of school completed, drawing on the results of the 1994 and 2002 surveys. Overall, the data suggest gains (essentially a doubling) in non-school qualifications at both the degree or above level and the certificate/diploma level over eight years. Less dramatic—but significant nonetheless—is the increase in individuals who report they have no non-school qualification, yet they have completed Year 12. Those figures increased from 6.8 to 9.9 per cent between 1994 and 2002. Interestingly, the 2002 figure is well below the 2001 Census figure of 17 per cent (non-Indigenous Australians reported about double that rate).

Table 14.3. Indigenous persons aged 15 years or over, main reason left school by remoteness, Australia, 2002

Main reason left school	Non-remote female	Non-remote male	Remote female	Remote Male	Significance test
	%	%	%	%	
Year 12 or equivalent not available	0.6	1.5	7.5	10.5	male[b] female[b]
Got/wanted a job/apprenticeship	14.5	17.5	15.7	15.8	
Other work-related reason	3.0	1.9	1.9	2.9	
Feel had done enough	3.6	9.0	9.5	12.5	
Did not do well	8.7	16.3	7.0	3.4	male[b]
Did not like school	28.3	24.2	24.7	30.9	
Other school-related reason	10.0	13.9	8.6	9.7	
Personal/family reason	25.8	11.5	20.7	8.4	non-remote[a] remote[a]
Other reason	5.6	4.3	4.3	5.9	

a. The difference between male and female respondents is statistically significant.
b. The difference between remote and non-remote respondents is statistically significant.
Source: Customised cross-tabulations from the 2002 NATSISS

Table 14.3 summarises the findings of a new question. It was asked of people who had not completed Year 12 and who were not, at the time of the survey, currently at secondary school, and it asked the main reason they left school. The table shows the responses from the 2002 NATSISS according to remoteness and sex. There is a great deal of interesting information in the table but I would like to highlight four points. First, the most common reason people gave for leaving school was that they did not like it. The rate was highest among remote males and non-remote females, at about 30 per cent. While one would expect to find some number of students in the general Australian population who left school because they simply did not like it, it is troubling that the proportion is so high among Indigenous people who already attend school at a lower rate than other Australians. Secondly, the survey shows that many people in remote areas (especially males) left because Year 12 was not available in their community. Thirdly, the responses show that in general terms, females were more likely to leave for personal/family reasons. This finding is slightly higher for non-remote females but in both remote and non-remote settings, females were more than twice as likely as males to have left school for that reason. Finally, the survey showed that between 14.5 and 17.5 per cent left school because they 'got/wanted a job/apprenticeship'. Unfortunately, the question was poorly worded and confounds the responses of those who actually left school because they got a job or apprenticeship with those who left because they hoped for such an outcome. Obviously, in terms of policy, there is a big difference between getting a job and wanting one.

Table 14.4. Unemployed Indigenous persons aged 15 years or over, main difficulty finding work by remoteness, Australia, 2002

Main difficulty finding work	Remote	Non-remote
	%	%
Transport problems/distance [a]	8.6	15.9
No jobs at all [a]	29.4	8.2
No jobs in local area or line of work [a]	18.3	10.5
Insufficient education, training or skills [a]	13.4	27.9
Own ill health or disability	3.9	6.0
Racial discrimination	0.7	2.4
Age	3.1	6.8
Other	10.3	13.0
Total with difficulties	87.8	90.9
No difficulties reported	8.6	6.1

a. The difference between remote and non-remote respondents is statistically significant at the 5 % level.
Source: Customised cross-tabulations from the 2002 NATSISS

Table 14.4 depicts respondents' perceptions of their main difficulty in finding work with a breakdown contrasting remote and non-remote people. Again, there is some useful data that emerges from the question. Transport problems to reach distant work sites was a more significant problem for people in non-remote areas, with nearly 16 per cent reporting this was their main difficulty in finding work. Most people in remote areas cited a lack of jobs over all other reasons as the primary problem in finding work. Nearly 30 per cent of remote respondents said there were no jobs at all and another 18 per cent said there were no jobs in the local area or in their line of work. This rate was significantly higher than among non-remote people, for whom a lack of education and training was the primary problem.

Table 14.5. Indigenous persons aged 15 years or over, whether attended vocational training, by remoteness, Australia, 2002

Vocational training (in last 12 months)	Non-remote female	Non-remote male	Remote female	Remote male	Significance test
	%	%	%	%	
Attended	26.6	34.1	16.7	23.4	non-remote [a]
					remote [a]
					female [b]
					male [b]
Did not attend	73.4	65.9	83.3	76.6	non-remote [a]
					remote [a]
					female [b]
					male [b]

a. The difference between male and female respondents is statistically significant.
b. The difference between remote and non-remote respondents is statistically significant.
Source: Customised cross-tabulations from the 2002 NATSISS

Obviously, the lack of education and training is a critical problem, particularly among non-remote people. Table 14.5 provides a closer look at people's experiences with vocational training, focusing on people in non-remote and remote areas who attended vocational training of some sort in the previous 12 months. Overall, the data show that men in both remote and non-remote areas were more likely to attend training than women, and people living in non-remote areas were more likely to attend training than those in remote areas. Higher attendance in vocational training in non-remote areas is coupled with the strong perception that individuals in non-remote areas lack the necessary education and training or skills. One might ask whether there is a gap between the training offered and the training needed, but the data do not provide evidence to answer this question.

It is often said with a large degree of sarcasm by both Indigenous and non-Indigenous people that Indigenous Australians are the most highly trained workforce in the country, yet that training rarely leads to jobs. Nevertheless—as Table 14.6 shows—when asked, people overwhelmingly affirm that their training experiences are useful. More than four out of five remote and non-remote males and females indicated they have used their training. It is worth noting, as well, that the question asked was not a vague question like 'was the training useful', but instead it asked 'Have you used the information or skills you got from the training'?

Table 14.6. Indigenous persons aged 15 years or over, whether used vocational training, by remoteness, Australia, 2002

Was training used?	Non-remote female	Non-remote male	Remote female	Remote male	Significance test
	%	%	%	%	
Used training	86.8	82.8	86.0	92.3	remote [a]
					male [b]
Did not use training	13.2	17.2	14.0	7.7	

a. The difference between male and female respondents is statistically significant.
b. The difference between remote and non-remote respondents is statistically significant.
Source: Customised cross-tabulations from the 2002 NATSISS

The 2002 survey enables an even closer exploration of the reported value of vocational training. In the survey, people who said they used the training were asked the follow-up question, 'How was the training used?'. Table 14.7 summarises the responses to that question and shows that the vast majority, again in both non-remote and remote areas, male and female, said they used it 'for work'. Another 10 per cent or so said they used it 'to get a job'.

Interestingly, the 1994 NATSIS recorded other responses to questions about the use of training, including 'for personal growth' and 'hobby'. Those options do not appear on the 2002 survey and presumably are subsumed under 'other'. This streamlining, if that is what accounts for the change, is unfortunate, since more people cited personal growth as a use in the 1994 survey than 'to get a job' (about 30 per cent). It appears that the changes to the 2002 survey provide disappointingly less insight into the myriad reasons people engage in study.

Table 14.7. Indigenous persons aged 15 years or over, how vocational training was used, by remoteness, Australia, 2002

How training was used	Non-remote female	Non-remote male	Remote female	Remote male	Significance test
	%	%	%	%	
Used for work	75.5	69.1	74.7	81.4	male[b]
Used to get a job	12.2	11.3	10.7	8.0	
Used, other	3.5	1.7	3.7	1.4	

a. The difference between male and female respondents is statistically significant.
b. The difference between remote and non-remote respondents is statistically significant.
Source: Customised cross-tabulations from the 2002 NATSISS

The real power in surveys like NATSISS, from my perspective, is the possibility of exploring the patterns of relationships between variables. Table 14.8 cross-tabulates a handful of social characteristics with data on school completion and qualification. The questions that animate the analysis presented in this table, are the following: 'What is the place of the western-educated person in Indigenous communities?'. 'Are they different as a result of their western education and, if so, how do they compare to their friends, family and other 'non-western-educated' or 'less-western-educated' members of their community? The notion of 'western-educated' obviously requires some explanation. Firstly, given that there are few—if any—Indigenous individuals in Australia today who have not come in contact with western schools or other education and training institutions, the notion of a 'western-educated' individual assumes a continuum of experience and knowledge. Individuals who are 'western-educated', as I am using the term, have typically successfully engaged with and reached a recognised level of competency (completion of primary or secondary schooling and/or a qualification) in a government or non-government school or other educational institution. For the purposes of this analysis, I am defining a 'western-educated person' as a person towards one end of the continuum: from those with some secondary education through to those who have gained a non-school qualification. These are people who have some degree of facility with English literacy and numeracy, who are recognised by other Indigenous people as having particular skills, and who can thus move with some confidence within the structures and expectations of mainstream institutions.

Table 14.8. Indigenous persons aged 15 years or over, non-school qualification by highest year of school completed by selected characteristics, Australia, 2002

	Non-school qualification?				
	No				Yes
	Year 9 or below	Year 10 or 11	Year 12	Total	Total
	%	%	%	%	%
Current daily smoker	57.3	53.2	40.7	53.3	45.7
Employed CDEP	13.2	15.8	13.4	14.2	9.5
Employed non-CDEP	17.4	32.4	52.7	28.2	53.5
Not in the labour force	57.5	36.1	22.7	44.3	23.9
Unable to raise $2000 within a week for something important	71.6	56.5	40.0	61.3	38.8
Cannot, or often has difficulty, getting to the places needed	16.7	10.4	9.5	13.3	8.4
Had undertaken voluntary work in last 12 months	16.7	23.3	33.1	21.6	42.0

Source: ABS (2004c: Table 7)

Acknowledging that this analysis is exploratory, note first that the contrasts between those with and those without non-school qualifications are statistically significant for each of these social characteristics. In other words, statistical analysis suggests that individuals in these two categories are different from one another in each of the characteristics. There is a great deal to ponder in these results but space is limited here, so I would simply like to highlight some of the findings and put forth some propositions. Firstly, there is a clear association between smoking and lower educational attainment. Indeed, the proportion of smokers declines as the number of completed years of secondary education increases. If smoking has a social cost (e.g. in economic terms or in terms of an individual's or a family's health), then it appears that educated individuals are less likely to suffer those costs as, presumably indirectly, are members of their families.

The next three characteristics in Table 14.8 reveal an association between employment and education. First, relatively lower levels of education correlate with a greater likelihood of employment in CDEP while higher levels of education and non-school qualification are strongly associated with employment outside the CDEP scheme. Similarly, those with the least education are by far the most likely to have withdrawn from the labour force. Education is strongly associated with the ability to alleviate financial stress. When asked if they could raise $2000 within a week for something important, over 70 per cent of respondents who had completed only Year 9 or below said they could not. On the other hand, over 60 per cent of individuals with a non-school qualification said they could. Increased levels of education appear to be associated with increased domestic financial agility, in the sense that these people are most likely to be able to raise money at short notice. Similarly, people who are more educated would appear to be better placed to draw on whatever resources they must to get to places they need to reach. For example, only about 8 per cent of people with non-school qualifications said they cannot or often have difficulty getting to places they need to go, while double that proportion of individuals who had attained education at Year 9 or below had such difficulties.

Finally, and most intriguingly, there appears to be a strong association between education and undertaking voluntary work. In the context of the survey, the notion of voluntary work appears in a question about volunteering work for a range of different organisations, not in caring for family or other such 'voluntary work'. While only about 17 per cent of individuals with education levels at Year 9 or below engaged in voluntary work, 42 per cent of individuals with non-school qualifications did so. While this result deserves more analysis than is possible here, one could interpret this to mean that people who have invested in education—who, as economists might say, are engaged in 'human capital acquisition'—have an increased tendency to re-invest in their communities through voluntary work. That proposition is interesting, in that it flies in the face of the often-stated perception that as Indigenous people gain education (and income) they are in increased danger of disconnecting from the Indigenous community. Such are some of the important questions the NATSISS raises and might allow us to answer or at least explore.

NATSISS, education and training beyond 2002: is it worth the trouble?

Unlike 1994, today we have mountains of increasingly good data depicting the experience of Indigenous Australians with education and training. For example, relevant data are collected by:

- the Commonwealth Department of Education, Science and Training (DEST)
- the Ministerial Council on Education, Employment, Training and Youth Affairs (MCEETYA)
- the Commonwealth Department of Employment and Workplace Relations (DEWR)
- the National Centre for Vocational Education Research (NCVER)
- the Australian Council for Education Research (ACER), and
- all the various States and Territories.

Many of these are administrative data sets useful in assessing and monitoring the delivery and outcomes of programs. The DEST data are particularly detailed, with omnibus statistics now assembled annually in the National Report to Parliament. While there are always issues about access and data quality, there is currently more education and training data available today than ever before.

Data are generally of higher quality now, and some agencies and departments appear to be opening up and welcoming collaboration and analysis in ways that were in the past inconceivable. But in regard to NATSISS, some important data have been lost in streamlining the education and training questions. Much of what we now have in NATSISS (and presumably will get again in 2008) is identical to what we can get from DEST or NCVER on an annual basis.

Comparing the 1994 NATSIS with the 2002 NATSISS, it appears the focus has shifted and the education and training components of the survey seem to favour easy questions over difficult ones. What was lost was the attempt to acquire a range of attitudinal data. It should be noted that many of these questions were included in the 1994 survey at the request of Indigenous people who were concerned with a range of important social justice and cultural issues. These questions included things like, 'Are you happy with your child's education? Has your child þeen taught Indigenous studies or an Indigenous language? Do they have an Indigenous teacher? Do you feel welcome in your child's school? Are you involved in decision making at the school? Would you prefer a community-controlled school?'. From one perspective these are all, one could argue, questions which are politically unsavoury. They highlight what the education system is (or is not) providing Indigenous families while the 2002 questions focus on what may be the outcome of that provision: the level of participation. Yet, without insight into the former, it is too easy to blame the Indigenous families for low levels of participation.

One must acknowledge that the attitudinal questions did not work well in 1994 and I was one who criticised the questions and the data they yielded (Schwab 1996). Yet I'm sorry to see the baby tossed out with the bathwater and I would hope the ABS will have another look at how to get at those important aspects of the Indigenous experience. When the most common reason given for leaving school is 'I didn't like school', it seems we should be redoubling our efforts to understand why. And perhaps we should be asking more about the experience of school and not just about patterns of participation.

Even in its pared down form, the 2002 NATSISS provides a rich data set that enables an exploration of the linkages among a broad range of variables that simply are not available in other surveys. It is in that way that the survey is such a powerful tool. While education and training data are increasingly accessible from a wide range of sources, there is no other robust data set that enables researchers to explore, for example, how education might relate to health, or training to internet use, or non-school qualification to arrest. NATSISS is—and should be—much more than a device to track outcomes. There are enough of those sorts of data sets already. It should, rather, assist movement beyond the 'what' questions to the more difficult and ultimately important 'why' questions. It should be a tool for helping researchers, policy makers and communities explore what we can and should do to address the needs of Australia's most disadvantaged citizens. In that way, it would be worth the trouble.

15. Indigenous Australians and transport—what can the NATSISS tell us?[1]

Sarah Holcombe

The 2002 NATSISS is the first national survey of Indigenous Australia that includes a transport module and, as such, provides a unique opportunity to examine Indigenous transport needs. The analysis of the transport data in this chapter is approached from an anthropological perspective, comparing this survey data with the ethnographic record. In doing so, I focus primarily on the findings of the remote area NATSISS but I also (as relevant) highlight comparative findings using the data from the non-remote NATSISS and the GSS. My focus reflects the concentration of ethnographic research in remote areas. It also reflects the fact that the issue of adequate transportation is magnified for Indigenous people in these areas because of their low socioeconomic status, the large distances, poor roads and relatively low access to vehicles.

In a short paper such as this, the obvious complexity and multi-dimensional nature of the topic also necessitates a narrowing of the analytic frame. The questions that drive this analysis are: Does the data adequately fix the scale of the issue of transport availability? Do we get a picture of vehicles per capita in remote regions? Can any distinction be drawn between types of vehicles, such as those purchased by government or those purchased privately? Finally, does the data reveal any relationship between transport and equity of access, including by gender?

Although it is far from comprehensive, the ethnographic record on transport in remote Indigenous contexts is rich in local and regional detail. This literature indicates that vehicle access is significantly more restricted in remote areas than the 2002 NATSISS data suggests. While the 2002 NATSISS data offers statistically significant evidence of a divide between remote and non-remote areas in terms of access to a vehicle to drive (of approximately 15%), the ethnographic evidence indicates that this divide is far greater. Recent evidence suggests, however, that transport disadvantage for Indigenous people in non-remote areas in NSW, for instance, is also significant—perhaps more so than is indicated by the 2002 NATSISS (Wadiwel 2005). The ethnographic research for remote areas suggests that the figure for Indigenous access to a vehicle 'to drive' may be as low as 5–10 per cent in some areas. According to the NATSISS data, 44.2 per cent of

[1] I would especially like to thank John Taylor and Ben Smith for their editorial insights and Boyd Hunter for his unrelenting humour.

Indigenous people in remote areas have such access. This compares to 85 per cent of the general population. How could these remote area figures be so different? I suggest that cultural factors have played a major role in the interpretation of the question and, thus, the data generated.

The question that elicited this figure in both the 2002 NATSISS remote and non-remote surveys and in the GSS is similar, though the phrasing and contextual information varies. This headline question reads: *'(Including community vehicles you can drive at any time) is there a car, 4WD or truck that you can drive if you want to?'*. This question conflates the issue of *access* to transport with the availability of a vehicle *to drive*. In the remote context, one does not imply the other. This confusion of issues sets the tone for the remaining three sets of questions: all modes of transport used in the last two weeks; main reason for not using public transport; and the perceived level of difficulty with transport. The questions could be re-focused and re-sequenced, while several of them are too general to gain meaningful data. The fundamental issue concerns *access* to transport and this is addressed in the final set of questions: *'Can you get to places you need to go?'*.

I will return to this crucial question. But first, in order to make a prognostic assessment of how the next NATSISS could be improved, the assumptions and issues that the headline question (see above) raises require unpacking. This question assumes that having access to a vehicle 'to drive if you want to' indicates ready availability of access and is thus implicitly indicative of a form of vehicle 'ownership'. Unlike the data produced from this question in the GSS, the results in this remote NATSISS survey give us little indication of vehicle numbers per head of population. This is one of the key areas in which the survey questions impact on the quality of data that is available.

Another significant issue that impacts on the quality of the data is the aggregation of the ARIA classifications 'remote' and 'very remote' areas. This collapsing of geographic regions in the data compounds the lack of regional data, so that the NATSISS results provide a blunt instrument of comparison. I also note here that the ethnographic data called on is from very remote regions, according to this five-fold classification system, although like the 2002 NATSISS I simply use the term 'remote'.

The issue of transport is fundamentally tied to Indigenous spatial mobility, which depends on access to a vehicle. A focus on transport, rather than mobility, refers to *how* people travel, what options they have and how these options are influenced by infrastructural factors, such as availability of access to public transport and socioeconomic factors which impact on private vehicle ownership. *Why* people travel and how frequently they do so, is addressed in the mobility section of the survey (in the remote survey, this follows the transport module).

There is value in understanding the relationship between the two data sets, as they reflect the broader social dimensions of transport.[2]

Remote area ethnography of transport

The ethnographic literature on remote area Indigenous vehicle usage examines the social dimensions of vehicles, providing a grounded analysis of this complex issue. Known variously as 'Toyotas' or 'trucks', vehicles have come to play an essential role in the livelihood practices of remote Indigenous peoples. Research has shown us that vehicles are a necessity, not a luxury, for Indigenous people in remote areas. Research has been undertaken in desert settlements in the Northern Territory and South Australia (Hamilton 1987; Peterson 2000; Stotz 2001; Young 2001; Young & Doohan 1989), the Western Desert (Holcombe 2004; Lawrence 1991; Myers 1989), Arnhem Land, (Altman & Hinkson 2005; Fogarty 2005; Gerrard 1989), the Kimberley in Western Australia (Kolig 1989) and Cape York in Queensland (Smith 2000a, 2004). The tyranny of distance is sharply drawn in this literature.

Settlement decentralisation was actively encouraged in the early 1970s across Central and Northern Australia, occurring during the same period as citizenship rights and social service benefits, enabling Indigenous Australians to purchase vehicles for the first time. This policy shift from assimilation also saw the establishment of various federal funding regimes to assist in purchasing vehicles to enable this 'homelands movement' to occur (Cane & Stanley 1985; Coombs, Dexter & Hiatt 1982; Nathan & Japanangka 1983). As Stotz noted 'it was only the Toyota that could actually replace the loss of mobility people had suffered since they were institutionalised [from] the late 1940s' (2001: 227). Without access to a reliable vehicle, people cannot now reside on homelands or in outstations, which limits their participation in customary economy and land management activities. 'Looking after country', ceremonially and through foraging, fire regimes and other land use and management activities, requires living on country, and today people will say this is not possible without access to a vehicle (Payne 1984; Young 2001: 38). This is more apparent in desert regions than in tropical and sub-tropical regions, due to environmental constraints.

During the 1990s there was considerable criticism about government expenditure on such vehicles for homelands or outstations (documented in Altman 1996; Altman 1999; Cooke 1994) and a resultant limitation on funding vehicles for mobility purposes (Smith 2004). The vehicles now purchased via government funds tend to be driven only for specific purposes, such as aged care support

---

[2] The Royal Commission into Aboriginal Deaths in Custody emphasised the importance of mobility, and thus access to transport, as an Indigenous mechanism for social control. The report noted that 'the option to resolve conflicts by simply moving away was one of the deepest and most significant freedoms of Aboriginal society' and that this 'practice remains important in urban and rural as well as remote areas' (Commonwealth of Australia 1991: 104).

or community policing ('night patrol').[3] They are not freely available for general purposes, such as visiting the nearest service centre, for shopping, banking etc. Such vehicles also tend to be monopolised by certain individuals and are not shared across the settlement 'community'. Perhaps as a result, one hears anecdotally that privately purchased vehicles are becoming more common. The focus on private and 'community-owned' vehicles also underlines the lack of public transport options in very remote regions. There remains a dearth of data about the forms and costs of public transport availability, such as charter aircraft, commercial buses, etc.

The anthropological literature is unambiguous on the issue of vehicle availability: vehicles are a scarce resource, with severely limited access to vehicles common for significant proportions of the remote Indigenous population. In the late 1980s, Gerrard observed a 'chronic shortage' of vehicles in the Arnhem Land settlement of Maningrida where there were 'roughly 60 functioning vehicles at any given time for a population of approximately 800 people' (Gerrard 1989: 101). Of these, Gerrard estimated one Aboriginal-owned car for every 44 Aboriginal people, a ratio of less than 2 per cent, in broad NATSISS terms, of access to a vehicle 'to drive' at any given time. In 2005, fifteen years after Gerrard's study, Altman and Hinkson (2005) found that for a sub-set of this population in the Maningrida region, there was 'a ratio of one truck (or 4WD) for up to 30 Kuninjku'. Even though this group is more affluent (in cash terms) than ever before, the community of 300 still shares between only 10 and 20 functioning vehicles at any one time (2005: 6). A similar figure is evident in an ethnographic example from the Tanami Desert, where the family from one outstation (which includes 25 permanent and 25 non-permanent residents) shared a single Toyota Land Cruiser and a tractor (Stotz 2001). In the Central Australian settlement of Papunya in the late 1980s, Myers (1989) found that only a very small percentage of the population had cars. In the mid-1990s Smith found that in the township of Coen, in Cape York Peninsular (which has a majority Indigenous population of approximately 200), 14 'community' and 'private' vehicles were available for relatively general access (2000a: 152–3).

There appears to have been only marginal growth in actual vehicle numbers across the 20-year time period in which these studies were conducted. These figures all suggest approximately 5–10 per cent of the Indigenous population having access to a vehicle 'to drive' at any given time.

This data clearly raises questions about the veracity of the 2002 NATSISS remote area figure of 44.2 per cent of access to a vehicle to drive. That people in remote areas do not have ready access to a motor vehicle, either as a driver or passenger, is also suggested by the 2002 NATSISS figure of 74.4 per cent of people who

---

[3] Night patrol is community policy.

used walking to 'get around'. This differs very significantly from the non-remote figure of 49.2 per cent of people who walked as a form of transport.

Another consistent theme that emerges in the ethnography, and which is corroborated by the 2002 NATSISS data, is the issue of women's limited access to vehicles relative to men. Men have more ready access to vehicles to drive, both private and 'community'. According to Altman and Hinkson, there are no Kuninjku women who drive (2005: 8), though they may be recognised as co-owning a private vehicle with their spouse. The issue of lack of women's access to vehicles was recognised as a special needs case in the Pitjantjatara Lands in SA, NT and WA where, with the assistance of the Ngaanyatharra, Pitjantjatjara, Yankunytjatjara Women's Council (NPYWC), an active program of both purchasing women-specific vehicles and teaching women to drive was implemented in 1988 (NPYWC 1990: 9). This government-sponsored program sought to address the paucity of vehicles for women's own use. Nevertheless, this program only resulted in one such vehicle per settlement, limiting vehicle access to senior women associated with the Women's Council. There is also considerable evidence to suggest that women have limited access to private vehicles, either as owner or driver (Holcombe 2004; Stotz 2001; Young 2001).

A key aspect of the value of vehicles in Aboriginal Australia is in their capacity to generate and sustain relationships between people. The social importance of vehicles means that the chronic vehicle shortage creates an intense demand, often referred to as 'humbugging' and an expectation that the person in possession of a car has a duty to share the resource (Gerrard 1989). Vehicles are a dynamic resource, flowing into and out of remote settlements at a far greater rate than non-remote areas. There are several other interrelated reasons for this short lifespan. The lack of remote area road infrastructure ensures that the driving conditions are difficult, and often dangerous, on unsealed and often unformed and un-maintained roads (see Lawrence 1991). This lack of infrastructure, combined with poor vehicle maintenance on vehicles that are already generally secondhand, ensures that the lifespan of cars is extremely short. They may last from several weeks (Myers 1989: 25) to several years (Gerrard 1989:101). The sheer use wear of vehicles also attests to the fact that 'if faced with a choice between caring for their property or for their relatives, they prefer to invest in people rather than things' (Myers 1989: 24). Vehicles are often regarded as necessarily expendable and the loss of a vehicle compounds the demands made on other vehicles owned by, or accessible to, others in a 'community'.

Vehicles are now essential to the production of remote Indigenous culture. In the Arnhem Land region, the purchase of vehicles is a crucial motivator in Kuninjku production of art for sale (Altman & Hinkson 2005: 5). The importance of having access to a vehicle is also often cited by people as a means to avoid social conflict (Hamilton 1987; Young 2001: 37), as noted in the findings of the Royal Commission into Aboriginal Deaths in Custody . Awareness of the cultural importance of decentralisation is embedded in an innovative mobile schooling program operating for outstations in the Maningrida region, which relies on 4WD vehicles to take schooling to the children in an adaptive classroom approach (Fogarty 2005). Similarly, Smith (2000a, 2004) argues that access to vehicles has formed an essential foundation for central Cape York outstation development and for the various social and economic benefits that flow from outstation use.

Does having access to a vehicle mean being able to drive it, if you want to?

The headline question in the remote survey, as for the 2002 NATSISS non-remote and the 2002 GSS, implies an individualistic form of transport use and mobility. In remote areas, the intersection of Indigenous values and practices with the welfare economy precludes such a standard from developing. Being able to drive a vehicle at 'any time' assumes vehicle ownership, or at least primary control. Thus, in a non-Indigenous context this may provide a relatively accurate figure of vehicle ownership, as suggested by the figure of 85 per cent of access to a vehicle (or vehicles) to drive. Ironically, it is this very phrasing that has likely caused the figure of almost one vehicle, available to drive, for every two Indigenous people to be so high.

An explanation for the remote Indigenous figure is the cultural norm of 'demand sharing' (Myers 1989) which is particularly prevalent in regard to vehicles. As vehicles are among the most valued objects in contemporary remote Indigenous life, demands to access a vehicle are 'difficult to refuse', and 'open rejection is impossible' (Myers 1989: 23). As Myers notes, 'to have a car, one might say, is to find out how many relatives one has' (1989: 23). When one considers demand-sharing in the context of the 2002 NATSISS headline question, we can assume that the respondent's pre-emption to vehicle access relates to *any* family member within the settlement who has vehicle access, as they will have a right to demand access to that vehicle. Thus, the NATSISS question does not tell us about actual vehicle numbers, but rather about perceptions of vehicle access and self-perceptions of rights to vehicles within the kin group. In order to provide meaningful data, this issue of *access* must be separate from the availability of a vehicle *to drive*. The 2008 NATSISS could usefully distinguish between rates of access to transport and enumerating vehicle numbers per head of Indigenous population.

Sample selection: remote, very remote and non-remote [4]

The conflation of selected ARIA classification areas in the publicly available data (e.g. the CURF)—the remote with the very remote areas—in the data sample selection has implications for assessing transport needs by geography. Hence, the issue of access to transport and locational disadvantage cannot adequately be drawn because of this blunt level of aggregation. [5]

There is a number of government and non-government bodies noting the dearth of evidence on the availability of transport and related infrastructure in remote and very remote areas (Brice 2000). Transport and associated infrastructural needs differ significantly between remote and very remote Australia. There is considerable locational diversity between these two classes of remoteness, a diversity that affects transport infrastructure, and thus access to employment and education opportunities. These, in turn, impact on socioeconomic status, which affects vehicle access. In very remote areas, where Indigenous people account for approximately 42 per cent of the population, detachment from services is more pronounced than the basic remoteness structure can portray, reflecting unique aspects of the Indigenous settlement pattern (Taylor 2002: 4).

Equity and access to vehicles

Both the remote and non-remote area 2002 NATSISS data and the available remote area ethnographic data suggest that gender is a key dimension in vehicle access (Table 15.1). There are two issues about this inequality of access that are worth raising here. These concern access to vehicle-related training, as these corroborate the ethnographic evidence on gender equity, and issues of equity of access within the Indigenous population generally.

[4] Comparison with the GSS, though useful on a national scale in identifying major trends, is limited by the scope of the survey, as it was only conducted in urban and rural areas, not in sparsely settled areas (as are to be found in NSW, Qld, SA, WA and the NT). The GSS notes that, with the exception of the Northern Territory, the population living in sparsely settled areas represents only a small proportion of the total population (ABS 2003b: 58). However, in terms of the Indigenous demographic profile by geographic classification, 69% of Indigenous people live outside of major urban areas. In 2001 approximately 1 in 4 Indigenous Australians lived in remote areas compared with only 1 in 50 non-Indigenous Australians (ABS 2003b: 1).

[5] Even if customised cross-tabulations were purchased from the ABS, the relatively small sample sizes in very remote areas are unlikely to withstand the substantial disaggregation into the various transport categories. That is, the resulting estimates might not be particularly reliable.

Table 15.1. Access and use of motor vehicles and walking (aged 15 and over) by sex and remoteness, 2002[a]

	Remote	Non-remote
	%	%
Male		
Has access to motor vehicles to drive	50.5	62.5
Used car/4WD/motorcycle/motorised scooter as driver	47.7	58.3
Used car/4WD as passenger	55.1	56.7
Walked	73.4	46.5
Female		
Has access to motor vehicles to drive	38.2	55.3
Used car/4WD/motorcycle/motorised scooter as driver	30.7	50.6
Used car/4WD as passenger	61.9	61.6
Walked	75.4	51.6
Population		
Has access to motor vehicles to drive	44.2	58.7
Used car/4WD/motorcycle/motorised scooter as driver	39.0	54.2
Used car/4WD as passenger	58.6	59.3
Walked	74.4	49.2

a. The estimates of access to motor vehicles are slightly larger than figures presented in ABS (2004c), as we exclude from the calculations those who did not answer the question. In their calculations, the ABS implicitly assumed that the 'not stated' (31 individuals) did not have access to motor vehicles to drive. Source: Customised cross-tabulations from the 2002 NATSISS CURF

Daly found a significant discrepancy between male and female rates of undertaking training in transport-related fields through non-CDEP training providers in the 1994 NATSIS, with 15.2 per cent of males undertaking training in this area compared with 2.6 per cent of females. In CDEP training, 9.6 per cent of males undertook training in transport-related fields, while no females undertook training in this area (Daly 1996:100). This male monopoly on vehicles is also suggested by the figures on transport-related accidents. In the Pilbara, transport-related accidents are the third-highest cause of death among Indigenous men (at 8.9% of the population), while among Indigenous women it is the seventh-highest cause of death (at 4.3%) (Taylor & Scambary 2005). In 1991–92 there were 661 Aboriginal and Torres Strait Islander men hospitalised due to transport-related accidents, compared with 329 Aboriginal and Torres Strait Islander women. Indigenous Australians are over-represented in road fatalities by approximately 3.5 times (Moller, Dolinis & Cripps 1996) .

Given the general vehicle shortage found in the ethnographic evidence, the 2002 NATSISS figure of a 12.3 per cent discrepancy between male and female access to a vehicle to drive in remote areas would appear to overstate both the extent of availability and access. A gendered discrepancy in access is also apparent in the non-remote NATSISS, where women have 7.2 per cent less vehicle access than men. In the GSS, women were found to have 9.3 per cent less access than men, suggesting that regardless of scarcity men will always have greater vehicle access.

Equity of access to vehicles within the settlement is also assumed in the presupposition of the headline question, that 'community' vehicles are equally shared among the group or household. As Gerrard notes, 'in reality, access to Aboriginal-owned vehicles [is] strictly limited by clan and family affiliation, meaning it [is] unevenly distributed' (1989: 101). Likewise, the notion that a 'community vehicle' can be driven at any time is not supported by the ethnographic evidence. Vehicles for general community or group purpose use are virtually non-existent; they tend to be allocated to specific purposes. And where government-sponsored vehicles are allocated for less specific purposes, such as for outstations, they are associated with an individual: the outstation 'boss' and his immediate family.

Forms of transport and questions of availability

The 2002 NATSISS survey also investigated all forms of transport used in the last two weeks and the main reasons for not using public transport. The vast majority of the data produced from this section highlight statistically significant differences between patterns of remote and non-remote transport use and availability. However, given the findings of the ethnographic research on the first or headline question, the veracity of a number of the figures from these other sections of the survey may also be thrown into some doubt.

The figures generated for the total number of Indigenous people who used transport for both the remote and non-remote categories seem very high (85.9% and 98.1% respectively). What can these figures tell us? They indicate that transport of some form or another is readily available. Yet, as the preceding discussion suggests, we cannot assume that this is motorised transport, particularly for remote areas. Rather, these figures beg the question of what 'transport' is and what the NATSISS wants to elucidate in terms of access to it. This figure is a conflation of all forms of transport used in the last two weeks, including walking. I query the value of including walking and more generally 'other modes of transport' that are unidentified in this total figure—might it also include horse riding, for instance? Walking is an important form of transport and it needs to be distinguished as separate from motorised transport, as indeed it is in the survey form. A total figure that stated clearly 'all forms of motorised transport' would be more useful.

The numbers of people who use walking as a form of transport in remote areas is high. The figure of 74.4 per cent of the remote area population who walked in the last two weeks 'to get around' compares with the non-remote figure of 49.2 per cent. This very significant difference seems anomalous. However, it may be reflective of remote area people regularly walking locally within the settlement and walking for customary economy activity. It might also be reflective of the severe lack of motorised transport that remote area respondents seem to be suffering and the lack of choices that this necessitates. In many rural areas in developing countries, for instance, walking is the major form of transport, and it is clear that in these contexts this transport method is closely associated with lack of access to economic and social services, transport infrastructure, and thus 'real' poverty (Starkey et al. 2002).

Interestingly, the GSS does not include any reference to walking as a form of transport. Rather, it is only included as a form of leisure, such as exercise. So, comparisons cannot be drawn here.

In remote areas, 63.3 per cent of respondents indicated that they did not use public transport, as no service was available, compared to 16.3 per cent in non-remote areas. This very significant difference does seem to reflect actual availability of public transport. However, it needs to be read closely in relation to the total figures for both the remote and non-remote respondents who did not use public transport. This total is qualified by the statement that it 'includes persons who did not use public transport for reasons of personal safety, cost, racial discrimination and time considerations'. In remote areas, a total of 85.6 per cent of people did not use public transport. I suggest that the specific qualification or reason for such a high figure would (principally) be because no service was available, as the figure of 63.3 per cent suggests, and the cost. Consider the cost of charter planes and commercially run buses, such as the 'fizzer bus' near Katherine in the Northern Territory: the cost for this 'public' bus, which is run commercially, is $350 per person for a 180km trip (Toohey 2000).

In non-remote areas, the total number of respondents who did not use public transport is 65.1 per cent, which may appear surprising given the apparent availability levels. However, the qualifications listed above may also well apply to this non-remote figure, as reported in a recent article in the NCOSS newsletter on transport disadvantage in NSW Aboriginal communities. Lack of coordination of public transport services between country towns means that a 208 km trip by sealed road can take up to eight hours (Wadiwel 2005: 7). Likewise, there is a real issue of affordability where CDEP participants lose their public transport concession entitlements, despite receiving an equivalent income to unemployment allowance. Discrimination is also a factor where 'across NSW Aboriginal people are routinely being refused bus, taxi or other services on what can only be said to be racial grounds' (Wadiwel 2005: 7).

Diversity of transport

The 2002 NATSISS category 'other modes of transport' could be further clarified (Table 15.2). In remote areas, other forms of transport include chartered aircraft (planes and helicopters) and horses. With the advent of the new rail link from Alice Springs to Darwin, trains are also a possibility in that region. Likewise, in other remote areas across Australia, trains have long been a transport possibility.

Establishing figures of particular types of transport usage could be useful in clarifying demand and costs. Air charter and trains tend to be significantly more costly than road transport, yet air charter, in particular, may be the only option at certain times of the year, given the impact of seasonal weather patterns on road access.

In remote areas, 'public' transport is not necessarily easily defined. Buses, such as the 'fizzer bus' mentioned above, are specific for Indigenous use, and thus not really 'public' in the general sense of the term. They tend to be private enterprise buses run for commercial gain by local non-Indigenous entrepreneurs. Likewise, there are no public or government subsidies for airline charters, even though in some remote regions (in Arnhem Land and Cape York, for instance) there may be no other way to access services, such as clinics or stores, during the wet.

Table 15.2. Indigenous people aged 15 years or over, modes of transport usage by remoteness, Australia, 2002

	Remote	Non-remote	Total
	%	%	%
All modes of transport used in last 2 weeks [b]			
Bus [a]	13.6	29.6	25.2
Car/4WD as passenger [a]	58.6	59.3	59.1
Car/4WD as driver [a]	38.6	53.4	49.3
Taxi [a]	10.6	19.0	16.7
Bicycle [a]	4.2	9.0	7.7
Walking [a]	74.4	49.2	56.1
Total used transport [a c]	85.9	98.1	94.7
Did not use transport [a]	14.1	1.9	5.3
Main reason for not using public transport in last 2 weeks			
Did not use public transport in last 2 weeks			
Prefer to use own transport or walk [a]	17.5	33.6	29.2
No service available at all [a]	63.3	16.3	29.1
No service available at the right/convenient time [a]	2.3	5.0	4.3
Total did not use public transport [a d]	85.6	65.1	70.7
Used public transport [a]	13.8	34.6	28.9
Indigenous persons aged 15 years or over	100.0	100.0	100.0

a. An asterisk denotes that the difference between remote and non-remote data is statistically significant at the 5% level.
b. Respondents may have indicated more than one response category.
c. Includes other modes of transport.
d. Includes persons who did not use public transport for reasons of personal safety, cost, racial discrimination and time considerations.
Source: ABS (2004c, Tables 1 and 20)

Perceived level of difficulty with transport

The final set of 2002 NATSISS transport questions relate to the perceived level of difficulty in accessing transport (Table 15.3). Unfortunately, the key question about difficulty of access relates to perceptions rather than facts. The questions, which include, *'Do you have difficulty getting to places needed?'* and *'Do you have these problems often or sometimes?'*, offer a subjective self-assessment in terms of perceived level of difficulty. Such questions carry similar methodological baggage to self-assessed health measures (e.g. Anderson & Sibthorpe 1996: 124; Brady 2005: 133). Likewise, the concept of 'conditioned satisfaction' may play a role in the self-perception of difficulty and how this relates to needs (Nussbaum 2001).

This final set of questions is so generalised that the figures generated tell us little. The figures generated for people who 'can easily get to the places needed' are 65.6 per cent and 71.8 per cent for remote and non-remote respectively. This compares with 84.3 per cent of the general population who 'can easily get to the places needed'. This begs questions of where the 'places' are, how people are getting there, and how 'need' is defined? Clearly, the answers are different in all three cases, as are transport needs. As a result, these figures are not comparable.

Table 15.3. Perceived level of difficulty with transport—percentage of the population (aged 15 plus) by sex

	Remote	Non-remote	Non-Indigenous [a]
	%	%	%
Male			
Can easily get to the places needed	68.2	72.5	86.8
Sometimes have difficulty getting to the places needed	15.9	17.2	10.4
Often have difficulty getting to the places needed	2.8	4.5	2.4
Can't get to the places needed/never go out/housebound	13.1	5.8	0.4
Female			
Can easily get to the places needed	63.0	71.2	81.9
Sometimes have difficulty getting to the places needed	18.5	18.9	13.3
Often have difficulty getting to the places needed	3.8	4.8	4.1
Can't get to the places needed/never go out/housebound	14.8	5.1	0.7
Persons			
Can easily get to the places needed	65.6	71.8	84.3
Sometimes have difficulty getting to the places needed	17.2	18.1	11.8
Often have difficulty getting to the places needed	3.3	4.6	3.3
Can't get to the places needed/never go out/housebound	14.0	5.4	0.6

a. Non-Indigenous statistics refer to people aged 18 and over.
Source: Customised cross-tabulations from the 2002 NATSISS and GSS CURF

Suggestions for re-phrasing and re-sequencing the questions

Questions in the next NATSISS could be improved by proceeding from the general to the specific, by changing the sequencing of questions, and by separating out the issues. A less deterministic approach to structuring and framing the questions could be followed, starting with the questions on access to transport. The key question is, *' Can you get to places you need to go?'*, followed by questions on *'How do you get there?'*. By clarifying the *how* in transport, actual vehicle numbers and types could also be elicited, while further choices could be given, such as types of transport (as suggested above for the remote areas).

One approach to establishing approximate numbers of private vehicles would be to begin with the question, *'Do you own a vehicle?'*. If the answer is *no*, then the question could be followed with 'Do you have access to a vehicle?'. If the answer to this question is *yes*, it could be followed with, *'Who owns this vehicle: family member, the council, clinic, school, other?'*.

In asking *'What other forms of transport do you use?'*, a broader range of options could be listed. Finally, I have not discussed in detail the issue of drivers licenses in this paper. However, gaining and maintaining a license for Indigenous people in remote and non-remote areas is a significant issue, and has a major impact on the ability to drive and to teach others (see also Wadiwel 2005). Thus, it needs to be a separate question, rather than conflated with another question, as it is in the 2002 NATSISS. A specific question could be, 'D*o you hold a current drivers license?*' and if the answer is *no*, the next question could be, *'Have you ever held one?'*.

Conclusion

The aggregation of very remote areas with remote areas in the NATSISS data means that the relationship between access to transport and locational disadvantage cannot adequately be drawn. Yet, there is considerable locational diversity between these two classes of remoteness which impact very significantly on transport infrastructure, and thus on access to employment and education opportunities. These factors impact on socioeconomic status, in turn affecting transport access.

The 2002 NATSISS suggests that access to transport in remote Australia is lower than elsewhere, but that almost half of the Indigenous adult population does have access to some form of transport. This level is substantially higher than indicated by evidence (admittedly partial) from select ethnographic case studies, leading to some doubt regarding the validity of 2002 NATSISS findings. Thus, although the transport access patterns identified in remote areas are correct (for example, the limitations on access to motorised transport and women's lack of access relative to men's), it is likely that the headline question was misleading. If this is true, it is unfortunate: the importance of transport needs in remote areas are heightened, because of the dispersed population and services, the need to access lands, the vast distances and road quality. A recent World Bank paper noted that 'patterns of transport demand and supply are [always] linked to population density and income levels' (Starkey et al. 2002: 3). Gaining an accurate fix on the scale of the issue is fundamental to developing an informed policy approach to remote transport needs. Despite the survey, we still do not have a fix on the proper scale of the issue. Nor do we get a picture of vehicles per capita in this remote region, nor gain any sense of the predominant vehicle types.

16. Information and Communication Technology

Peter Radoll

This chapter presents data from the 2002 NATSISS as it relates to Information and Communication Technology (ICT) use, namely computers and the internet. While a number of determinants of ICT use have been well established, like education and income, there are other factors that have a similar impact on ICT use in the Aboriginal and Torres Strait Islander community. Presented here for the first time, these factors include health status, the impact of being on the CDEP scheme, the impact of the justice system, access to online services, and the overall digital divide between remote and non-remote Indigenous communities.

Firstly, it should be noted that this data is dated, in that since its collection there have been many ICT projects targeting both Indigenous and non-Indigenous communities at both the State and federal government levels. Such programs include the Telecommunications Action Plan for Remote Indigenous Communities (Department of Communications, Information Technology and the Arts [DCITA] 2002) and Networking the Nation Program (DCITA 1999).

Statistical analysis of the data has been conducted in this paper and where the term 'significant' is used, it refers to statistical significance. This is the first time this data has been collected, so a comparison with previous NATSISS data is not possible.

Usefulness of the data

Capturing ICT use in the 2001 Census was well founded, as it not only provided an overall picture of ICT use by individuals but it also allowed us to see whether there were segments of the population that were lagging behind others in ICT use or suffering from a 'digital divide'. The results of the national census clearly demonstrated that a 'digital divide' existed between the Indigenous and non-Indigenous communities. Capturing similar data in the NATSISS has enhanced the understanding of the 'digital divide' facing the Indigenous community. Unfortunately, only national data has been released by the ABS (2004c) which prevents a more localised region-by-region analysis of ICT use in Aboriginal and Torres Strait Islander communities. However, on a positive note, we can now compare remote and non-remote Aboriginal and Torres Strait Islander communities, which give us a better understanding of where government ICT policy and resources can be targeted.

Digital divide defined

The term digital divide has many connotations and there has been much research attempting to define the term. Fink and Kenny (2003) define the digital divide as:

- a gap in access to ICT—measured primarily by the number of users or individuals with access per one hundred of population
- a gap in the ability to use ICT—measured by an individual's skills set
- a gap in actual use—measured by the amount of time an individual goes online
- a gap in the impact of use—measured by economic returns or savings attributed to ICT.

This multi-faceted approach to defining the term 'digital divide' is drawn from a decade of literature that has emerged in the area, so it is a good reference point to understanding the term. The ABS collected data on only one aspect of digital divide, namely 'use', which excludes the other three aspects of digital divide. Data on use is by far the easiest to collect but leaves a large gap in truly understanding the extent of the digital divide.

Australia's digital divide

To understand the depth of the existing digital divide within Australia it would be beneficial to first examine the differences between the Indigenous and non-Indigenous communities from the NATSISS data then more closely examine data as it relates to remote and non-remote Indigenous communities.

Figure 16.1. Home computer use and home internet use by Indigenous status, 2002

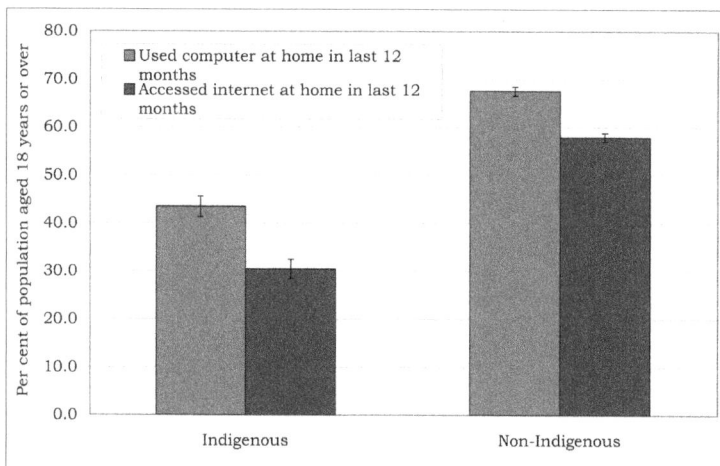

Source: ABS (2004: Table 5).

Figure 16.1 highlights the digital divide between Indigenous and non-Indigenous communities. It shows that at the time of the data collection 67.6 per cent of the non-Indigenous population had used a computer at home in the previous 12 months compared with only 43.5 per cent of the Indigenous population. Similarly, internet use at home was significantly different between the Indigenous and non-Indigenous populations, at 30.4 per cent and 57.9 per cent respectively. This significant digital divide between the Indigenous and non-Indigenous populations has been highlighted by Daly 2005 and shows that Indigenous people are two-thirds as likely as non-Indigenous people to use computers at home and only half as likely to use the internet at home (ABS 2004c).

Previous research has demonstrated that there are two key factors in determining ICT use in Australia, namely income and education level (Daly 2005; Lloyd & Bill 2004; Lloyd & Helwigg 2000). These determinants are no different in the NATSISS, as figures 16.2 and 16.3 clearly show a positive relationship between income and ICT use and education and ICT use.

Income

Figure 16.2. Computer use and internet use by income quintile, 2002

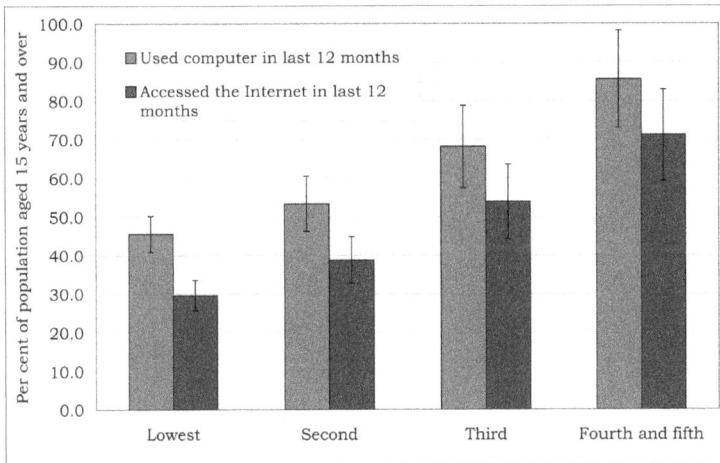

Source: ABS (2004: Table 9).

Specific individual income levels were not published in the NATSISS, so it is difficult to say which level of income would fall within which quintile. However, as a rough comparison using available data, the national mean equivalised gross household income for the non-Indigenous community is $665 compared with $394 for the Indigenous community. These figures clearly demonstrate that there is a significant difference in the mean household incomes between Indigenous and non-Indigenous communities. Therefore, we can say that the first quintile would be quite low and even the highest quintile would still be relatively low. While the mean household income of the Indigenous population is low in comparison to the non-Indigenous community, ICT use still increases as income increases, making income a very strong determinant of ICT use.

While we may expect this, as previous research has highlighted this fact, we should still be aware of international research on how income and the digital divide is perceived. Mark Malloch-Brown from the United Nations Development Programme stated there is a 'growing digital divide between rich and poor' (quoted in Fink & Kenny 2003: 3), suggesting that we can only make inroads to the digital divide between Indigenous and non-Indigenous communities when Indigenous and non-Indigenous incomes are equal.

Education

Figure 16.3. Computer use and internet use by education level, 2002

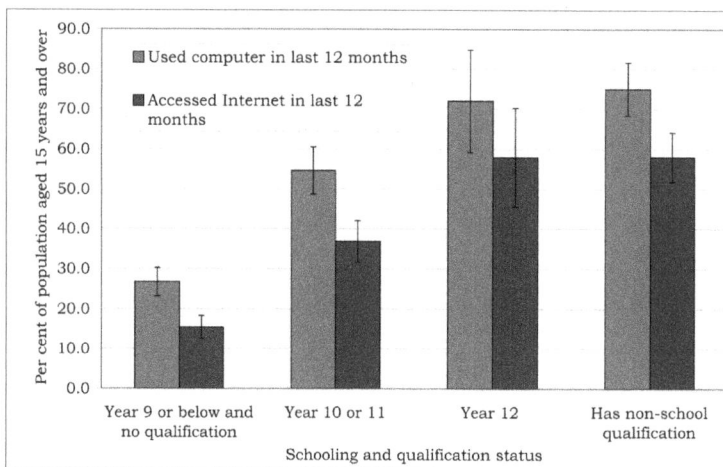

Source: ABS (2004: Table 7).

As we know from previous research, education—or at least the level of education—plays a role in ICT use. Fig. 16.3 highlights that Aboriginal and Torres Strait Islanders with low or no formally quantifiable educational qualifications were found to have lower use rates of ICT, whereas those with Year 12 and post-school qualifications have a higher rate of computer and internet use. These findings that education qualifications are a strong determinant of ICT use are not surprising, as Lloyd and Helwig (2000) also found that the level of education was a strong factor in determining computer use in Australia. Post-secondary education was found by Lloyd and Helwig to be the strongest determinate of computer use for the overall Australian population. However, post-Year 11 education is the most prominent determinate of computer use for the Indigenous community, with Year 12 and post-school qualifications making no significant difference. One point of interest from the data on internet use is that both Year 12 and post-school qualifications have the same rates of use, suggesting that while computer use may increase (albeit nominally) as a person's educational qualifications increase, internet use may, in fact, stabilise.

While income and education do play a major role in determining ICT use for the Aboriginal and Torres Strait Islander population, other factors affecting ICT use have arisen from the NATSISS that I will go on to discuss. But firstly, it would be useful to examine the digital divide that exists between remote and non-remote Aboriginal and Torres Strait Islander communities.

Figure 16.4. Computer use and internet use by remoteness, 2002

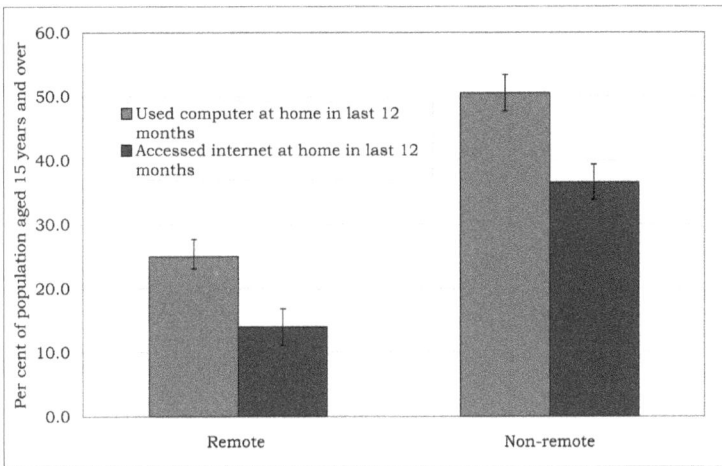

Source: ABS (2004: Table 5), persons aged 18 and over.

It has already been demonstrated that a significant digital divide exists been the Indigenous and non-Indigenous communities in Australia, and Figure 16.4 above highlights a significant digital divide been remote and non-remote Indigenous communities. Remote communities use computers and the internet less than half as much as their non-remote counterparts. There are many factors that point to why this may be and I will now examine the available data from the NATSISS to attempt to determine this.

Location

Figure 16.5. Computer use by location and remoteness, 2002

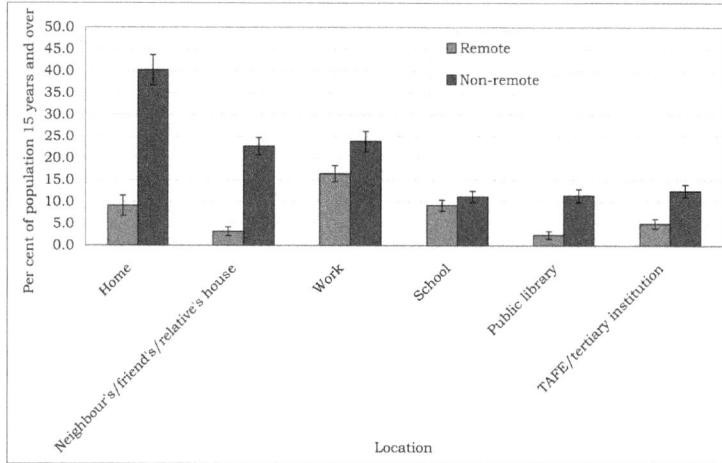

Source: ABS (2004: Table 22).

One important component of ICT use is the location at which you are able to use the technology. Figure 16.5 shows the locations that remote and non-remote communities use computers, with some interesting results.

All areas of computer use in the NATSISS are significantly different when comparing remote and non-remote computer use, with the exception being use in schools. This further highlights the digital divide between remote and non-remote Aboriginal and Torres Strait Islander communities and demonstrates the importance of schools in remote communities in bridging the digital divide.

There are three main computer access points for non-remote communities (see Fig. 16.5). These are in the home, at another person's home and at work. However, in remote communities the main access points are work, school and home. It is interesting that remote and non-remote communities are contrasting in the locations where computers are accessed. There is no one factor that indicates why this is the case; rather, there seems to be a number of factors that contribute to such low ICT use in remote communities overall.

Before discussing the additional factors, we should first examine the issues associated with the above findings; in particular, the main access point to computers in remote communities being in the work place or at work. While digital divide literature would argue that any use of computers is bridging the digital divide, using computers at work only enables or permits use of, and access to, the technology mainly for work purposes and associated tasks. By this, I mean that using ICT provided by the employer in the workplace is often restricted. Examples of this are the various e-mail and internet 'appropriate use' policies utilised in all levels of the public service and other employing entities.

These 'appropriate use' policies are designed to address issues of security and productivity, both of which are vital to organisations. Using computers at work drastically inhibits other types of use; namely personal and private use.

Home use is one key to bridging the digital divide. NATSISS shows that home use of ICT lags well behind in remote communities, at 9.1 per cent, while in non-remote areas it is 40.2 per cent. It is interesting to note that use in schools in remote communities is 9.2 per cent which is not significantly different to home use in remote communities. Home access has many benefits over other types of access in that there are no restrictions on when, why, and what you access. In addition, by having a computer at home, you can install personal software and programs.

Purpose

Figure 16.6. Purpose of internet use by remoteness, 2002

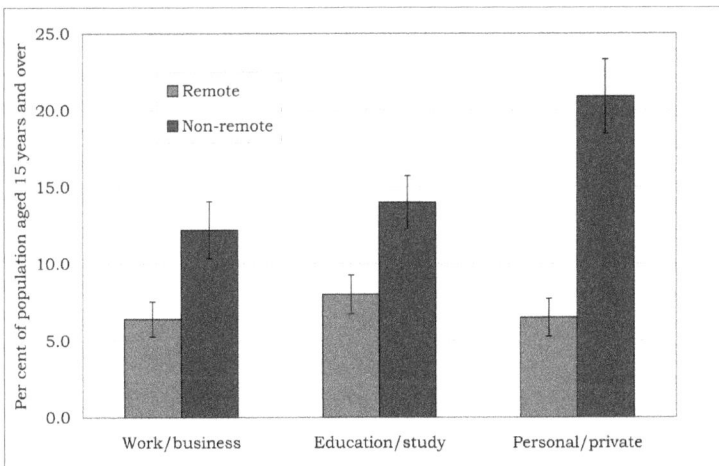

Source: ABS (2004: Table 22).

The purpose for using the internet in remote and non-remote Indigenous communities highlighted in Figure 16.6 shows that the level of internet access for specific purposes is also significantly different. Non-remote communities tend to use the internet for personal/private reasons. In remote communities, work/business, education/study and personal/private are the main reasons for accessing the internet, with no significant difference between the three purposes.

Frequency

Figure 16.7. Frequency of internet use by remoteness, 2002

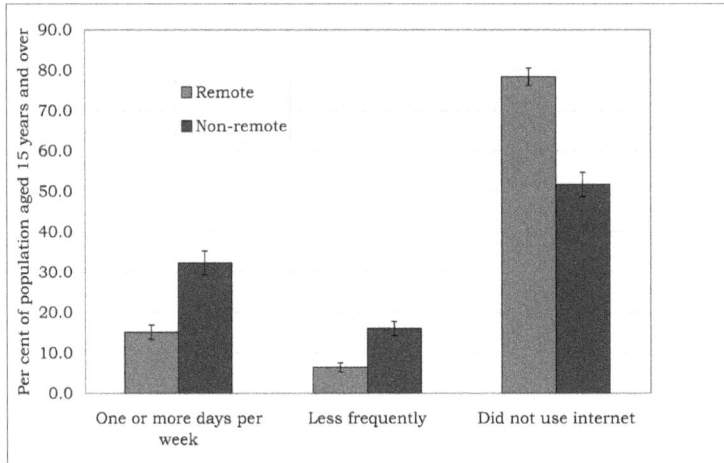

Source: ABS (2004: Table 22).

As we have seen previously in the definition of the digital divide, an individual's skill set is a sub-component of the digital divide. While we do not have data on the time individuals spend online, we do have data on the frequency of internet use. Therefore, examining the rates or usage would also be a further indicator as to which segment of the Indigenous population suffers from the digital divide. Figure 16.7 clearly shows that internet usage rates are very low overall, demonstrated by two different sets of data. Firstly, 78.4 per cent of Indigenous people in remote communities did not access the internet at all compared with 51.7 per cent of non-remote Indigenous people. Secondly, 32.3 per cent of non-remote Indigenous communities accessed the internet one or more days per week compared with only 15.1 per cent in remote Indigenous communities.

Other inhibitors of ICT use

As we know, Fink and Kenny (2003) define the digital divide as both infrastructure access and individual capabilities. It would be ideal to have data that demonstrate—or at least go some way towards understanding—the depth of the digital divide in Aboriginal and Torres Strait Islander communities. Unfortunately, only a few statistics were collected on ICT use in the NATSISS, leaving many aspects of the digital divide unable to be examined. Therefore, we cannot speculate too much on an individual's ICT skill set, nor can we gauge whether the Indigenous community derives any economic advantage by using ICT as outlined by Fink and Kenny (2003), because that data is simply not available. What we can discuss further is the additional inhibitors to ICT that have come out of the NATSISS that impact on Aboriginal and Torres Strait Islander communities.

Employment

Figure 16.8. Information and Communication Technology use by employment status, 2002

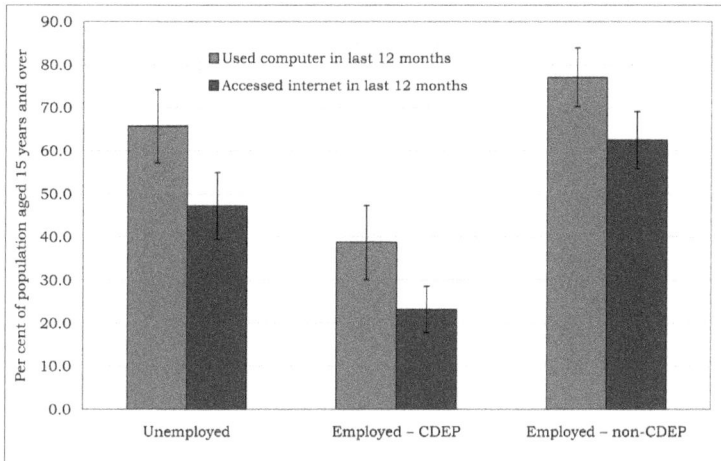

Source: ABS (2004: Table 8).

Employment status and the type of employment have an impact on ICT use. Being employed makes Aboriginal and Torres Strait Islanders significantly more likely to use a computer and access the internet; this may be closely correlated with income. However, what is striking is that being on the CDEP scheme makes you less likely to use computers and the internet than those who are unemployed. Moreover, only 38.7 per cent of those employed under the CDEP used computers, whereas 65.7 per cent of the unemployed used a computer. Similarly, 23.2 per cent of CDEP participants accessed the internet compared with 47.2 per cent of the unemployed participants. While this might be surprising to some, it really should not come as a great surprise as there has been research suggesting that CDEP is not an effective program for employment opportunities (Langton 2002). Also, the data presented here further demonstrate that CDEP programs perpetuate the digital divide. It should be noted that those on unemployment benefits are required to seek employment opportunities, and ICT is usually provided by job search agencies for this purpose.

Age

Figure 16.9. Information and Communication Technology use by age, 2002

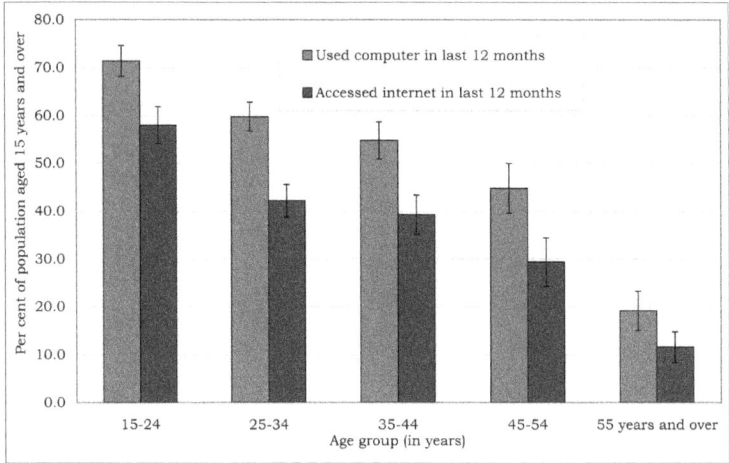

Source: ABS (2004: Table 3).

We have seen previously (see Fig. 16.2 and Fig. 16.3) that higher incomes and higher education rates lead to higher ICT use. Therefore, you would expect that ICT use would be predominantly an older age group activity. However, this is simply not the case. Surprisingly, the youngest age group has the highest ICT use. This may demonstrate that age could play an equal role along with income and education in determining ICT use, or it may in fact show that education is the stronger determinant rather than income and the fact that income and education are correlated may only be coincidental. This discrepancy between age groups could also be partly explained by the 'digital natives—digital immigrants' concept. This theory asserts that technology is more accepted the earlier it is introduced to a person ('digital natives'), while there is some resistance if the technology is introduced later in life ('digital immigrants') (Prensky 2001).

Health status

Figure 16.10. Information and Communication Technology use by health status, 2002

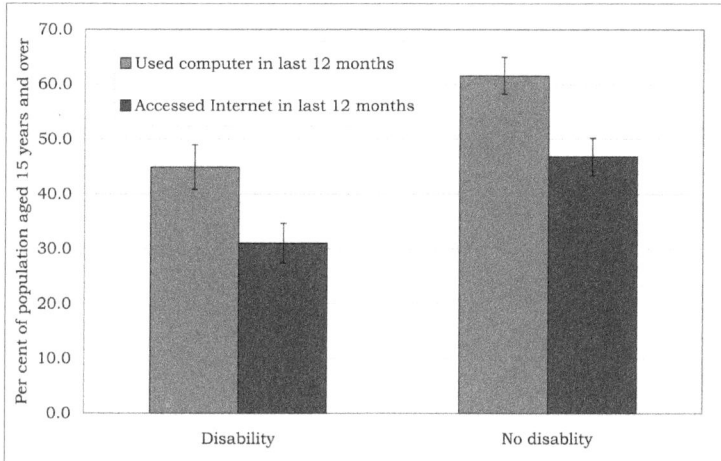

Source: ABS (2004: Table 10).

Health of an individual plays a significant role in the use of ICT. It is well known that the overall health of the Aboriginal and Torres Strait Islander community is lower than that of the non-Indigenous population. However, just because you have a disability or ailing health does not exclude you from your citizenship responsibilities. Neither should your condition exclude you from the digital era, but it quite obviously does. Of those with no disability, 61.6 per cent used computers compared to 44.9 per cent of those with disabilities. Once again, internet access is similar, with 46.8 per cent of those with no disabilities having accessed the internet and only 31 per cent of those with a disability having accessed the internet in the same period. This leaves those with disabilities and poor health suffering a digital divide.

Justice system

Figure 16.11. Computer use by whether ever charged, 2002

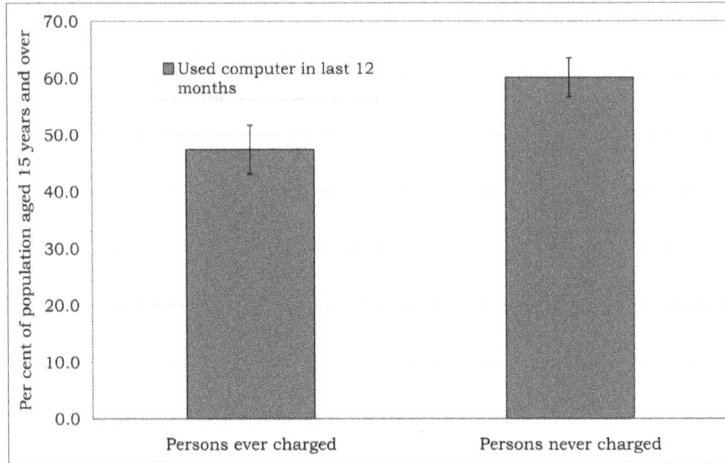

Source: ABS (2004: Table 11).

Being charged appears to have an effect on ICT use, with a significant difference in computer use between people having been charged and people who have never been charged represented by 47.4 per cent and 60.1 per cent respectively. This may also be closely correlated with education and income. Unfortunately, internet access data were not available for charged and non-charged people but we can assume that the percentage of internet access for people 'ever' charged would also be significantly different from the percentage of people 'never' charged.

Telephone status

Figure 16.12. Telephone status by remoteness, 2002

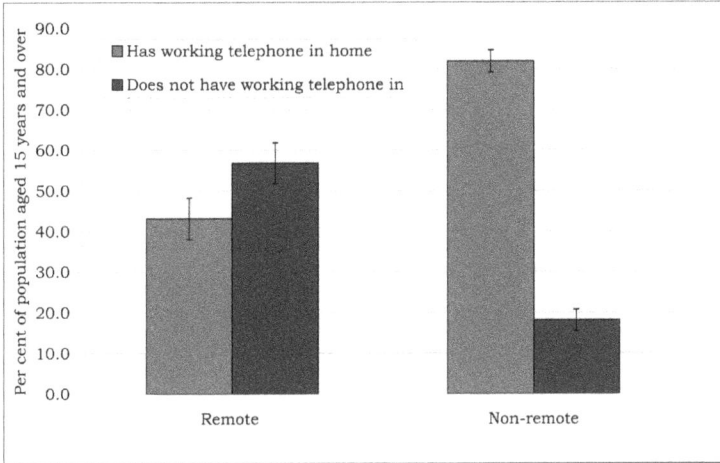

Source: ABS (2004: Table 22).

Unlike computer use—where only electricity is required to use the technology—internet access relies on other infrastructure for connectivity, namely a telephone line, whether it is for a dial-up service or for broadband services (excluding direct satellite-based technologies). The NATSISS has highlighted a significant internet barrier for remote communities, namely the connection point. Only 43.2 per cent of the remote homes surveyed had a working telephone, which is approximately half that of the non-remote population at 81.9 per cent. This suggests that the potential to have the internet connected at home in non-remote communities is higher than that of remote communities, which may be an indicator of low internet use.

Accessing money

Figure 16.13. Money access mode by remoteness, 2002

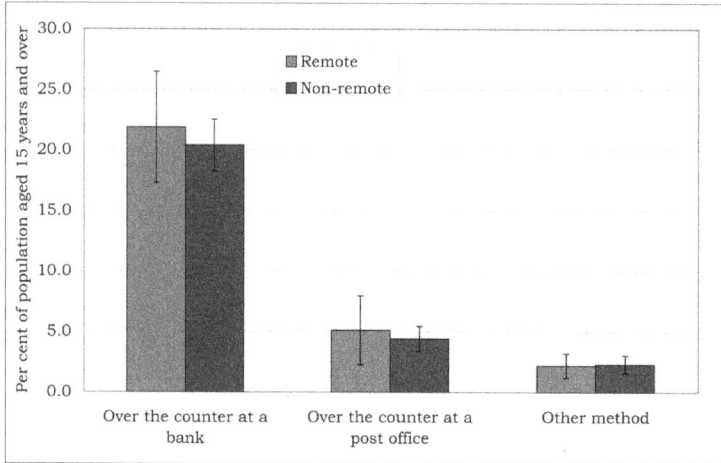

Source: ABS (2004: Table 18).

Accessing money is one of the data items collected in the NATSISS. This provides an opportunity to examine how Aboriginal and Torres Strait Islanders access their money and to see if more modern technologies are being used by this segment of the Australian community. The NATSISS data highlighted that more conventional methods of accessing money—like face-to-face—are relatively low (see Fig. 16.13) but nevertheless there was no significant difference in using these modes between remote and non-remote Indigenous communities.

Figure 16.14. Electronic money access by remoteness, 2002

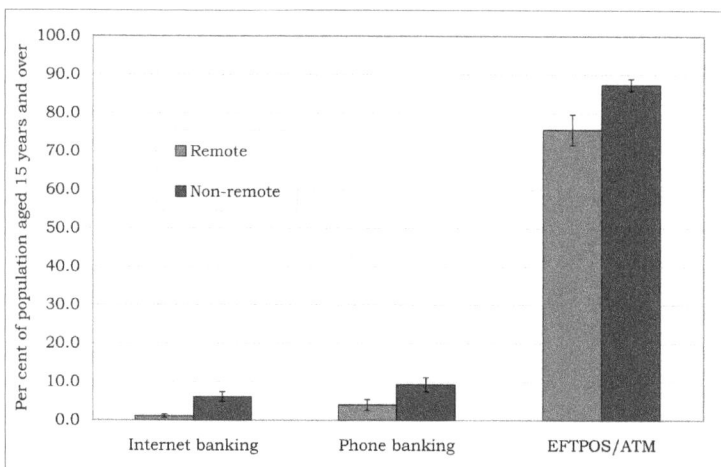

Source: ABS (2004: Table 18).

In contrast to Figure 16.13, Figure 16.14 shows that some electronic forms of money access have become quite popular in both remote and non-remote communities. EFTPOS and ATMs were the most used of the electronic technologies, with 75.7 per cent of people living in remote communities using this mode compared with 87.4 per cent of people in non-remote communities—a significant difference. Phone banking was the next most popular method of electronic money access, leaving internet banking the least popular method of accessing money. Only 6.1 per cent of people in non-remote communities and just 1.1 per cent of people in remote communities were making use of this mode.

Conclusion

We have seen that there is a significant digital divide been the Indigenous and non-Indigenous community, and the NATSISS has provided data showing that there is a significant digital divide between remote and non-remote Indigenous Australia as well. While there are many aspects of ICT use collected in the NATSISS, there are other aspects of the digital divide that have not been collected; namely, a user's skills set, the number of hours an individual spends online and the economic benefit derived from ICT. This is a weakness of the NATSISS data.

There is no doubt that, given such low ICT use rates and the fact that numerous inhibitors to ICT use are faced by the Aboriginal and Torres Strait Islander community, targeted policy is required to overcome the digital divide. However, the complexities of ICT use inhibitors (namely, low incomes, low education, being charged in the justice system, poor health and being on CDEP) suggest that policy targeting these key areas will not see the digital divide being bridged.

It must be stated that while remote Indigenous communities are in most need of ICT, the Indigenous community as a whole lags well behind the rest of Australia in ICT use and any Australian ICT policy should target Indigenous communities as a whole.

To conclude, it is worth noting that the above discussion focuses on the bi-variate analysis of the relationships of various socioeconomic factors to ICT issues. While a multi-variate analysis of the factors underlying ICT use might eliminate some of the duplication arising from the correlation between these factors, this is left for future research. This chapter has provided one of the first overviews of ICT issues in the Indigenous community, providing a first step towards a more structured empirical analysis that might include a multi-variate modelling exercise.

17. Health

Russell Ross

The beginning point of this analysis is the established fact that Indigenous health outcomes are recognised to be very poor both relative to those for the non-Indigenous population and in absolute terms—see, for example, Gray, Hunter & Taylor (2004) and Booth & Carroll (2005).

Life expectancies for Indigenous Australians are some 20 years lower [1] than for non-Indigenous Australians, an unacceptable statistic. Equally unacceptable is the fact that this life expectancy appears not to have risen in recent times. Of particular interest is to ascertain whether there have been significant improvements in Indigenous health outcomes; not only in an *absolute* sense but also *relative* to the non-Indigenous population standards.

Whatever the conclusion is about the aggregate picture for Indigenous peoples, it is also important to ascertain how balanced the picture is within the Indigenous population. For example, we can compare outcomes by factors such as location (defined by the remote/non-remote distinction in this context); geographic regions (States and Territories); age; gender; education; and labour force status. Finally, it is also important to ascertain what progress has been made between 1994 and 2002; that is, between the two NATSIS surveys.

A central question must be: how much of the health gap can be explained by socioeconomic factors, and how much can be explained by different levels of access to health services? Further, where there is access to health services, what is the take-up rate? These are very important questions. Depending on the answers, the policy responses should be very different. If socioeconomic factors can explain the entire gap, then policies must be directed towards improving Indigenous socioeconomic outcomes. However, the policy response should be quite different if it is a problem of availability and access to health services.

There is another caveat to be made upfront. That is that the coverage of 2002 NATSISS was limited to people living in private dwellings. While this may not be a major limitation for many aspects of the 2002 NATSISS analysis, I would argue that in the context of health, it is potentially a significant limitation. As the 2002 NATSISS did not canvass individuals who were in institutions, it is potentially overstating the general health levels in the Indigenous population. I'm thinking in particular of hospitals, nursing homes, hostels and prisons. By definition, people in hospitals are on average going to be suffering below average

---

[1] Over the period 1999–2001, the life expectancy at birth for an Indigenous male was 59 years, and for an Indigenous female, 64 years. Comparable life expectancies in the total Australian population were 77 years for males and 82 years for females. All these figures are from ABS (2004c).

health. Although many people in hospitals are there for very short spells and may be experiencing very temporary health setbacks, many are hospitalised for treatment for ongoing health conditions. A similar pattern occurs in relation to nursing homes and hostels, where it is likely that the health status of residents is lower than the average among residents of private dwellings. It is also well known that prison inmates are on average less healthy than those outside prisons. Given the disproportionately high incidence of incarceration for the Indigenous population, this is expected to bias the health measure upwards. According to ABS (2004e), the Indigenous population is over-represented in prisons by a factor of 10 to 1 and on the night of 30 June 2004, 20 per cent of all prisoners were Indigenous. The ABS (2005c: 3) estimates that at the time of the 2002 NATSISS, there were 19 320 Indigenous Australians living in non-private dwellings. This is equivalent to 7 per cent of the estimated total Indigenous population aged 15 and over of 282 200.

A survey such as the 2002 NATSISS cannot be expected to answer all the big questions about Indigenous health. For example, it is not a suitable instrument for assessing factors such as life expectancies which require externally measured, scientific data. It is not an appropriate instrument for collecting any epidemiological or public health information which requires the arms-length, objective collection of raw data.

The data

What information was asked?

The focus of the questions was on the respondent's self-perception of the state of their general health, and the disabilities they experienced. This was gathered via a *single* question seeking the person's own evaluation on a five-point scale: excellent, very good, good, fair or poor.

Considerably more information was gathered on disabilities and long-term health conditions. For the purposes of the 2002 NATSISS, disabilities were defined as 'conditions which you may have, that have lasted, or are likely to last, for six months'.

Participants were asked if they had any such conditions. For those who indicated they did have such a condition(s), a sequence of supplementary questions followed. These questions were designed to elicit considerations as to whether the conditions:

- meant they were restricted in any way in everyday activities as a consequence of the condition(s)
- necessitated any help or supervision with a range of tasks
- resulted in any difficulties with some specific tasks, and
- led to any difficulties with undertaking education or employment.

Most of the information obtained in the 2002 NATSISS was common to both sub-surveys,[2] but there were also several specific questions asked only in one sub-survey. The remote area questionnaire included some questions about medications and visits to clinics/doctors—information not requested in the non-remote area questionnaire. Conversely, included in the non-remote area questionnaire but not the remote area one was reference to some specific conditions such as arthritis, asthma, heart disease, Alzheimer's disease and dementia.

Although much of the information obtained in the 2002 NATSISS was common to both sub-surveys, there were significant methodological differences in the way the information was obtained. These methodological differences are highlighted in Biddle and Hunter (in this volume). Not only were the questions asked differently between the remote and non-remote areas, there were also differences in the amount of detail sought for some of the health-related questions; see especially the explanatory notes section in ABS (2004a). The issues specific to health will be addressed below.

What information should have been asked

It is unfortunate that several other questions were not included in the 2002 NATSISS. In particular, it would have been very useful to be able to assess whether people's perceptions of their individual general health level had changed over time. This would have enabled a better feel for whether individuals see themselves as getting healthier, getting less healthy, or otherwise.

It would also be useful to have data on *access* to health services, and the rate of *use* of such services (where available).[3]

However, against this is the reliance on self-perception. I see a problem with using questions that are loaded with words such as 'may', 'likely to', and so on. The responses gained are very subjective, and subject to considerable variation in people's interpretation of both their own health and exactly what the question is seeking. Further, it relies heavily on the person's willingness to answer openly.[4]

Finally, I believe that the questions are too general in nature. It would be better to have included some questions which sought to gauge the depth of the

---

[2] By this I mean the distinction between the remote and non-remote sub-surveys. See the papers by Webster, Roger & Black, and Biddle & Hunter in this volume for further discussion of this difference in the two sub-surveys. Although the ABS refers to the CA and NCA questionnaires, the questionnaires themselves are titled remote and non-remote.

[3] This is an important distinction, as there is a difference between people choosing not to use health services when they are accessible and being forced not to use them because the services are not accessible.

[4] The ABS is aware of this problem, at least in relation to disabilities and long-term health conditions, stating that 'there may be some instances of under-reporting as a consequence of respondents being unwilling to talk about a particular subject when interviewed' (ABS 2004c: 58).

problems. That is, where health problems are identified, there is no measure of depth or seriousness of the condition(s).

Of course, there will be a trade-off between the number of health questions and the overall length of the questionnaire, but in order to collect useful information, it is necessary to have a critical mass of questions, to include questions such as those referred to here.

Issues with the way the information was collected

It is important that we are able to relate the context of the questionnaire and the survey to the information in the data. The major methodological differences in the way the data were asked between the two sub-surveys raises serious questions about the comparability of the responses. In addition, the coverage of some questions and the amount of detail elicited in the two sub-surveys is of concern for analysts.

A problem I see is the over-reliance on self-perception of health status. Sibthorpe et al. (2001) and Crossley and Kennedy (2002) have both addressed this issue in relation to earlier surveys on Indigenous health. Crossley and Kennedy analysed data from the 1995 National Health Survey. That survey asked some respondents to categorise their own health status twice. A disturbing 28 per cent of people gave different responses to the two questions. Although most of these were only one category away from their previous answer, 3 per cent of the total sample gave answers more than one category different. This implies that the robustness of the answers is of some concern, and the sequencing of questions is very important.

Sibthorpe et al. (2001) have suggested that there are important differences in the link between self-perception of health and more objective measures (of health), especially for people living in very remote areas and/or for those for whom English is not their first language. Booth and Carroll (2005) also considered this aspect of reliance on self-perception measures but concluded that it probably is not a major concern.

Nevertheless, we do need to be aware of this issue and how it may impact on interpretation of the results.

A second issue with the data relates to significant differences between the questions asked, and the way they were asked, as there were major differences between the two sub-samples.[5] For example, in the remote area questionnaire, there were fewer questions asked. Unlike their non-remote counterparts, respondents in remote areas were not asked for information about factors such as disfigurements, deformities, mental illness, and restrictions on physical activities/work due to conditions such as back pain and migraines. They were

---

[5] This information is taken from Explanatory Notes 43–7 in ABS (2004c: 58–9).

also not asked about any psychological disabilities. However, the remote area questionnaire did include some questions on visits to medical practitioners and medications—questions not asked in the non-remote questionnaire. These questions could, in some cases, allow the ABS to identify disabilities. The importance of this is seen in the explanatory note 47 to ABS (2004c: 59), which states: [6]

> In tables showing disability data from the 2002 NATSISS only, the disability populations are limited to the set of criteria used to identify disability in remote areas. In the table comparing the disability status of Indigenous people in non-remote areas and non-Indigenous people (Table 5), more extensive criteria have been used to identify disability.

It is also important to keep in mind how the data was organised once it had been collected. The primary limitations on the usefulness of the data are generated by the format of the questions asked, rather than by the way the data were coded. Although some of the tables provided in ABS (2004c: 59), the ABS's published data output from the 2002 NATSISS, do collapse data into more aggregated categories, this should be less of a problem for those able to access the data via RADL. [7]

This is especially true in the health data, where often the 'good' category is omitted from the tables in ABS (2004b). Although it is straightforward for the analyst to reconstruct the missing data, it is simply inefficient for every researcher to have to do so.

Analysis of the 2002 NATSISS data

Examination of the existing ABS publications can give only very basic information; and even some of that is subject to interpretation. The following comments are based on analysis of the health components of selected tables in ABS (2004c: 59). It should be stressed that this analysis is based only on one source of data. There are a number of other sources of data on health. These include the general social surveys which ask some questions on health (e.g. ABS 2003b), the national health surveys (ABS 2002b), and publications from the ABS and Australian Institute of Health and Welfare (e.g. ABS/AIHW 2005).

Where applicable, the comparison also includes corresponding information from either the 2002 GSS or the 1994 NATSIS. The former is used for comparing the Indigenous figures with those of the non-Indigenous population, and the latter is used for comparisons of Indigenous figures at the two points in time (1994

---

[6] The table referred to in this quote, table 5, is in ABS (2005b). It is not Table 5 of this paper.

[7] RADL is an ABS service which allows web-based analysis in much greater detail than is possible with published data. It permits analysis based on individual-level data, subject to a number of restrictions which protect the confidentiality of NATSISS respondents. This enables researchers to structure their analysis to very specific issues.

and 2002). In the comparisons with other data sets, some compromises are necessary. For example, the 2002 GSS had a slightly more restricted age coverage (18+ years of age) than the 2002 NATSISS and also did not survey in remote areas. Also, 1994 NATSIS did not ask questions on disabilities.

Throughout the following discussion, two main indicators of health are used: firstly, the distribution of the responses to the self-perception of health question; and secondly, the overall indication of the incidence of disabilities (as defined above). Each table has these figures for the target group, and in each table is an indication of which statistics are significantly different from others. The standard of statistical significance is at the 5 per cent level. Occasional reference is made to 'weak' statistical significance—this is used to indicate where significance is at the 10 per cent level but not the 5 per cent level.

The analysis discussed here can only identify linkages between the variables. It does not identify the direction of causality of the linkage. That is, when two factors are connected, it is often important to know which factor is causing the other factor. For example, when we see a link between poor health and unemployment; is the unemployment leading to poor health or is the poor health meaning that the person cannot work? This issue, of lack of indication of the direction of causality, is common to all the tables presented here. For a complete picture, it is important to be able to determine this direction of causality, especially if we are to make policy implications, but that analysis is beyond the scope of this chapter. For further discussion of the general issue of causality, see Kawachi et al. (1999: xi–xxxiv).

Summary health indicators, Indigenous Australia, 2002

The broad aggregate comparisons within the Indigenous population are shown in Table 17.1. That table shows that at the aggregate levels of remote versus non-remote, and Aboriginal versus Torres Strait Islander, there are virtually no discernible differences. The only statistically significant difference is between remote and non-remote, and even then only for the 'fair/poor' categories. Although Table 17.1 also appears to show a real difference in the 'good' category as well, that difference is only weakly significant. More will be made of this implication later.

This overall 'sameness' of outcomes suggest that further analysis is needed at a less aggregated level. Throughout this analysis, no further distinction is made between Aborigines and Torres Strait Islanders, as there are no statistically significant differences between the two populations in any of these health categories.

The first two rows of Table 17.1 do not contain information on the breakdown between excellent and very good health categories separately for Aborigines and Torres Strait Islanders, as this was not available in ABS (2004c). In the remainder of the tables, these two categories are re-collapsed into the one category 'excellent/very good'. This has been done as there are no discernible variations in the composition of these two categories between remote and non-remote areas.

Table 17.1. Summary health indicators by remoteness and Indigenous status[a]

	Remote	Non-remote	All Indigenous	Aboriginal	Torres Strait Islander
	%	%	%	%	%
Self-assessed health status					
Excellent	16.9	18.0	17.7	n/a	n/a
Very good	27.3	26.0	26.4	n/a	n/a
Excellent/very good	44.2	44.0	44.1	44.1	44.9
Good	35.1	31.4	32.4	32.4	32.3
Fair/poor	20.0	24.5	23.3	23.4	22.8
Has a disability or long-term health condition	35.4	36.9	36.5	36.3	34.8

a. There is a significant difference between the remote and non-remote data for people with 'fair/poor' health status. The difference between the remote and non-remote data for people with 'good' health status is weakly significant.
Source: ABS (2004c: Tables 1 & 13)

Selected disability characteristics by remoteness, Indigenous Australia, 2002

The second comparison looks at the types of disabilities reported in the survey. This is shown in Table 17.2, which presents the disability component disaggregated into a list covering:

- those without a disability and/or long-term condition
- those with a disability or long-term condition, and
- for those with a disability, some broad categories of the type of disability/condition.

This information shows that there are no significant differences between the remote figures and the non-remote figures for any type of disability except for the 'intellectual' option. Even there, the figures, although statistically significantly different from one another, are very low at 7.7 per cent (non-remote) and 4.9 per cent (remote). Consequently, in the latter tables, only the aggregate level of 'disability or long-term health condition' is used.

It should also be remembered that there were significant differences in the methods used to collect the data; see discussion in Biddle and Hunter (in this volume).

Table 17.2. Selected disability characteristics by remoteness

	Remote	Non-remote	Total
	%	%	%
Disability status			
Has a disability or long-term health condition			
Has profound core activity restriction	4.0	2.9	3.2
Has severe core activity restriction	4.9	4.4	4.5
Disability/restriction not defined	26.5	29.6	28.7
Total with a disability or long-term health condition	35.4	36.9	36.5
Has no disability or long-term health condition	64.6	63.1	63.5
Disability type			
Sight, hearing, speech	14.2	13.5	13.7
Physical	24.0	23.4	23.6
Intellectual [a]	4.9	7.7	7.0
Disability type not specified	16.7	16.1	16.3
Total with a disability or long-term health condition	35.4	36.9	36.5

a. There is a significant difference between the remote and non-remote data for people with a declared intellectual disability.
Source: ABS (2004c: Table 13)

Health variables by quintile of weekly equivalised gross household income, Indigenous Australia, 2002

The first factor examined is income. Table 17.3 and Figure 17.1 present health status by a very broad measure of household incomes; that is, by weekly equivalised income quintiles. The numbers in Table 17.3 indicate that there is a strong, positive relationship between income and self-assessed health, at least at this level of aggregation. Figure 17.1 also shows that the incidence of disabilities/health conditions falls as household incomes rise. Of course, this table does not indicate the direction of causality; is poor health causing low incomes, or are low incomes causing poor health? It makes a big difference to the policy implications of this analysis if it is poor health that is leading to (i.e. causing) people to have low incomes, or the reverse, that is, if it is the fact that people have low incomes which causes them to have poor health.

Table 17.3. Self-assessed health variables by quintile of weekly equivalised gross household income

	Income quintiles			
	Lowest	Second	Third	Fourth and fifth
Self-assessed health status	%	%	%	%
Excellent/very good	38.0	42.8	49.2	56.7
Good	32.3	34.2	33.3	32.9
Fair/poor	29.7	23.0	16.5	12.4

Note: There is a significant difference between the first quintile and the highest two quintile groups for groups with the highest and lowest health status. There is also a weakly significant difference between the second quintile and the first quintile group for those with excellent or very good health and those with fair/poor health.
Source: ABS (2004c: Table 9)

Figure 17.1. Disabilities by quintile of household income

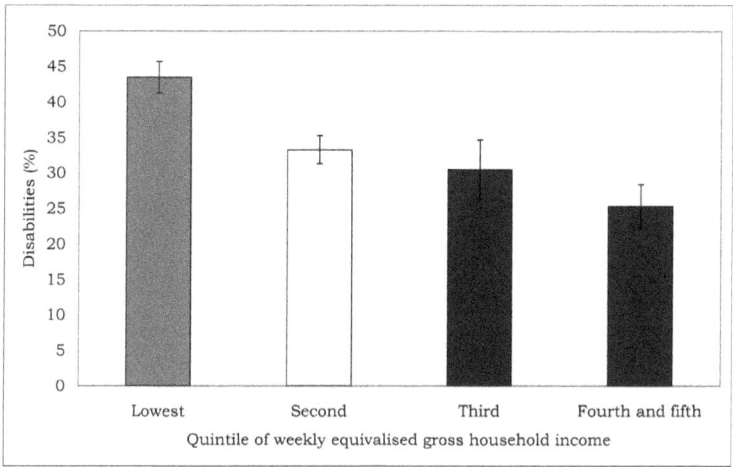

Source: ABS (2004c: Table 9).

Health indicators by age and gender, Indigenous Australia, 2002

Another important factor to consider is the age and gender profile of the population and hence of health status. These are shown in Table 17.4 and Figure 17.2. The figures show clearly that health is a declining function of age. The figures also show that there are some significant gender differences in health status.

The dispersion of health status varies with age. Whereas there is no difference in the proportion of each age group who reported their health as 'good', the proportions reporting better than 'good' declines with age and conversely the proportion reporting worse than 'good' rises with age. This is true for both males and females.

This is also true for the incidence of disabilities, where the incidence rises with age throughout the entire age range.

In addition, there are some differences between the genders within some age groups, especially at the younger end of the range. In the 15–24 age group, males' self-assessed health status is significantly better than that for females, although there is no statistical difference in the reported incidence of disabilities. Of young males, 63.9 per cent reported themselves to be in better than good health, compared with only 53.4 per cent of females. Conversely, a higher proportion of females rated themselves to be good or worse.

Table 17.4. Self-assessed health status by age and gender

	Aged 15–24	Aged 25–34	Aged 35–44	Aged 45–54	Aged 55 and over	All respondents
	%	%	%	%	%	%
Males						
Excellent/very good	63.9	53.0	41.1	33.2	15.5	47.1
Good	29.7	29.9	33.1	26.3	30.7	30.1
Fair/poor	6.2	16.6	25.7	40.1	53.7	22.5
Females						
Excellent/very good	53.4	46.9	37.7	31.1	18.0	41.3
Good	36.6	36.3	34.9	33.1	27.6	34.6
Fair/poor	9.9	16.8	27.3	35.8	54.2	24.0

Source: ABS (2004c: Table 3)

Figure 17.2. Disabilities by age and gender

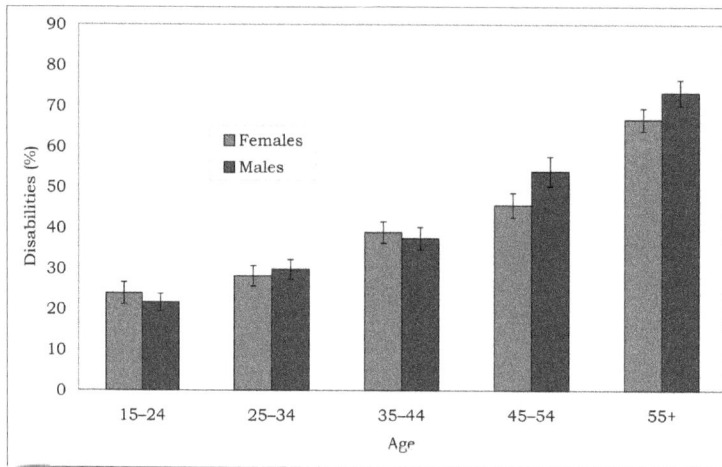

Source: ABS (2004c: Table 3).

Comparing health indicators, Indigenous and non-Indigenous Australia, 2002

The statistics in Table 17.1 above did not provide a comparison with the non-Indigenous population. This is done in Figure 17.3, which compares NATSISS figures with those from the 2002 GSS. In order to provide a meaningful comparison, it is necessary to take into account the differences in the age structure of the Indigenous and non-Indigenous populations. The NATSISS data in Figure 17.3 have been recalculated to provide this comparison. The Indigenous figures in this figure differ from those in the earlier tables because the 2002 GSS covered people aged 18 and over, whereas the 2002 NATSISS included ages 15 and over.

Figure 17.3 clearly shows two important differences. Firstly, these figures confirm that Indigenous health is still way below that for the non-Indigenous population.

The figures in the last two columns indicate that a majority (58.9%) of the non-Indigenous population enjoys 'excellent' or 'very good' health, but only one in three (35.2%) Indigenous Australians report their health thus. At the other extreme, twice as many Indigenous people report their health as 'fair' or 'poor' compared to the non-Indigenous population (32.7% and 16.1% respectively).

Secondly, when the populations are standardised for age, the difference between the remote and non-remote Indigenous populations for the 'good' category becomes statistically significant, while it is only weakly significant in Table 17.1 for the non-standardised figures.

Due to the differences in the way the disability information was collected for the two sub-surveys in the 2002 NATSISS, it is not possible to compare disabilities data from the remote sub-survey with those from the GSS. This limits any comparison with the non-Indigenous population to non-remote areas only. The incidence of disabilities is almost twice as high among the Indigenous population (at 56.6%) as it is among the non-Indigenous population (40.1%). [8]

Figure 17.3. Self-assessed health status by Indigenous status, non-remote areas

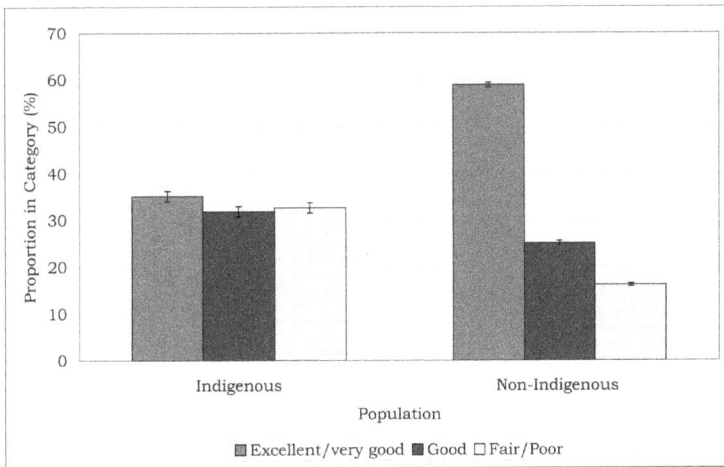

Source: ABS (2004c: Table 4). Non-Indigenous data are from the 2002 GSS.

Indigenous health, Australia, 1994 and 2002

Has Indigenous health improved between 1994 and 2002? Table 17.5 presents a comparison of these broad indicators with the corresponding figures derived from the 1994 NATSIS. This comparison is only presented for the self-assessed health status, as it is not possible to derive a similar measure for disabilities from

[8] It seems redundant to state that this difference is very statistically significant.

the 1994 NATSIS. The figures in Table 17.6 indicate that the situation has actually worsened, at least if any credence is to be given to these figures.

Although the same proportion reported health as 'excellent' or 'very good' (45.3% and 44.1% with no statistical significance between these two figures), there was a shift from the 'good' category to the 'fair' and 'poor' categories. The good category fell from 37.1 per cent to 32.4 per cent, while the fair and poor categories rose correspondingly (17.5% to 23.3%), with all these differences being statistically significant.

Table 17.5. Indigenous health, Australia, 1994 and 2002[a]

	1994	2002
	%	%
Self-assessed health status		
Excellent/very good	45.3	44.1
Good[b]	37.1	32.4
Fair/poor[b]	17.5	23.3

a. The content of this table has been restricted to those items that are comparable between 1994 and 2002.
b. There is a significant difference between 1994 and 2002 estimates for those with good and fair or poor health.
Source: ABS (2004c: Table 6)

However, I caution against reading too much into these figures without further analysis. It is possible that these figures are hiding some improvements, and may be picking up a fallacy of composition problem.

For example, these figures could be reflecting improvements in health care that have saved lives but left the people in poor health. Alternatively, these figures could reflect a greater awareness of poor health issues over time. Further, the figures could be reflecting better sampling design and administration. Most likely, a combination of all three of these is important.

What the inter-year comparison should really be trying to ascertain is: Are individuals getting healthier over time? I'd like to know if people aged 'X' in 1994 were reporting better health positions in 2002 when they were aged 'X+8'. In any event, these figures give a very superficial picture of the changes over time. Considerably more analysis is required on this aspect.

I also recall that at the CAEPR conference held to discuss the results from the 1994 NATSIS, Anderson and Sibthorpe (1996) stressed that the use of self-assessed health status was a comparatively recent phenomenon. They highlighted some issues with this measure in a sample where over 80 per cent of the respondents declared themselves to be in good or better health, yet it was widely known that Indigenous health levels were low. This may simply reflect the fact that if people do not have good access to health services, they may well have undiagnosed health problems. Alternatively, their perception of what is good health may be different from what is considered the norm for 'good' in other situations. As one example, Anderson and Sibthorpe (1996) noted that a research paper based on the 1995 NHS reported that one half of all people with diabetes considered themselves to be in good health.

Health characteristics by non-school qualification and highest year of school completed, Indigenous Australia, 2002

We now turn our attention to education, measured by two broad categories of educational achievement: attainment of a post-secondary school qualification, and years of completed schooling. Table 17.6 presents the results for the post-school qualification, and for those without such a qualification, by years of completed schooling. Schooling here excludes tertiary level education; that is, it only refers to primary and secondary schooling. The qualifications measure is very heterogeneous, covering anything from a trade certificate to a university higher degree.

Table 17.6. Non-school qualification by highest year of school completed by selected health characteristics[a]

	Does not have a non-school qualification				Has a non-school qualification
	Year 9 or below[b]	Completed Year 10 or Year 11	Completed Year 12	Total without qualification	
	%	%	%	%	%
Self-assessed health status					
Excellent/very good	30.7	47.1	57.9	40.9	46.1
Fair/poor	35.1	17.7	12.7	25.2	22.9
Has a disability or long-term health condition	49.7	29.9	22.1	38.2	35.3

a. Excludes people who were attending secondary school.
b. Includes people who never attended school. Year of schooling is only shown for those who do not have a non-school qualification.
Source: ABS (2004c: Table 7)

Table 17.6 shows that although there is no statistical significance at the 'aggregate' level—that is, between those with and without a non-school qualification—there are some very significant differences *within* the group without non-school qualifications, as follows:

- there is a clear trend among those without post-schooling qualifications
- those who have completed Year 12 are far healthier than those who have not, and have fewer disabilities
- in turn, those who have completed Years 10/11 are healthier than those who have only completed Year 9, and have fewer disabilities, and
- those who completed Year 11 or Year 12, but have no non-school qualifications, are healthier than those with non-school qualifications, and have fewer disabilities.

It is likely that this trend is really capturing the fact that more and more Indigenous people are completing high school (i.e. Year 12). By default, these people are younger than those who previously only completed Year 9, and as younger Indigenous people tend to be healthier, these figures follow. Biddle (2005) tested a number of hypotheses concerning the inter-relationships between age, education and health. He shows a clear positive (negative) relationship between poor (good) health and age which is mitigated by higher educational attainment; see, in particular, his Figure 4 (Biddle 2005: 24).

Health characteristics by labour force status, Indigenous Australia, 2002

It is also widely acknowledged that there is a link between health status and employment prospects in the labour market; see, for example, Booth and Carroll (2005). Table 17.7 and Figure 17.4 present the evidence from the 2002 NATSISS. As is now standard with Indigenous employment statistics, the employment figures are disaggregated between CDEP and non-CDEP. The figures presented tell a very interesting story about the link between health and employment. However, as with the other tables, it does not indicate in what direction the causality is directed.

Figure 17.4. Labour force status by self-assessed health

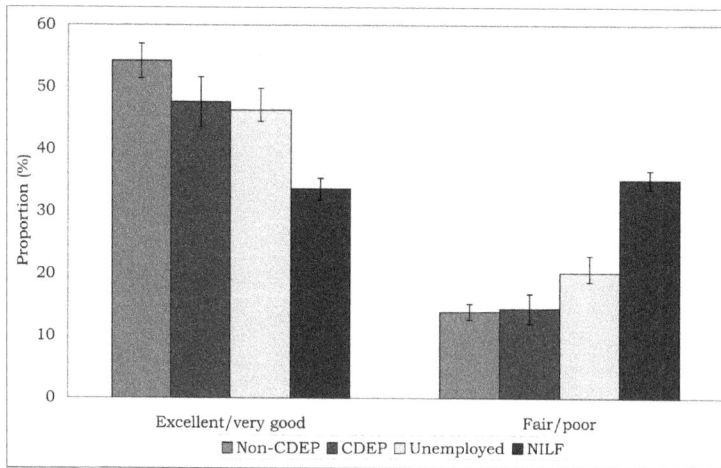

Source: ABS (2004c: Table 8).

Table 17.7. Labour force status by disabilities[a]

	Employed			Unemployed	Not in labour force
	CDEP	Non-CDEP	Total		
	%	%		%	%
Has a disability or long-term health condition	31.2	24.4	26.2	35.2	48.8

a. The differences between 'non-CDEP employed' and 'unemployed' is statistically significant at the 10% level, whereas the difference between 'total employed' and 'unemployed' is statistically significant at the 5% level.
Source: ABS (2004c: Table 8)

These figures indicate a significant difference in health status between those employed and those unemployed for the 'fair/poor' category (see Fig. 17.4) and for those with disabilities (Table 17.7). However, the gap for the 'excellent/very good' categories (i.e. 52.5% against 46.3%) is not even weakly statistically significant.

Within the employed groups, there are significant differences in each category. Those in non-CDEP employment are significantly healthier and have fewer disabilities than those in CDEP employment. Indeed, and not surprisingly, those in CDEP employment have health statuses identical to those for the unemployed for all practical purposes. That is, the gaps between the figures for CDEP employment and for unemployed are not statistically significant for any row.

Further, the gaps between the unemployed and those outside the labour force are statistically significant. The unemployed are healthier and have fewer disabilities than do those outside the labour force.

Disability status and self-assessed health status by age

The relationship between health status and the incidence of disabilities for the older population is also revealing, as is demonstrated in Table 17.8. The emphasis in that table is on the impact of age on the two measures of health.

Comparing those without disabilities—that is, columns 2 and 4—it is clear that health status deteriorates with age even without disabilities. For those aged under 50, the majority (59%) rated themselves in better than good health, and only 8 per cent considered themselves to be in fair or poor health. Conversely, only 39 per cent of those aged over 50 with no disability rated themselves as in better than good health, and 16 per cent rated themselves in fair or poor health.

Similarly, for those with disabilities, columns 1 and 3, a significantly larger percentage of the older group considered themselves to be in bad health. For those aged 50 or over, the vast majority (68%) are in fair or poor health—up from 39 per cent for those under 50. Conversely, only 11.6 per cent of those aged over 50 with a disability rated themselves as in better than good health, compared to 25.1 per cent of those under 50.

Table 17.8. Self-assessed health status by age and disability

	15–49 Years		50 Years or over		All respondents	
	Disability[a]	No disability	Disability[a]	No disability	Disability[a]	No disability
	%	%	%	%	%	%
Excellent/very good	25.1	59.4	11.6	39.4	20.9	57.4
Good	35.6	32.2	19.9	44.7	30.7	33.5
Fair/poor	39.1	8.2	68.3	15.9	48.3	9.0

a. Includes people with a long-term health condition.
b. Statistics in italics indicate that the difference between 'disability' and 'no disability' data is statistically significant.
Source: ABS (2004c: Table 10)

Indigenous health indicators by State/Territory and Australia, 2002

Also of interest are the regional variations, as measured by State and Territory boundaries. These figures are shown in ABS (2004c: Table 2). Compared to the national figures, there are significant differences, as follows:

- in the Northern Territory, Indigenous people are healthier and have fewer disabilities
- in Victoria, Indigenous people are less healthy and have more disabilities
- in the ACT, fewer Indigenous people are in poorer health, but there is no difference in disabilities
- in New South Wales, more Indigenous people are in poorer health, but this figure is only weakly significant
- in Tasmania, there are no differences in health status but more disabilities.

It is worth highlighting the fact that on all measures, Queensland reflects the national averages. None of the Queensland statistics are significantly different from the national averages. This supports the position taken earlier to not present separate figures for the Torres Strait Islander population, all of whom are in Queensland.

Concluding remarks

There is a wealth of information on health contained in the 2002 NATSISS. This chapter has provided an overview of the data and offered some insights into these data. One particular conundrum of interest is the comparison with the 1994 NATSIS. If taken at face value, that comparison indicated that Indigenous health has gone backwards in the intervening eight years. This raises doubts more about the reliability of the 1994 data than about the efficacy of health expenditures.

It is unfortunate that there were such differences in the methodology and coverage on the two sub-samples—the remote versus non-remote areas samples—which has resulted in some restrictions on the validity of comparisons on key health data between the two types of localities.

The analysis does highlight the complex interactions between a person's health and key socioeconomic factors such as education, employment, age and gender.

Nevertheless, the 2002 NATSISS provides strong evidence that there is still much work and effort required before it can be concluded that the gap between Indigenous and non-Indigenous health standards has been substantially reduced, let alone eliminated.

18. Substance use in the 2002 NATSISS

Tanya Chikritzhs and Maggie Brady

It is important at the outset to acknowledge with candour that questioning Aboriginal or Torres Strait Islander people about their use of alcohol and other drugs is always fraught with difficulty, whatever the circumstance. As Anderson and Sibthorpe (1996: 118–134) observed of the 1994 NATSIS, one wonders about the subjective meanings that might be attached to such questions, what are the perceptions of personal risk that might be attached to such questions, and what interviewees understand to be the purpose of such information. Questions of this sort can meet with resistance, underestimation and 'fudging' even if asked privately by a health professional and in an ostensibly 'safe' environment such as an Indigenous health service (Brady et al. 2002). Usually, health service providers will ask such questions only after having first put people at ease, opening the discussion on alcohol and other drugs by phrasing questions colloquially and/or by first having the results of preliminary screening to hand.

Reliable survey estimates of substance use that specifically address Indigenous populations are important for several reasons. They provide a measure of the prevalence of substance use among Indigenous people for males and females and for a range of ages; such information is not available from any other source. For example, Indigenous alcohol consumption cannot be identified from per capita alcohol consumption estimates. When surveys are conducted regularly and consistently, they can also be used to measure changes in use over time. Survey estimates of the population prevalence of tobacco, alcohol and other drug use are a vital component in the estimation of 'population aetiologic fractions' for these substances—the degree to which a substance can be said to cause various diseases and injuries in a certain population. Accurate prevalence estimates of alcohol consumption are therefore necessary for the estimation of alcohol-attributable deaths and hospitalisations. In turn, reliable estimates of levels of alcohol-related harms can be used to inform the allocation of funding and resources for prevention and treatment programs (Gray et al. 2002).

For a national survey to be of local relevance, it needs to:

- contain sufficient numbers of respondents from Indigenous populations throughout Australia
- include a broad range of ages and both sexes, and
- use appropriate methods for measuring drinking levels and patterns.

It should be kept in mind that although alcohol survey methods have improved in recent years (Stockwell et al. 2004), when compared to per capita consumption estimates from wholesale data, past surveys have always dramatically underestimated alcohol consumption.

Overview of the 2002 NATSISS methods in relation to substance use

In their chapter to this volume, Biddle and Hunter describe the survey methods used by the 2002 NATSISS in detail. This section will briefly review those methods with a specific focus on the collection of information in relation to alcohol, tobacco and other substances (e.g. illicit substances). It is worth pointing out here that we use the term 'substance use' to include alcohol and tobacco use, as is the normal practice in the alcohol and other drugs area, whereas in the NATSISS 'substance use' seems to refer only to other drugs, including illicit drugs.

Selecting a representative sample

The 2002 NATSISS used distinctly different sets of methods for sampling Indigenous populations. These methods were not applied at random or equivalently throughout the Indigenous population of Australia but were purposely targeted at specific sub-populations largely relating to geographic location. Thus, the sampled population was effectively divided into two parts:

1. A random sample of households from *discrete* Indigenous *communities* and the outstations associated with them in areas of Queensland, South Australia, Western Australia and the Northern Territory. This was known as the CA sample. Although it is not entirely clear, the ABS has suggested that the majority of respondents sampled using this method were resident in *very remote* areas as distinct from remote areas (i.e. using ARIA classification)—see Biddle and Hunter in this volume. Approximately 23 per cent of the total 9359 respondents were sampled in this way.

2. The remaining respondents were resident in NCAs and were selected using a stratified multi-stage area sample based on the 2001 Census. A random selection of dwellings within selected census CDs was then screened to assess Indigenous status of usual residents. According to the ABS, an insufficient number of households with Indigenous residents were initially collected, so additional CDs were sampled during February to April 2003. The majority of respondents to the 2002 NATSISS were sampled using this method.

Those who lived in remote regions but *not in* discrete Indigenous communities were sampled using only the NCA method.

Asking about drug use in the 2002 NATSISS

Two main methods were used for asking about drug use in the 2002 NATSISS:

- face-to-face interviewer questioning of the respondent requiring a verbal response, using either computer assisted or pen and paper methods, or
- voluntary, confidential (i.e. self-sealed), self-complete questionnaire.

The voluntary self-complete option was exclusively available to respondents of NCAs who agreed to answer questions on illicit drug use. For both alcohol and tobacco use, *only* face-to-face interviews were used. In addition, the self-complete questionnaire on illicit drug use was not available to residents living in CAs and these respondents were required to submit their responses verbally to an interviewer.

Tobacco

From the non-remote areas questionnaire, it can be ascertained that three questions were asked about tobacco use during face-to-face interviews. They were:

- Do you currently smoke?
- Do you smoke regularly, that is, one smoke a day or more?
- Have you ever smoked regularly (that is, one smoke a day or more)?

Residents of CAs were similarly asked about smoking behaviour.

Alcohol

The alcohol questions used in the 2002 NATSISS were different from those of the 1994 NATSIS. For the 2002 survey, respondents were asked to estimate how often they drank in the last year and the quantity and type of alcohol they usually consumed on a drinking day. These were later converted to standard drinks by the interviewer. The quantity/frequency method is adequate for estimating rates of abstainers but is known to result in underestimation of the volume of alcohol people drink per occasion (Stockwell et al. 2004).

Illicit drug use

The 2002 NATSISS asked respondents about their use of substances other than alcohol and tobacco, referring to this group of drugs as 'substance use'. The 'substance use' questions included the use of a range of drugs for non-medical purposes and were based on questions used in the NDSHSs. In the survey, 'substance use' included analgesics, tranquillisers, amphetamines, marijuana, heroin, cocaine, hallucinogens (both synthetic and naturally occurring), ecstasy and other designer drugs, petrol and other inhalants, and kava. Respondents were asked to report on their use in the last 12 months or if they had ever used any of these substances. Unfortunately, however, according to the ABS (2004c), information on substance use will not be released for residents of remote areas 'due to data quality concerns' (see p.79), largely arising from the manner in which the questions were asked (i.e. face-to-face interview requiring a verbal answer). In effect, therefore, the findings of the 2002 NATSISS in relation to illicit drug use are limited to non-remote populations and cannot be used to inform *overall* levels of use among Indigenous people (since use in over 20% of the sample is not accounted for).

Estimates of Indigenous drug use from the 2002 NATSISS

The results of the 2002 NATSISS in relation to substance use (i.e. tobacco, alcohol and illicit drugs) have been summarised by the ABS (2004c).

Table 18.1. Drug use among Indigenous Australians: results from the 2002 NATSISS

Risk behaviour/characteristics	Remote	Non-remote	Total
	%	%	%
Smoker status			
Current daily smoker	50.4	48.0	48.6
Occasional smoker	2.9	2.0	2.3
Ex-smoker	11.5	16.7	15.3
Never smoked	32.7	33.3	33.2
Alcohol consumption level in last 12 months			
Low risk	32.2	51.3	46.1
Risky	10.0	9.4	9.6
High risk	6.8	5.1	5.6
Any alcohol use	53.6	75.3	69.4
Did not consume alcohol in last 12 months	46.4	24.7	30.6
Type of substance used in last 12 months[a]			
Used substances in last 12 months			
Analgesics and sedatives for non-medical use	-	4.4	-
Amphetamines or speed	-	4.7	-
Marijuana, hashish or cannabis resin	-	19.1	-
Kava	-	0.6	-
Total used substances in last 12 months	-	23.5	-

a. Respondents may have indicated more than one response per category. Data only available for non-remote areas. 'Substances' includes heroin, cocaine, hallucinogens, designer drugs, petrol and other inhalants.
Source: ABS (2004c: 40)

According to the 2002 NATSISS, almost one half of Indigenous respondents were current daily smokers. In comparison, non-Indigenous current tobacco use has been estimated by the 1998 and 2001 National Drug Strategy Household Surveys to be around 25 per cent and 23 per cent respectively—less than half that for the Indigenous population (Higgins, Cooper-Stanbury & Williams 2000).[1]

---

[1] According to the 2004 NDSHS, smoking rates among the non-Indigenous population have declined since 2001 (AIHW 2005).

The ABS identified a 90 per cent response rate to its self-complete 'substance use' questionnaire (administered in NCAs but not CAs). However, as discussed earlier, the ABS has only reported on illicit substance use by respondents who resided in NCAs. With this qualification in mind, 19.1 per cent of Indigenous respondents claimed to have used cannabis in the previous 12 months. The 2001 NDSHS has estimated that among the non-Indigenous population, cannabis use in the last 12 months was about 13 per cent (AIHW 2002). More Indigenous respondents reported having used cannabis in the last 12 months than any other illicit substance—including amphetamines (4.7%) and kava (0.6%). Less than 5 per cent of respondents reported having used analgesics or sedatives for non-medicinal purposes.

In relation to the proportion of respondents who did or did not use 'substances' in the last 12 months, the summary table given in the ABS report presents some conflicting results. The last two lines of the table show: 'Total used substances in last 12 months' and 'Did not use substance in last 12 months', the proportions for which are given as 23.5 per cent and 16.2 per cent respectively. These two figures are irreconcilable, and it would appear that an error has been made in at least one of these totals. [2]

According to the 2002 NATSISS, 30.6 per cent of *all* Indigenous participants reported being abstinent from alcohol in the past 12 months. Respondents who lived in remote regions were almost twice as likely as those living in non-remote areas to report no alcohol consumption in the previous 12 months (46.4% versus 24.7%). In addition, 46.1 per cent of all respondents reportedly drank at levels that placed them at *low* risk for alcohol related harms. Of the remaining 23.3 per cent of the sample, 15.2 per cent were reported as having drunk at levels that placed them at risky or high risk for alcohol related harm. [3] However, a substantial proportion of the sample (8.1%) was not accounted for in the summary of drinking patterns provided in the ABS (2004c) (i.e. levels reported on p.40 do not add up to 100%). It is possible that these respondents either refused to respond to questions about alcohol use or they gave responses that were problematic.

Previous reviews that have compared Indigenous and non-Indigenous alcohol use in Australia have found that although Indigenous people are more likely to abstain from alcohol, they are also more likely to drink at risky/high-risk levels for alcohol-related harm than their non-Indigenous counterparts (e.g. Gray et al. 2004). The 2001 NDSHS estimated that among the non-Indigenous population Australia-wide, 17.8 per cent abstained from drinking alcohol (i.e. had not drunk

---

[2] The ABS confirmed that the summary table is subject to typographical error at CAEPR conference and that these values are incorrect. We await further information from the ABS.

[3] It is not clear from the ABS report whether the drinking patterns reported were in relation to the latest Australian drinking guidelines for acute or chronic harms (i.e. NHMRC 2001) or a composite of both.

in the last 12 months)—about half that estimated for the Indigenous population by the 2002 NATSISS. In addition, some 71.7 per cent of non-Indigenous respondents reported drinking at low-risk levels for alcohol-related harms, while the remaining 10.4 per cent drank at risky or high-risk levels. [4] The recently released results of the 2004 NDSHS confirm that levels of drinking have remained relatively stable among the non-Indigenous population (AIHW 2005b). In summary, therefore, according to the results of the 2002 NATSISS and the 2001 NDSHS, the comparative discrepancy between Indigenous and non-Indigenous risky/high-risk use of alcohol in recent years was less than 5 per cent (see Fig. 18.1).

Figure 18.1. A comparison of Indigenous and non-Indigenous drinking levels, from the 2002 NATSISS and the 2001 NDSHS

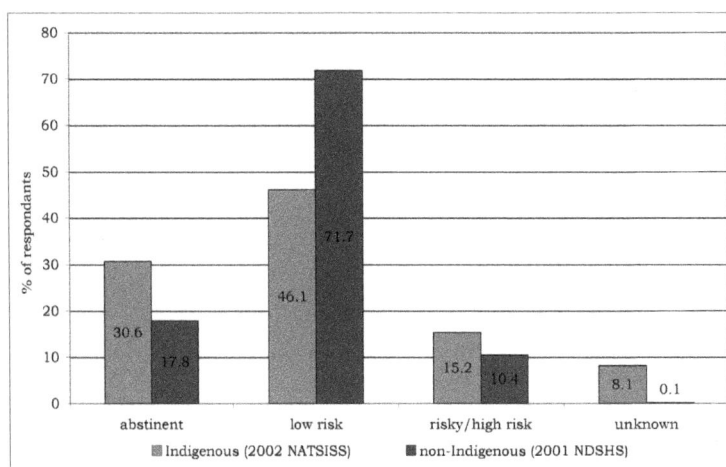

Source: ABS (2004c: Table 13) and AIHW (2005b).

The small apparent difference between Indigenous and non-Indigenous risky/high-risk alcohol consumption suggested by the 2002 NATSISS results is particularly surprising given recent comparisons of alcohol-attributable hospitalisations for the two populations. For example, data cited by the SCRGSP (2005) noted that in 2002–03 the rate of hospital admission among Indigenous males was between two and seven times greater than for non-Indigenous males (see Table 18.2). This was for conditions related to high levels of alcohol use, such as acute alcohol intoxication, alcoholic liver disease, harmful use, and alcohol dependence.

---

[4] These summaries are not available in the AIHW detailed findings report. They were calculated by the National Alcohol Indicators Project for the 2003 report on Australian Alcohol Indicators (Chikritzhs et al. 2003).

Table 18.2. Indigenous and non-Indigenous rates of hospitalisation (per 1000) for a selection of alcohol-attributable conditions 2002–03

Condition	Indigenous rate	Non-Indigenous rate	Rate ratio
	%	%	
Acute intoxication	3.5	0.5	7.0
Alcohol dependence	2.4	1.0	2.4
Alcoholic liver disease	1.5	0.3	5.0
Harmful use	0.4	0.1	4.0

Source: SCRGSP (2005: Table 8A.2.2)

Other studies have shown that the rate of death from *wholly* alcohol-caused conditions (e.g. alcoholic liver cirrhosis, alcohol dependence) among residents of WA, SA and the NT is almost eight times greater for Indigenous males than for non-Indigenous males and 16 times greater for Indigenous females (Chikritzhs et al. 2000). The level of alcohol-attributable death among young Indigenous Australians (aged 15–24 years) has also been shown to be almost three times greater than for their non-Indigenous counterparts—with the divergence between the two populations apparently increasing in recent years (Chikritzhs & Pascal 2004). The Indigenous National Alcohol Indicators Project has also estimated the crude rate of alcohol-attributable deaths among Indigenous Australians versus the general population from 1999 to 2002.[5] As shown in Figure 18.2, the rate of death from alcohol-attributable conditions is over two and a half times greater for Indigenous people than for the general population, and the relative proportions have not changed substantially in recent years.

[5] Alcohol-attributable crude death rates have been derived from the data held by the Indigenous National Alcohol Indicators Project at the National Drug Research Institute, and have been specifically produced for this report. Age standardised rates are not yet available.

Figure 18.2. Estimated national alcohol-attributable death rates for Indigenous people versus the general Australian population

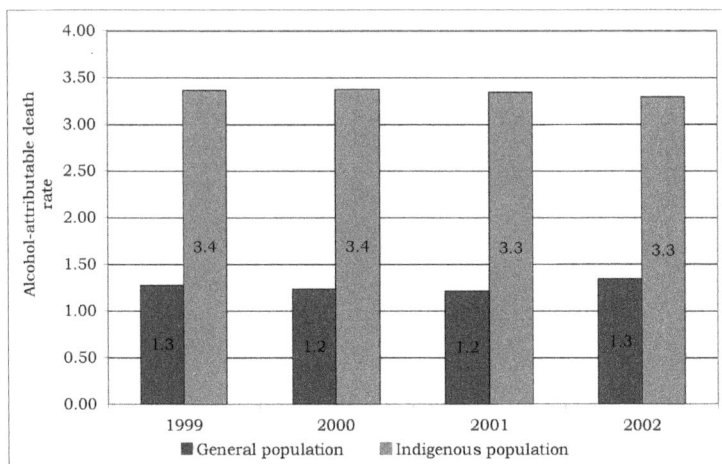

Source: Customised cross-tabulations from the National Alcohol Indicators Project alcohol-attributable mortality database, National Drug Research Institute, Perth.

How do the 2002 NATSISS results compare with other national surveys of Indigenous drug use?

To date, only five national surveys with moderate to large sample sizes have specifically addressed Indigenous alcohol consumption throughout Australia. They include the:

- *1994 National Drug Strategy Household Survey Urban Aboriginal and Torres Strait Islander Peoples Supplement* (NDSHS) conducted by AGB McNair on behalf of the Commonwealth Department of Human Services and Health (CDHSH 1996)
- *1994 NATSIS* (ABS, 1995b)
- *1995* NHS (ABS 1999)
- *2001 NHS* (ABS 2002a), and
- *2002 NATSISS* (ABS 2004c).

The latter four surveys were conducted by the ABS. A range of small local surveys have also been conducted but they are generally not suitable for comparisons between regions (see Saggers & Gray 1998). This section compares the results of the 2002 NATSISS with those derived for the Indigenous population by the 1994 NATSIS, the 2001 NHS and the 1994 NDSHS.

1994 NATSIS

The most obvious starting point for comparison with the 2002 NATSISS outcomes for substance use is the 1994 NATSIS. Unfortunately, the use of incompatible methods and different questions applied in the two surveys has limited the comparisons that can be drawn between them. The ABS has limited its own tabulated comparisons to estimates of current smokers and the proportion of respondents who drank alcohol in the past 12 months, with no comparisons possible for illicit substances.

Despite sampling differences, the two surveys estimated similar levels of current smokers—51.7 per cent in 1994 and 50.9 per cent in 2002. Likewise, the estimated proportions of Indigenous people who abstained from alcohol in the last 12 months were 32.4 per cent in 1994 and 30.6 per cent in 2002. It is not possible however, to compare the proportion of drinkers consuming alcohol at various levels (i.e. low, risky, high risk) between these two surveys. In summary, although the comparisons are severely limited, there is no strong evidence of change over time in tobacco and alcohol use—on these two measures, at least.

The 2001 NHS

It is also useful to compare the 2002 NATSISS with the 2001 NHS (ABS 2002b). Given the temporal proximity of these two surveys—and if both had adequately and reliably measured drug use—then all other things being equal, we would expect to find similar national estimates. [6]

The 2001 NHS included respondents from both remote and non-remote areas of Australia but did not include an especially large number of respondents (1853 adults) (ABS 2002b). (Indeed, due to the small sample size, the ABS limited the scope of all publications to reporting at a national level.) Nevertheless, among Indigenous respondents, the estimated proportion of current smokers from the 2001 NHS was about 51 per cent—a few points higher than that estimated by the 2002 NATSISS (48.6%).

With regard to patterns of alcohol use, the method used by the 2001 NHS was especially problematic. Although a quantity–frequency method was applied for measuring alcohol consumption, the survey only asked about drinking that had occurred in the week before the interview (those who did not drink in the previous week recorded no alcohol consumption). What is more, only details of the three most recent drinking days during that week were recorded. The shortcomings of this method have been well documented elsewhere (e.g. Rehm et al. 1999; Stockwell et al. 2004; WHO 2000)—especially in relation to underestimated levels of consumption and numbers of episodic drinkers (i.e. binge drinkers). In particular, since the drinking patterns of a large proportion of Indigenous drinkers tend to be intermittent and clustered around fortnightly payment of social security entitlements, the 'three day method' used was particularly prone to underestimation of total alcohol consumption and ill-equipped to measure Indigenous alcohol consumption.

---

[6] The 2001 NHS did not ask Indigenous respondents about illicit drug use.

Keeping in mind the shortcomings of the survey method applied in the 2001 NHS, it was reported by the ABS that 42 per cent of Indigenous respondents drank in the week before the interview (58% were abstinent in the previous week) and 29 per cent of those who drank consumed alcohol at risky or high-risk levels (ABS, 2001a). From this we can deduce that 12 per cent of all Indigenous respondents drank at risky/high-risk levels, while the remaining 30 per cent drank at low-risk levels in the previous week.[7]

1994 NDSHS Urban Aboriginal and Torres Strait Islander Peoples Supplement

The only major population survey that has focused *specifically* on substance use among Indigenous Australians is the *1994 NDSHS Urban Aboriginal and Torres Strait Islander Peoples Supplement* (CDHSH 1996).[8] Despite its age, this survey is arguably the most reliable to date (Gray et al. 2004). Indeed, a comparison between the 1994 NDSHS and the 2002 NATSISS is perhaps the most revealing measure of the shortcomings of the latter survey.

The 1994 NDSHS was conducted in response to recommendations by the Royal Commission into Aboriginal Deaths in Custody. It was conducted by AGB McNair on behalf of the then Commonwealth Department of Human Services and Health. As part of the survey, face-to-face interviews were conducted with 2993 Indigenous people aged 14 years and older residing in 'urban centres' of Australia. Urban centres were defined as either 'major' or 'other' but were required to include a minimum of 1000 people for inclusion in the survey (the majority of Indigenous Australians reside in such centres) (CDHSH 1996).

The 1994 NDSHS estimated that about 54 per cent of respondents were 'current' smokers (i.e. smoked in the previous 12 months)—about 4 percentage points higher than that reported by both the 2002 NATSISS (48.6%) and the 2001 NHS (51%). This may indicate a possible reduction in the proportion of Indigenous smokers over the period between surveys. Others have also noted that apparent declines had occurred in the proportion of current smokers between 1994 and 1998 (Gray et al. 2004). In relation to those who have 'ever smoked', there was also an apparent decline, with the 1994 NDSHS estimating about 77 per cent of the sample as having ever smoked, compared to about 67 per cent from the 2002 NATSISS. Nevertheless, the prevalence of smoking among Indigenous people remains about twice as high as that for non-Indigenous people (Gray et al. 2004).

---

[7] Although questions about drinking in the last two weeks were asked by the 2002 NATSISS, concerns regarding the methodological approach have restricted publication of the results. However, personal communication with the ABS has identified that the proportion of respondents reporting risky/high-risk use in the last two weeks of the 2002 NATSISS was substantially higher than the proportion of respondents reporting the same level of drinking in the last 12 months.
[8] With the exception of the—as yet— unpublished 2004 NHS.

Some 48 per cent of Indigenous participants in the 1994 NDSHS reported having ever used cannabis and 22 per cent had used in the last 12 months. According to the 2002 NATSISS, about 19 per cent of respondents residing in NCAs had used cannabis in the previous 12 months. In the 1994 NDSHS, 19 per cent of Indigenous participants reported using an illicit drug other than cannabis. The summary of results from the 2002 NATSISS regarding 'substance use' indicates that the overall estimated proportion of residents in NCAs that had used a substance in the last 12 months was 23.5 per cent (ABS 2004c). Nineteen per cent had used cannabis in the last 12 months and a total of 9.7 per cent had used other drugs listed in the report (i.e. analgesics, amphetamines, sedatives, kava). Given this, it would appear (although it is difficult to tell definitively from the report because not all drugs were listed) that somewhere in the order of 10 per cent of respondents had used drugs other than cannabis in the recent past. This is a much smaller proportion than that estimated by the 1994 NDSHS. However, comparisons between the two surveys—and therefore any consideration of changes over time—are complicated by the fact that the 2002 NATSISS findings in relation to illicit drug use are specifically limited to non-remote Indigenous populations. In addition, it is worth noting that the standard errors in relation to reported kava use were considerable (between 25% and 50%) (ABS 2004c).

The 1994 NDSHS used a standard quantity-frequency method for estimating 'usual' alcohol consumption—as did the 2002 NATSISS. This method is adequate for estimating rates of abstainers but will result in underestimation of the volume of alcohol people drink per occasion (Stockwell et al. 2004). Most importantly, questions about alcohol in the 1994 NDSHS were contained in a confidential sealed section for self-completion by the respondent, so no direct questioning from the interviewer was required.

In relation to alcohol, results from the 1994 NDSHS showed that about 38 per cent of Indigenous people living in urban centres had not drunk alcohol in the previous year, compared to 28 per cent of the general population (CDHSH 1996). Thus, Indigenous people were more likely to be non-drinkers than members of the general population. In fact, the 1994 NDSHS reported a greater proportion of Indigenous respondents as being abstinent in the last 12 months than did the 2002 NATSISS (30.6%). Nevertheless, based on amounts usually consumed, the 1994 NDSHS estimated that 82 per cent of all Indigenous current drinkers were considered at risk or high risk compared to 28 per cent of non-Indigenous drinkers (CDHSH 1996). Overall, therefore, an estimated 51 per cent of all Indigenous respondents to the 1994 NDSHS were considered to be drinking at risky or high-risk levels for alcohol-related harm—*over three times greater than that estimated by the 2002 NATSISS*. Reported levels of alcohol consumption from the 2002 NATSISS, 2001 NHS and 1994 NDSHS have been summarised in Figure 18.3.

Figure 18.3. Estimates of alcohol use by Indigenous Australians: a comparison of three surveys

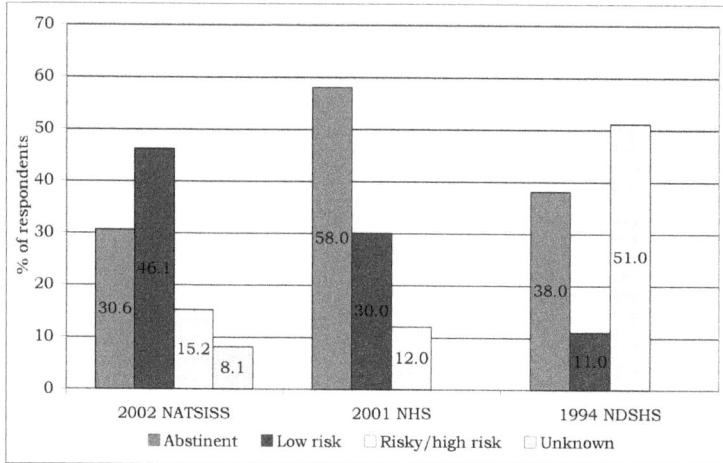

Source: ABS (2004c: Table 13) and CDHSH (1996).

Another surprising finding was that a higher percentage of Torres Strait Islander people than Aboriginal people were found to have risky or high-risk alcohol consumption (see Fig. 18.4).

Figure 18.4. 2002 NATSISS, Aboriginal and Torres Strait Islander risky/high-risk alcohol consumption

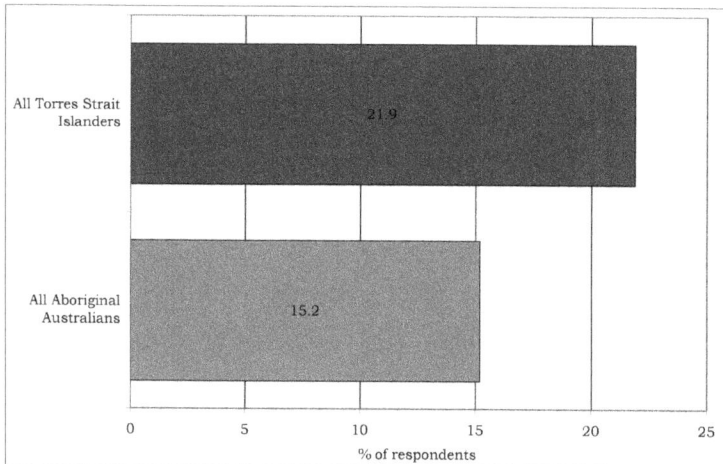

Source: ABS (2004c: Table 1).

What can be said about the results of the 2002 NATSISS?

All the surveys identified above have sought to inform on Indigenous substance use, yet they are all different and each of them has its limitations—some more serious than others. It is difficult to reliably compare results between surveys that do not have very similar methods, particularly where they vary in relation to sample selection, survey design and administration (e.g. questionnaire). Errors related to sampling can often be estimated and due consideration given to interpretation of the results. However, non-sampling error—such as may arise in the actual process of questioning respondents about their drug use—is less likely to be immediately obvious and the extent of it can rarely be measured. Variation between surveys also makes it difficult to determine whether differences which may appear over time or between populations are, in fact, real, rather than aberrations of the methods applied.

So, what can be said about the results of the 2002 NATSISS in relation to Indigenous drug use, and with what level of confidence? In short, this is a thorny question and one which is best served by examining some of the likely sources of error in relation to substance use from the NATSISS. These are discussed below.

Exclusion of residents living in non-private dwellings

Collecting information only from those living in private dwellings will arguably have a greater impact on responses to questions on the topic of alcohol and other drug use than on many other topics. Numerous Indigenous people living in hostels, short-stay caravan parks, prisons and other correctional facilities, hotels and hospitals were excluded. Alcohol-dependent people and problem drinkers in any population will be under-represented in household surveys, as Stockwell et al. (2004) have pointed out. They are more likely to be of no fixed abode, to be in an unfit state to participate, more likely to be incarcerated, and reluctant to discuss their drinking. Larson (1996) among others, has pointed out that many Aboriginal drug users (especially injectors) have insecure accommodation, are itinerant, or are living in supported accommodation and rehabilitation centres. One can conclude then, that the NATSISS would have missed numerous users of illicit drugs during the 2002 survey, as well as many problem drinkers for the same reasons.

Insufficient data on drinking patterns

Two 'big' questions that are important to alcohol researchers working on Indigenous data are the extent of drinking to the point of intoxication (i.e. binge drinking), and whether more people are becoming abstainers. Binge drinking is proving to be increasingly important as a predictor of physical harm, and is usually defined as drinking in order to get drunk in a short time. The number of drinks consumed per occasion is an important risk factor for death from injury, whereas frequency of consumption is not. There is a developing consensus (based largely on studies of alcohol-related mortality in the former Soviet Union) that repeated binge drinking is linked to sudden cardiac death, increased risk of thrombosis and high blood pressure (Hemstrom 2001; Room 2001). The 2002

NATSISS asked 'when you drink, how much do you usually drink in a day', rather than 'how often do you have six or more drinks on one occasion', which is the preferred 'binge' question. On the second big question—that is, the proportion of individuals who had quit drinking—the 2002 NATSISS did not have a question on being an ex-drinker (although they did have a question on being an ex-smoker).

Lack of confidentiality leading to unreliable responses

Despite its lack of intrinsic appeal, the decision to avoid the use of self-completed questionnaires in CAs for illicit 'substance use' questions was prompted by results from pilot testing. According to the ABS, pilot testing indicated that poor literacy levels in these very remote areas posed significant problems in relation to the use of self-complete questionnaires. Unfortunately, however, the lack of confidentiality imposed on such a sensitive topic diminished the reliability of the responses to such an extent that it led the ABS to abandon the use of the information. The abandonment of these evidently problematic findings on illicit substances begs the question of whether the alcohol consumption data was similarly diminished in quality. It is perplexing that while questions on illicit substance use were considered too sensitive for face-to-face questioning in NCAs, despite the potential discomfiture, at no time were respondents given the opportunity to reply to questions on alcohol use in a confidential fashion. We believe that alcohol consumption is as sensitive a topic as illicit drug use.

The use of facilitators may also have resulted in varying degrees of error, although it is hard to predict the potential differential impact of questioning in the presence of, or without, an Indigenous facilitator. Whether the Indigenous facilitator is known to the respondent could influence responses. Social proximity to an Indigenous facilitator could result in the respondent minimising their consumption rates of alcohol and other drugs (Brady et al. 2002).

Insensitivity to geographical and cultural diversity

Surveys that include questions on illicit drug use should ideally be part of specialised national drug use surveys (rather than a general 'social' survey), such as the Australian Needle and Syringe Program surveys. Otherwise, they should be conducted regionally by less threatening organisations that have more intimate contact with the subjects. National surveys do not take into account geographical and cultural variations in the prevalence, patterns or availability of illicit drugs among so many diverse Aboriginal groups (Shoobridge 1997). They find that more regionally specific and targeted surveys using rapid assessment methodologies are of more use (Aboriginal Drug and Alcohol Council 2004).

The inclusion of a question on kava use is puzzling, as Indigenous users of this soporific drink (imported from the Pacific) are confined almost entirely to one region of the country (Arnhem Land), and probably number in the hundreds.

No findings are available for CAs (remote areas) and the estimates for the non-remote sample are unreliable. This is unsurprising, as the use of kava by Indigenous Australians in these settings is unknown.

Limited corroboration with other sources of information

The ability to corroborate findings between various sources of information is indicative of the level of confidence that can be placed in the results of any particular study. In the case of the 2002 NATSISS and its findings on substance use, the corroborative evidence is weak.

Population estimates of risky/high-risk alcohol use drawn from the 2002 NATSISS indicate that about 15 per cent of the Indigenous population drink at risky/high-risk levels. This can be compared to other surveys which have estimated that about 10 per cent of the non-Indigenous population drink at the same level (AIHW 2002). This does not concur with the exceptionally high rate of death and hospitalisation from alcohol-attributable conditions which arise from risky/high-risk drinking (e.g. alcoholic liver cirrhosis, alcohol dependence) among the Indigenous compared to the non-Indigenous population. Indeed, the resultant drinking patterns from the 2002 NATSISS more closely resemble a *non*-Indigenous drinking pattern than any other previously conducted surveys of Indigenous drinking. That is, the majority of Indigenous respondents are consuming at low-risk levels; a smaller proportion are abstinent; and the smallest proportion drink at risky/high-risk levels. With the exception of the 2002 NATSISS, all of the national surveys described here indicated much larger proportions of Indigenous abstainers than low-risk drinkers. (Even the 2001 NHS, with all its problematic methodology, found a larger proportion of Indigenous abstainers than low-risk drinkers.)

Moreover, data on other alcohol-related issues, such as homicide, do not corroborate the suggestion in the 2002 NATSISS that a large proportion (46%) of Indigenous drinkers are consuming alcohol at low-risk or relatively safe levels. For example, between 1999 and 2003, 71.3 per cent of Indigenous homicides occurred in situations where both perpetrator and victim were drinking (as opposed to 19.5% of non-Indigenous homicides) (SCRGSP 2005).

The 2002 NATSISS results are vastly dissimilar to the only national survey specifically relating to substance use conducted in Australia to date—and arguably the most reliable—the 1994 NDSHS. The 1994 NDSHS indicated over three times the number of risky/high-risk drinkers than did the NATSISS. Given that all surveys of drinking underestimate overall levels consumption when compared to what is known about per capita consumption (Stockwell et al. 2004; WHO 2000), it is reasonable to assume that surveys which yield the greater overall estimated levels of consumption are, in fact, closer to actual levels of drinking (Stockwell et al. 2004). What is more, it is difficult to imagine that such a large proportion of respondents to the 1994 NDSHS routinely overestimated their consumption to such a large extent. It is more plausible that there was a significant degree of underestimation of actual consumption in the 2002 NATSISS. Adding to this, ABS reports on the results from its various surveys are often forthcoming about the limitations of the surveys. For instance, the 2002 NATSISS report states that 'results from previous ABS surveys and administrative data collections suggest a tendency for respondents to under-report substance use and alcohol and tobacco consumption levels' (ABS 2004c: 58). Perhaps, in this instance, some greater acknowledgment of the atypical nature of the methods and results should be at the forefront of any further discussion of the 2002 NATSISS.

Finally, the findings of the 2002 NATSISS are particularly questionable when Aboriginal risky/high-risk drinking is compared with Torres Strait Islander risky/high-risk drinking. The NATSISS found that a greater percentage of Torres Strait Islanders (nearly 22%) were drinking in this way than were Aboriginal respondents (15%). Although there are no specific alcohol and other drug surveys that have detailed Torres Strait Islander consumption patterns, observational and other data suggest that social drinking is the norm, rather than explosive, high-risk drinking (Torres Strait Health Workshop Working Party 1993: 15).

Conclusions

The questioning in the 2002 NATSISS on alcohol and tobacco use was neither confidential nor self-completed. Respondents were asked whether they would like a one-on-one interview but, in practice, they often answered questions in the presence of other family members. With alcohol and drug use being highly personal, potentially embarrassing and inevitably sensitive, and with family members perhaps listening in, it is hardly surprising that many respondents would have been unwilling to provide accurate estimates.

The findings on alcohol consumption do not only contradict the most reliable survey to date (the NDSHS), they run counter to qualitative understandings of Indigenous alcohol consumption, to smaller scale surveys, and counter to what we might call 'anthropological intuition' based on intimate daily participation in, and observations of, Indigenous life. Smaller scale studies such as these show that Indigenous people who drink do so at levels variously described as 'harmful', 'very heavy regular' or 'binge drinking'. These definitions have been used to describe remote communities (Hunter 1993; Martin 1998), rural towns (Perkins et al. 1994) and urban areas (Lake 1989; Tharawal Aboriginal Corporation 1994). All these report that moderate consumers were very much in the minority. They also consistently show an all-or-nothing pattern of consumption. Not only is 'moderate' consumption much less likely to be the chosen style, but there is active resistance to the idea of moderate or 'social' drinking among large sections of the Indigenous community. For the most part, we know that people tend to give up drinking altogether (i.e. become abstainers) rather than attempting (against huge social pressures) to consume moderately, and there are a number of good social and cultural reasons why Indigenous people make this choice (Brady 2004). It is highly unlikely that Indigenous drinking patterns would have changed so dramatically in the relatively short years since these studies were published.

What is of most concern, however, is the way in which findings from a survey such as the 2002 NATSISS quickly assume the mantle of authority. The new Productivity Commission report, *Overcoming Indigenous Disadvantage* (SCRGSP 2005: 22), has already cited the alcohol and other drug findings of the 2002 NATSISS, with its ink barely dry. Reproduced and reiterated in subsequent government reports and other documents, these statistics can take on the aura of authenticity, and fail to include the necessary caveats suggested by Biddle and Hunter in this volume—that is, to describe the findings on alcohol consumption specifically as referring to the 'Indigenous population living in private dwellings'. All future discussion of the 2002 findings should also acknowledge the atypical methods and results. The lack of corroborative evidence greatly reduces our confidence in these 2002 NATSISS findings and increases our concern that poor data have the potential to adversely affect funding priorities, program planning and future policy making.

19. Crime and justice issues

Mick Dodson and Boyd Hunter

The over-representation of Indigenous Australians in prison continues to be a serious problem, more than a decade after the recommendations of the Royal Commission into Aboriginal Deaths in Custody (RCADC) were handed down (Baker 2001; Williams 2001). [1] For example, Baker (2001) finds that the over-representation stems initially from the higher rate of appearance at court by Indigenous Australians, but is amplified at the point of sentencing, with Indigenous offenders sentenced to imprisonment at almost twice the rate of non-Indigenous people. The violent nature of the offences for which Indigenous people are convicted and the greater likelihood of Indigenous people having prior convictions were also found to contribute to their higher rate of imprisonment.

The RCADC recognised that '...too many Aboriginal people are in custody too often', and recommended a strategy of imprisonment as the last resort to reduce the level of over-representation of Indigenous people in custody (Commonwealth of Australia 1991). More recently, Baker (2001) concluded that reducing the rate of court appearances provides the greatest leverage for reducing Indigenous imprisonment rates. Obviously, one clear way of achieving lower court appearance rates and in diverting people away from court is to reduce the rate at which Indigenous people are arrested. This requires police in the first instance to opt for alternatives to arrest.

This sort of analysis has been conducted using police and court data. However, insights into the socioeconomic forces underlying Indigenous interaction with the criminal justice system can only be obtained by interrogating omnibus social surveys like the 1994 NATSIS and the 2002 NATSISS that include a reasonably comprehensive set of potential explanatory factors.

While the 1994 NATSIS has provided some valuable insights into the processes underlying the disproportionate level of Indigenous arrest (Carcach & Mukherjee 1996; Hunter 1998, 2001; Hunter & Borland 1999), several important research questions remain unanswered. Why do Indigenous people appear in court at a rate five times higher than the rest of the population? Why are Indigenous people more likely to appear for (and be convicted of) certain types of offences? (Baker 2001). Clearly, factors such as the over-representation of Indigenous people in

---

[1] The recommendations emphasised the need to reduce the disproportionate levels of Aboriginal people in custody, rather than the need to directly prevent their deaths. This emphasis arose out of the Royal Commission's conclusion that the 99 Aboriginal deaths in custody which occurred during the 1980s were not a result of Aboriginal people being any more likely than others to die in custody, but were a result of their gross over-representation in prison.

prisons and other stages of the criminal justice system, the nature of Indigenous offending and re-offending, and the differential treatment of Indigenous people by the criminal justice system will all have a part to play.[2]

Hunter (2001) asserted that there was a need to reassess the evidence when the data from the 2002 NATSISS was released. One of his arguments for emphasising this survey was that analogous data would be collected for the non-Indigenous population, thus providing a national benchmark against which to compare the Indigenous analysis. Unfortunately, the scope for benchmarking Indigenous results is circumscribed by the failure to collect analogous crime and justice data for the non-Indigenous survey, the GSS. One exception is the limited range of variables relating to the individuals' experience of crime.

The 1994 NATSIS did permit some unique analysis of social interactions (Borland & Hunter 2000; Hunter 2001). However, it was probably too easy for NATSIS-based research (and future research based on the NATSISS) to be dismissed as being specific to the Indigenous population, simply because there was no general omnibus survey at the time that collected a similar range of data on arrest and incarceration.

The ABS's decision to leave out such variables from the GSS could be rationalised on the grounds that any survey estimates of incarceration (and most other aspects of involvement with the criminal justice system) for the non-Indigenous population would have high relative standard errors, especially given the current sample size. However, the failure to collect such information for the majority of Australians limits our capacity to understand Indigenous disadvantage, as many socioeconomic outcomes have been shown to be behaviourally related to arrest, at least for the Indigenous population. That is, while it is well established that socioeconomic factors explain crime rates, criminal activity also partially drives socioeconomic outcomes. This means that we need information on both factors for the whole population in order to understand the origins of Indigenous disadvantage (Borland & Hunter 2000; Hunter 2001).

Even if there is not sufficient number of GSS respondents to identify the extent of interactions between socioeconomic outcomes and crime for the non-Indigenous population, pooled estimates (i.e. a combination of the GSS and NATSISS data sets) would allow researchers to estimate more robust estimators of the factors underlying overall Indigenous disadvantage, and hence allow us to appreciate the overall importance of interactions with the criminal justice system in the cycle of social exclusion.

The 2002 NATSISS, like the 1994 NATSIS before it, is designed to provide a broad range of information across key areas of social concern and is ideal for exploring inter-relationships between these socioeconomic factors and crime

---

[2] See Broadhurst et al. (1994) and Cunneen & McDonald (1997).

and justice issues. It is also possible to identify broad trends in crime and justice issues—although the scope for any time series analysis is diminished by the removal and addition of questions between the 1994 and 2002 surveys. For example, no information was collected in 2002 as to whether an offence was for theft, assault, disorderly conduct and/or drink driving, and outstanding warrants and breach of orders. We will return to this in the discussion.

However, 2002 NATSISS does have two major advantages over the earlier Indigenous survey. Firstly, the 2002 survey collects information never attempted before in a social survey context—namely, whether respondents had been formally charged by police, the age they were first formally charged by police, and whether they had been incarcerated in the last five years. The 'age first formally charged by police' is potentially important, as it may be interpreted as introducing an implicitly longitudinal dimension to what would otherwise be a cross-sectional analysis.

This chapter discusses the range of strategies used to collect crime data with a particular focus on omnibus surveys such as the GSS, the NATSIS and the NATSISS. The chapter then provides a selective overview of what we already know from the analysis of administrative data and the 1994 NATSIS data collections. After identifying the holes in existing data collections, we then examine the 2002 NATSISS data for any new insights that research might shed some light on. The data analysis is by no means comprehensive and is more illustrative of potentially productive avenues for future research. The concluding section reflects on how the omnibus surveys might be improved for future collections.

Diversity in existing crime data collections

Data from prisons, juvenile corrective institutions, courts and police lock-ups confirm a disproportionate level of contacts between the Indigenous Australian population and the criminal justice system. Given that administrative data from the police and courts generally provide the most accurate information into the extent of Indigenous interaction with the criminal justice system (because they are, in effect, censuses of the relevant population), it is important not to place too much emphasis on such statistics in collecting survey data. In addition to the demographic characteristics usually adequately represented in administrative data sets, there is a need to also collect a diverse set of data on socioeconomic and cultural factors that allow analysts to explore inter-dependencies between socioeconomic factors and crime. However, it is not possible to explore these interdependencies unless some data on arrest and incarceration is collected in the survey.

Crime and justice issues are asked in a number of ongoing surveys and administrative databases. In addition to the police statistics and data collected in the administration of the criminal justice system, there is the National Crime and Safety Survey, the Personal Safety Survey, the GSS, the NATSISS, the International Crime Victims Survey, and the National Survey of Community Satisfaction with Policing. Each of these surveys asks about the experiences of crime or the criminal justice system. However, it is also important to collect information about the experiences outside the criminal justice system.

The impact of differing methodologies on the measurement of crime victimisation is the confidentiality of the surveys and data instruments. Variations in the level of confidentiality may result in substantial non-sampling errors and reporting bias. For example, the different methodologies between the various crime victimisation surveys, and between survey and administrative data, seem to produce substantially different results.

The following discussion briefly outlines alternative survey data sources and reflects on them as potential sources of information for the Indigenous population.

The 2002 NATSISS was collected using face-to-face interviews, but some were computer-assisted interviews (in NCAs) and some were pen and paper interviews (in CAs). A self-administered questionnaire on the non-medical use of legal and illegal substances was used to ensure that sensitivity of subject matter did not affect responses.

There is little information in the 1994 NATSIS and the 2002 NATSISS with which to gauge trends in crime or ascertain perceptions of police and criminal justice systems. The six-yearly frequency of the NATSISS data collections commencing in 2002 limits the ability of the NATSISS to respond to emerging issues. In any case, the existence of sampling error in survey data means that its comparative advantage is in identifying the inter-relationships between crime and justice and related factors rather than gauging trends. Note that the lack of data on the perceptions of police in the 2002 NATSISS can be contrasted to the 1994 NATSIS which collected an extensive range of Indigenous attitudes to police. However, this may not be a significant loss, since some analysts questioned the value of this data in the 1994 survey (Carcach & Mukherjee 1996).

As alluded to above, the GSS and NATSISS are broadly comparable, although the lack of arrest and incarceration data limits the ability to validly compare the experiences of Indigenous and non-Indigenous peoples. For example, employment decompositions—a common tool used by social scientists to examine the scope for labour market discrimination—will not be able to control for the independent effect of arrest on employment. This will have a tendency to mean that the unexplained component of the differential between Indigenous and non-Indigenous employment is more likely to be attributed to potential discrimination (Hunter 2004a). While the higher rates of arrest could be said to reflect historical or systemic discrimination, it would be desirable to provide analysis where the source of the problem can be better identified.

Prisoners were excluded from the 2002 NATSISS but not from the 1994 NATSIS. While surveys are obviously not the best ways to get information about the total numbers of prisoners, at a point in time prisoners represent about 2 per cent of the total Indigenous population (i.e. in the 1994 NATSIS), and the omission of prisoners may introduce selectivity issues into any analysis of the NATSISS data.[3] Over a lifetime, a significant number of Indigenous people will go to jail, so it is important to understand the long-term social dynamics of people who move into and out of prison. Notwithstanding apparent minor falls in the rates of deaths in custody over the 1990s (Williams 2001), it is still vitally important to understand the over-representation of Indigenous people in gaols. Given that non-private dwellings are not within the scope of the 2002 NATSISS, official prison data remains the most important source of information for this important segment of the population.

The National Crime and Safety Survey (NCSS) is designed to provide a snapshot of the prevalence of certain crimes in the Australian community, as well as provide some detail about victims, offenders and incidents. The stated aims of the survey are to inform the development of crime prevention and community education strategies by providing information on the incidence of selected crimes and reporting to police, and perceived levels of crime. The NCSS is unique in employing a mail-out and mail-back methodology. The surveys are completed on a three-yearly basis with respondents and are basically confidential in that they are a self-administered questionnaire. However, data on Indigenous people are not available in the NCSS data.

The Personal Safety Survey provides details on experiences of female and male victims of violence that are not available in other data collections. The mixture of face-to-face and telephone interviews used in the PSS is likely to lead to a substantial non-sampling within this survey and makes some of the results potentially non-comparable with other survey results.

The National Survey of Community Satisfaction with Policing (NSCSP) is designed to provide detailed information at a local level of public perceptions of policing and feelings of safety. Telephone interviews were used to collect information on respondents aged 15 and over. The low response rate is likely to be a

---

[3] Biddle & Hunter (in this volume) explore the relevant methodological differences between the 1994 NATSIS and the 2002 NATSISS; the most salient difference in the context of this chapter is that the dwellings sampled in the latter survey was constrained to Indigenous people aged 15 and over living in private dwellings—that is, it omits prisoners. Given that the Indigenous prison population did not fall significantly between 1994 and 2002 (Australian Institute of Criminology 2004: 97), the latter survey omits a small but potentially important component of the indigenous population. Biddle and Hunter provide some description of the unique characteristics of those living in non-private dwellings. In the context of this chapter, it is relevant to point out that residents of non-private dwellings were more likely than other Indigenous Australians to have been arrested in the last five years and to have been taken from their natural families. This is not surprising, since around 44% of NATSIS respondents (aged 15 and over) from non-private dwellings were in prison.

particularly pronounced issue for a mobile population such as Indigenous Australians, especially when surveys are based on telephone interviews (Hunter & Smith 2002). While information on Indigenous status may be collected in the NSCSP and the PSS, the Indigenous sample is likely to be too small and selective to be confident about the robustness of any resulting analysis.

The International Crime Victims Survey (ICVS) was initiated in order to facilitate international comparative research. It is designed to provide an international benchmark against which to compare crime and safety information in a standardised way. It is conducted on a four-yearly basis using a telephone interview with respondents aged 16 and over. It contains a detailed range of questions on crime victimisation, with the 2004 module asking several extra questions (including the number of assaults that were racially motivated, fear of racially motivated violence, and Indigenous status). The small sample size limits its usefulness. Hunter and Smith's (2002) critique of telephone interviewing technique in mobile populations is again relevant.

Therefore, while a number of surveys collect information that might shed light on the nature and extent of Indigenous interaction with the criminal justice system, the small sample of Indigenous respondents in most surveys rule them out for detailed analysis. The variation in survey methodologies means that most data will only be comparable in broad terms. Effectively, future analyses of Indigenous crime and justice issues are confined to the analysis of the 2002 NATSISS.

Crime, safety, and 'justice' questions in omnibus Australian surveys

While much of the 2002 NATSISS data is comparable with that in the 2002 GSS, the GSS does not include data on arrest, whether charged, and imprisonment. This is presumably because the proportion of the GSS sample who were arrested or charged is relatively small and hence it is difficult to get reliable estimates.

The GSS does, however, collate information on 'Crime and safety' rather than the 'Crime and justice' issues referred to in the official NATSISS publication (ABS 2003b, 2004c)—apparently, non-Indigenous people need information about 'safety' and Indigenous people need information about 'justice'. It would be nice if the need for justice for Indigenous Australians was really being recognised in public debates. However, the use of language probably indicates the tendency for the media to focus on 'crisis' and play on the (largely non-Indigenous) public's concerns about safety and victimisation that feed the 'law and order' campaigns that are a common feature of almost every State and—increasingly—federal election campaigns.

The GSS collects several questions on victimisation among respondents: feelings of safety at home alone after dark, whether they have been the victim of physical or threatened violence in the last 12 months, and whether they have been victims of an actual or attempted break-in in the last 12 months. Note that the only question that was also collected in the 2002 NATSISS is whether respondents were a victim of physical or threatened violence in the last 12 months. The richer set of Indigenous data on justice issues means that the NATSISS potentially provides insights into how interactions with the criminal justice system affect the socioeconomic outcomes, and vice versa.

There are more difficult issues of comparability when comparing the 1994 NATSIS and the 2002 NATSISS, as the question changed in subtle and important ways that may explain (partially at least) the significant increase in the rate of victims of physical or threatened violence in the last 12 months. In 1994, respondents were asked: 'In the last year, has anyone attacked or verbally threatened you?'. The 1994 questionnaire goes on to ask how many times a respondent was attacked, the broad characteristics of the violence, whether the respondent told the police, and the reason for not telling the police. In 2002, the questionnaire module, the respondents were asked about crimes that may have happened to them in the last 12 months: 'Did anyone, including people you know, use physical force or violence against you?'; 'Did anyone, including people you know, try to use or threaten to use physical force or violence against you?'; 'Were any of those threats made in person?'. While the questions were broadly similar, the questions about the involvement of the police may have lead to substantial under-reporting of the crime. If this is the case, then the decision to leave out the questions on attitudes to police in the 2002 NATSISS would be warranted.

In general, the remote questionnaire for CAs involved questions that were simplified versions of the form used in NCAs. While this should not affect the data significantly, there is more room for non-sampling error in the question about victimisation. The CA form asked two questions about crimes that may have happened to a respondent—firstly: 'In the last year, did anybody start a fight with you or beat you up?', and secondly: 'In the last year, did anybody try to or say they were going to hit you or fight with you?'. The use of plain English in such a sensitive area might be seen as more emotive by some respondents, and open to subjective interpretation of what a 'real' fight is by other respondents. While there are likely to be substantial variations in the cultural standards of what constitutes a 'fight' (or a 'beating') in the Indigenous and other Australian communities, there is no GSS data collected in remote areas. As a result, this issue is only relevant for making comparisons between remote and non-remote areas within the NATSISS.

Another 1994 variable that changed between 1994 and 2002 was whether a respondent felt that family violence was a common problem in their area. In 2002, respondents in NCAs were asked to identify social problems in their neighbourhood from a list of 12 items on a prompt card which included family violence as one possible option. [4] This obviously differs from the 1994 question in two important ways: firstly, it is specific about the geographic area in referring to a neighbourhood which is one of the smallest areas that can be analysed; secondly, and more importantly, family violence is just one of a number of local problems listed. In CA areas, respondents were asked about the 'problems in this community'. While the geographic area was also specified in the CA form, it is always difficult to get agreement as to what 'community' means among academics, policy-makers and, often, in the community itself—although the notion of community may coincide with some notion of 'neighbourhood' for some people. Again, the CA form listed a similar set of reasons that were not directly comparable to those provided for the questionnaire used in the NCA areas. Notwithstanding the lack of clarity in the definition of areas, the questions on the CAs and non-CAs are broadly comparable. The 2002 question is less 'leading' than the 1994 question, and places less focus on family violence. While the 1994 question may play down the incidence of family violence, the 2002 question is less likely to lead to non-sampling error and hence is probably preferable to the 1994 question.

Carcach and Mukherjee (1996) argued strongly that the 1994 NATSIS placed too much emphasis on the respondents' perceptions about police performance and the relationship between Indigenous people and the police. [5] There were no questions in 2002 about whether a person had been hassled by the police or attitudes to police—maybe because the 1994 question was seen as too leading, too vague, or too open to subjective interpretation. [6] The 2002 questions on police contact focus on more factual events, like whether a person had been charged and at what age a person was charged, as well as how many times a person was arrested. The arguments made by Carcach and Mukherjee obviously

---

[4] The prompt card listed several options, including: theft, problems involving youths such as youth gangs/lack of youth activity, prowlers or loiterers, vandalism or graffiti, or damage to property, dangerous or noisy driving, alcohol, illegal drugs, family violence, assault, sexual assault, problems with neighbours, levels of neighbourhood conflict, level of personal safety day or night.

[5] They were particularly concerned about the absence of a question on perceptions about the amount of crime in the area to give some background qualifier to the question on police performance in dealing with crime. Moreover, the 1994 questionnaire included among the reasons for the police not doing a good job in dealing with crime, that they 'Don't understand Aboriginal/Torres Strait Islander people/culture', which suggests that the question on police performance was referring to crime in Indigenous areas or perhaps among Indigenous people. However, the question was worded in such a way that it did not mention crime by Indigenous people or among Indigenous people at all.

[6] Carcach & Mukherjee (1996) also argued that the highly sensitive questions regarding victimisation of Indigenous people by police, such as hassling and physical attack, were potentially problematic. The NATSIS lacked additional details on incidents involving hassling or police attack, such as whether they were related to an arrest and, if so, whether or not the person resisted arrest or tried to avoid the arrest of another person. Perceptions about quality of the relationship between Indigenous people and the police and about the treatment received from police were in general positive. Again, there was no question on whether the relationship with police was good or bad five years ago. What is the meaning of a relationship being qualified as the same? Does 'same' mean good or bad?

had an impact, as the 2002 NATSISS no longer collects information on attitudes to or relationships between Indigenous people and the police.

The data on the 'stolen generation' was based on three questions in the 2002 questionnaire, as opposed to the one question used in the 1994 survey. The basic question used was almost identical in both surveys, asking: 'Were you taken away from your natural family by a mission, the government or welfare?'. The 2002 survey also asked whether their relatives were taken from their natural family and, if so, which relatives.

Both the 1994 NATSIS and 2002 NATSISS asked whether respondents used or needed legal services in the last 12 months and, if so, the type of legal services used. Given the variation in geographic access to legal services, and the Indigenous-specific nature of particular services, it is probably best to confine comparisons within the Indigenous surveys. Even then, it may be difficult to separate access from usage issues.

In summary, the 2002 NATSISS did not interrogate respondents about their attitudes to police or ask specific—and potentially personally sensitive—questions about the reasons for contact with the criminal justice system. To the extent that specific interactions with the police are sensitive matters, especially for people who do not trust government agencies such as the ABS, the omission of questions on attitudes to police may have led to more accurate responses to the relevant questions (i.e. with smaller scope for non-sampling error) than was possible in the 1994 survey. Having said that, Borland and Hunter (2000) conducted a benchmarking exercise on the arrest data in the 1994 NATSIS by comparing it with WA police data, and found that the NATSIS data was broadly consistent with the administrative data sources. Consequently, it seems reasonable to assume that the more modest 2002 questionnaires will have elicited similarly accurate responses about Indigenous arrest.

Socioeconomic factors underlying Indigenous arrest

Before outlining the analysis of published data, we first examine the extant analysis of the 1994 NATSISS, with a particular focus on the social and economic factors underlying the disproportionate Indigenous interaction with the criminal justice system. The emphasis in the following discussion is on data that are either not available or are limited by instrumental factors in the 2002 NATSISS.

Carcach and Mukherjee (1996) show that most of the arrests in the NATSIS were for disorderly conduct and/or drink driving, and outstanding warrants and breach of orders. Data show that alcohol consumption might have been associated with the reason(s) for arrest; a result consistent with findings from the National Police Custody Survey (Australian Institute of Criminology 1996). The links between alcohol and crime (violence, disorder and acquisitive crime) are well documented (see Ramsay 1996). Previous research would suggest that alcohol might have been involved in incidents of violence both in and outside the family, and in cases where the arrest was due to property crimes (e.g. Tuck 1989).

Hunter (2001) analysed NATSIS and found that the major factors underlying the high rates of Indigenous arrest were sex, labour force status, alcohol consumption, whether a person had been physically attacked or verbally threatened, various age factors, and the cluster of education variables. The top six factors underlying the various categories of arrests (drinking-related, assaults, theft and outstanding warrants) are basically the same as those identified above. However, alcohol consumption and being a victim of physical attack or verbal threat are particularly important factors underlying arrests on drinking-related and assault charges. This would seem to confirm the suspicion that there is a cycle of violence and abuse in Indigenous communities which is probably related to alcohol consumption. The overall results were robust, with the basic findings not changing substantially when the analysis was conducted separately for minors (under 18-year-olds), for each sex, or after prisoners were included in the analysis. This last finding points to the fact that the omission of Indigenous people under 15 years old and people in non-private dwellings should not distort future analysis of the 2002 NATSISS. The similarity in the results by reason for last arrest give us some confidence that the failure to collect information on the nature of the offence that led to arrest and imprisonment in the 2002 survey is not that important—or, rather, will not alter the overall analysis.

Borland and Hunter (2000) argue that at least some of the correlation between Indigenous arrest and labour force status is driven by a causal relationship, with arrest driving many of the poor employment outcomes experienced by Indigenous youth. Given this interaction, understanding the unique nature of Indigenous arrest is likely to be a key dynamic underlying ongoing Indigenous disadvantage and poverty. The lack of any arrest data in GSS is likely to impede our understanding of Indigenous disadvantage—or, rather, to control for the impact of arrest and interaction with the criminal justice system on economic outcomes.

Hunter and Schwab (1998) argued that the interaction with the criminal justice system may explain poor school participation rates among Indigenous children as young as 13. Hunter (1998) presented formal econometric tests that demonstrated that the direction of causality is from arrest to educational participation. Given that the 2002 NATSISS is constrained to those aged 15 and over, it will probably not be possible to replicate this earlier research.

Crime and justice issues in Indigenous social surveys

Broad trends in crime and justice between 1994 and 2002

Crime and justice issues were major components of both the 1994 NATSIS and the 2002 NATSISS. This section draws together the recently published data to provide a statistical overview for Indigenous Australia. Notwithstanding the substantial changes in the crime and justice data collected in the 1994 NATSIS and the 2002 NATSISS, and the limitations of the respective survey methodologies, it is possible to make some broad observations about the trends for Indigenous Australia (see Table 19.1).

Table 19.1. Indigenous people aged 15 years or over, selected law and justice issues in Australia, 1994 and 2002

	1994	2002
	%	%
Arrested once by police in last 5 years [a]	9.1	6.7
Arrested more than once by police in last 5 years	10.7	9.3
Total arrested in last 5 years [a]	20.2	16.4
Victim of physical or threatened violence in last 12 months [a]	12.9	24.3
Persons removed from natural family	8.3	8.4

a. The change between 1994 and 2002 is significant at the 5% level.
Source: ABS (2004c: Table 6)

The law and justice variables exhibited some significant changes between 1994 and 2002. The overall proportion of Indigenous adults who were arrested in the previous five years declined from 20.2 per cent to 16.4 per cent. The main driver here was the significant reduction in the number of people with only one arrest in the previous five years. This is a positive development, although it should be acknowledged that Indigenous people still have excessively high rates of interaction with the criminal justice system relative to other citizens. A less positive development is the increase in the proportion of the population who were a victim of physical or threatened violence in last 12 months. One-quarter of Indigenous people in 2002 reported that they had been a victim of physical or threatened violence in the previous 12 months—nearly double the rate reported in 1994 (12.9%). ABS (2004c) speculates that some of this increase may reflect under-reporting by respondents to the 1994 NATSIS. This is consistent with the above analysis of the way in which the questions were asked in 1994. The final observation from Table 19.1 is that the proportion of the population who were taken away from their natural family was basically unchanged. [7]

Describing selected crime and justice issues in 2002

While there were some changes over time at a national level, the main interest for most variables is how crime and justice issues vary by remoteness—a more meaningful measure of geographic accessibility than was available for the earlier survey (see Tables 19.2 & 19.3).

[7] This finding was robust to confining the analysis to being the same age cohort in the respective surveys. See ABS (2004c: Table 6).

Table 19.2. Selected law and justice issues by remoteness, 2002

	Remote	Non-remote	Total
	%	%	%
Arrested by police in last 5 years	16.9	16.2	16.4
Incarcerated in last 5 years	8.5	6.6	7.1
Used legal services in last 12 months	17.9	20.5	19.8
Victim of physical or threatened violence in last 12 months	22.7	25.0	24.3
Person or relative removed from natural family [a]	30.1	40.4	37.6
Risky/high-risk alcohol consumption in last 12 months	16.8	14.5	15.1

a. The difference between remote and non-remote statistics is significant at the 5% level.
Source: ABS (2004c: Table 1)

Arrest and incarceration rates are equally high in both remote and non-remote areas, but the usage of legal services in the last 12 months is slightly higher, albeit not significantly higher. Also, there was no significant difference between the incidence of high-risk alcohol consumption in remote and other areas in the last 12 months. The only significant difference between remote and other areas among the law and justice issues in Table 19.2 is whether a person or relative were removed from their natural family. The incidence of relatives (including oneself) being removed from their natural families is over 10 percentage points higher in non-remote areas. This pattern is consistent with the fact that many people were removed from remote communities and placed with families in cities or regional centres (Hunter, Arthur & Morphy 2005).

Table 19.3. Interactions with the justice system by remoteness and sex, 2002

	Remote	Non-remote	Males	Females	Persons
	%	%	%	%	%
Age first formally charged					
Less than 15 years [a,b]	4.0	7.0	10.1	2.6	6.2
15–16 years [a,b]	5.1	7.5	9.5	4.4	6.8
17–18 years [b]	6.8	8.5	12.6	3.8	8.0
19–24 years [b]	7.5	7.1	10.8	4.0	7.2
25 years or over [a]	8.6	5.4	6.7	5.8	6.3
Total formally charged [b]	32.8	35.8	50.4	20.8	35.0
Never charged [b]	67.2	64.2	49.6	79.2	65.0
Use of legal services in last 12 months					
Used legal services [a]					
Aboriginal Legal Services [a,b]	14.1	9.1	12.2	8.9	10.5
Legal Aid [b]	-	7.0	-	-	-
Other [a]	2.1	7.1	5.6	5.9	5.7
Total used legal services	17.9	20.5	21.1	18.6	19.8
Did not use legal services					
Needed legal services	2.9	3.1	2.9	3.2	3.1
Did not need legal services	79.2	76.4	76.0	78.2	77.1
Total did not use legal services	82.1	79.5	78.9	81.4	80.2

a. The difference between remote and non-remote is significant at the 5% level.
b. The difference between males and females is significant at the 5% level.
Source: ABS (2004c: Table 19)

Administrative data gives detailed information on all those who are involved in the criminal justice system but cannot provide direct information on what happens in the Indigenous population as a whole. As argued above, the inclusion of information on the age at which a person is formally charged allows an implicitly longitudinal dimension to what would otherwise be a cross-sectional analysis. Table 19.3 shows that over half of males (50.4%) have been charged at some time in their life, about 30 percentage points higher than the equivalent statistic for females (20.8%). There is no significant difference in the overall incidence of charging in remote versus non-remote areas. However, there is one potentially significant geographic difference evident: those living in non-remote areas are more likely than remote respondents to be charged before 16 years of age. Residents in remote areas tend to have more of their population being charged first at age 25 or older. This pattern of charging may result from the fact that some discrete communities do not have police in the local areas, so people are not charged until they move away from remote communities. Alternatively, they may have to commit a very serious offence before it would be brought to police attention so that charges could be laid. The higher incidence of charging among males is probably driven by charges being laid at an earlier age for Indigenous males compared to Indigenous females.

The usage of Aboriginal legal services is significantly higher in remote as opposed to non-remote areas, and such services are more likely to be used by males than by females (see Table 19.3). While the geographic differences may be related to the location of such services, the gender differences must be related to either the organisation of Aboriginal legal services or the power structures in relevant Indigenous communities, as there is no gendered pattern in the usage or need for legal services. That is, there is no significant difference in usage and need for legal services between sexes or by remoteness status. The lack of a gendered pattern in legal services use is in stark contrast to the fact that males are significantly more likely to have been arrested or imprisoned.

Table 19.4. Law and justice by age first formally charged, 2002[a]

Law and justice	Age first formally charged		
	Less than 17 years	17 to 24 years	25 years or over
Arrested by police in last 5 years	54.5	37.7	31.3
Incarcerated in last 5 years	28.6	13.6	14.4
Victim of physical or threatened violence in last 12 months	44.5	33.4	26.7
Indigenous persons aged 15 years or over (%)	100.0	100.0	100.0
Indigenous persons aged 15 years or over (#s)	36.7	43.0	17.6

a. The difference between being charged before the 17th birthday and being charged later in life is significant at the 5% level for all statistics in this table.
Source: ABS (2004c: Table 11)

The addition of 'age first formally charged' has been a substantial addition to the NATSISS vis-à-vis the earlier 1994 survey and deserves to be examined in greater detail (see Table 19.4). Having been arrested before 17 years of age leads to significantly higher arrest and incarceration rates than for other Indigenous respondents to NATSISS who were charged. For example, respondents who were charged before 17 years of age are 23.2 percentage points more likely to have been arrested than someone who was first formally charged at age 25 or older (54.5% and 31.3% respectively). However, being charged before your majority is also associated with a significantly greater likelihood of being a victim of physical attack or threatened with violence in the last 12 months (44.5% and 26% respectively).

Table 19.5. Socioeconomic factors underlying formal interactions with the criminal justice system by ever formally charged, 2002

	Total persons ever charged			
	Males	Females	Persons ever charged	Persons never charged
Law and justice				
Arrested by police in last 5 years [b]	44.7	38.9	42.9	2.2
Incarcerated in last 5 years [a b]	21.9	13.9	19.4	0.6
Victim of physical or threatened violence in last 12 months [a b]	32.6	44.7	36.4	18.0
Health				
Self-assessed health excellent/very good [b]	39.3	34.4	37.8	47.5
Self-assessed health fair/poor [b]	28.7	31.2	29.5	19.9
Risk behaviour/characteristics				
Current daily smoker [b]	63.7	71.0	65.9	39.3
Risky/high-risk alcohol consumption in last 12 months [b]	24.1	22.8	23.7	10.5
Education				
Attending post-school educational institution [b]	8.7	8.7	8.7	11.8
Has a non-school qualification	26.4	24.5	25.8	26.4
Does not have a non-school qualification				
Completed Year 12 [b]	6.9	5.0	6.3	11.9
Completed Year 10 or Year 11 [a]	25.4	35.4	28.6	29.0
Completed Year 9 or below [b]	41.3	35.1	39.3	32.7
Total with no non-school qualification	73.6	75.5	74.2	73.6
Labour force status				
CDEP [a b]	17.6	8.7	14.8	10.7
Non-CDEP [a b]	32.6	19.6	28.6	37.2
Total employed [a]	50.2	28.2	43.4	47.9
Unemployed 1 year or more	6.5	4.6	5.9	2.2
Total unemployed [b]	21.1	19.3	20.5	10.2
Not in the labour force [a b]	28.7	52.5	36.1	41.9
Income & financial stress				
Equivalised gross household income — second and third deciles	39.9	42.5	40.7	35.3
Unable to raise $2000 within a week for something important [a b]	58.5	70.8	62.3	49.8
Transport access				
Difficulty with transport				
Can easily get to the places needed [b]	65.0	60.1	63.4	73.7
Has difficulty, getting to the places needed [b]	14.1	14.9	14.4	10.2
Information technology				
Used computer in last 12 months [b]	45.5	51.8	47.4	60.1

a. The difference between males and females who were charged is significant at the 5% level.
b. The difference between those who were charged and never charged was significant at the 5% level
Source: ABS (2004c: Table 11)

Clearly, the data on the age at which a person is first charged contains a lot of useful information and should be analysed in detail by researchers and policy-makers. The introduction of data on whether a respondent was ever charged allows us to explore the likely factors underlying Indigenous engagement with the criminal justice system and the effects of that engagement. Family and culture have no significant correlation with the incidence of being charged, at least in the bi-variate analysis of cross-tabulations. Consequently, family and culture variables are omitted from Table 19.5.

It is understandable that very few people who were not formally charged were arrested or incarcerated. Given that it is rare to be imprisoned without being charged, it is probable that some in this category were errors in response. Among those who were charged, males were more likely than females to be imprisoned, and females were more likely than males to be a victim of physical or verbal violence in the last 12 months (44.7% and 32.6% respectively).

There is a significant association between health status and being charged. For example, people who were charged at some stage were 10 percentage points less likely to report having excellent or very good health (37.8% and 47.5% respectively). The converse of this is that such people are more likely to report fair or poor health to a similar extent. One reason for this relatively poor health status is that the respondents who were charged were significantly more likely to exhibit risk behaviour than those who were never charged. They were 26.6 per cent more likely to be a current daily smoker (65.9% and 39.3% respectively) and were over twice as likely to have engaged in high-risk alcohol consumption in the last 12 months (23.7% versus 10.5%).

The direction of causality between health and crime could be either way. However, the analysis of 1994 NATSIS data on arrest indicated that at least some of the correlation between arrest and education and labour force status is driven by arrest affecting the respective outcomes (Borland & Hunter 2000; Hunter 1998). People who were not charged are significantly more likely than those who were formally charged to be attending a post-school educational institution. This is consistent with the fact that people who were formerly charged were less likely than those who were never charged to have completed year 12 and more likely to have only have completed year 9 or below. Among those who were charged, there was no notable pattern of educational attainment between males and females.

The association between labour force status and being charged is significant, with those who are formally charged being less likely than other Indigenous people to be employed outside the CDEP scheme but more likely to be employed in the CDEP scheme. Unemployment rates are twice as high among those who were formally charged. The net effect of these employment and unemployment patterns is that those who were formally charged are slightly more likely to participate in the labour force.

In terms of the income and financial stress variables, there is no significant difference in the propensity to be in the ABS's low income group. However, those who were formally charged are significantly more likely than other

Indigenous people to experience financial stress (at least as measured by the inability to raise $2000 within a week for something important). The difference between the correlations between income and financial stress and having been charged probably reflects poorly on the adequacy of the ABS's measure of the low income group (see Hunter in this volume).

Two new variables introduced into 2002 NATSISS for the first time are transport access and information technology, both of which are strongly correlated with having been charged. Those who had previously been charged were more likely to have transport difficulties than other Indigenous people, and are much less likely to have used a computer in the last 12 months.

All these factors point to a cycle of Indigenous disadvantage and interaction with the criminal justice system that will be hard to break. Future research needs to put some more econometric structure around these statistics so that the mechanisms underlying these processes can be identified. Unfortunately, it is too difficult to do this until the RADL is developed so that it allows users to write computer programs with statistical packages that facilitate the relevant econometric tests. Stata is one such program that does not have the intellectual and financial setup cost of doing such analysis using SAS, which is one of two statistical packages currently supported in the RADL for NATSISS. The other supported package is SPSS, which is relatively cheap but is not flexible enough to allow analysts to easily construct relevant diagnostic statistical tests.

Concluding remarks

The United Nations Permanent Forum on Indigenous Issues has had some difficulty in securing submissions into the quality of Indigenous statistics (Economic and Social Council 2005). It appears that the quality of Indigenous statistics is best in post-colonial nations of Australia, NZ and the US. [8] For Australia's part, this is the legacy of the Royal Commission and the original NATSIS. The 2002 NATSISS offers the opportunity for researchers and policy makers to take an international lead in understanding the dynamics of Indigenous disadvantage and dispossessions, especially the factors underlying the disproportionate representation in the criminal justice system.

The policy implications are complicated by the fact that 'feedback mechanisms' have been identified where arrest reinforces disadvantage in several of these factors (especially employment prospects and educational attainment). Any

[8] The Economic and Social Council (2005) recommended that Latin America and the Caribbean countries continue and strengthen their efforts regarding the production, elaboration and use of relevant information from population censuses, household surveys and other adequate sources, in strong interaction with Indigenous peoples, aiming at improving the socioeconomic conditions and active participation of indigenous peoples in the development process throughout the Latin American region. The council goes on to recommend that member countries give full respect to the principle of self-identification and develop Indigenous participation in collections that take account of the full diversity and demographic/socioeconomic profile of Indigenous communities.

attempt to substantially reduce the high rates of unemployment among Indigenous people also needs to make inroads into Indigenous arrest. Education policy needs not only to improve the marketability of the Indigenous workforce, but to facilitate the citizenship skills required to operate in both the Indigenous and non-Indigenous domains. Notwithstanding such feedback, improving the labour market options of Indigenous people should markedly reduce the arrest rate.

The links between alcohol and crime, especially violent crime, are well documented. Substantial progress needs to be made on substance abuse problems before the cycle of violence in Indigenous communities can be broken. Restrictions on liquor supply are consistently nominated as producing the most tangible results in terms of reducing alcohol-related harm among Indigenous Australians.

Family and social factors are less amenable to direct policy intervention. Indeed, the misconceived policy interventions that led to the 'stolen generation' appear to be a major factor underlying Indigenous arrest rates. The negative effects of such policies are likely to be driven by the traumatic disruption to family life and the loss of culturally appropriate parenting skills. Early intervention approaches to dealing with risk factors associated with anti-social and criminal behaviour appears to offer a promising avenue for policy action. It is important that Indigenous people have some control over how family services are provided (e.g. the need for Indigenous carers for Indigenous clients is often identified as an issue). The needs of children of Indigenous prisoners, especially those from country areas, should also be taken into account if the risk of delinquent behaviour is to be minimised.

Overall, the NATSISS provides a valuable source of information on crime and justice issues. Apart from finessing the wording of particular questions to ensure that the survey data maximises its information content, the main recommendation relates to the data for the future GSS. The analysis of Indigenous disadvantage will continue to be hamstrung and dismissed until data on non-Indigenous interaction with the criminal justice system are collected in the general omnibus surveys for the whole Australian population. One may be concerned about the unreliability of population estimates for low frequency events such as arrest and imprisonment (at least in the non-Indigenous population). However, there should still be a sufficient number of the Australian population who were charged at some time to warrant its inclusion in future GSSs. Irrespective of the reliability of population estimates in the GSS, the ability to explore the interactions between social and economic factors is a key objective of these omnibus surveys. The omission of information on crime in existing general surveys means that it is not possible to effectively analyse this important dimension of Indigenous disadvantage.

Chapman and Gray (in this volume) provide an example of how social scientists can demonstrate the implications of failing to control for important factors underlying labour force status. However, they can only do this because of the existence of alternative sources of information for the total Australian population. While the NATSISS, and its earlier manifestation, includes information not available in other surveys, there appears to be a resistance in Australian public debates to acknowledge that Indigenous experience may inform the analysis for the population at large. Indigenous-specific research is dismissed as being just that because of the undeniably unique circumstances facing Indigenous Australians. Consequently, it is imperative that the GSS expands its scope to encompass the more general experience of the criminal justice system.

20. Culture

Nicolas Peterson

It was unclear in the original 1994 NATSIS survey what the purpose and significance of the 'Culture' questions were: in some respects this still remains true for the 2002 survey. A number of possible reasons for collecting cultural data were canvassed in the assessment of the previous survey. These included the need to recognise regional cultural variation, to help in designing culturally appropriate policies, to assist in formulating policies on cultural maintenance, and to help in identifying cultural issues or practices that may stand in the way of the achievement of policy goals.

Some insight into the thinking relating to the cultural questions can be gained from the planning of the original survey, when an early step was to identify nine topic headings under which information would be sought. Culture was one of these, and under this topic 'location and mobility' were proposed as important, as were matters relating to family. This suggests that culture was seen as referring to those aspects of mobility and family that differ substantially among Indigenous people from the rest of the population. As it turned out, the request for certain information about mobility was dropped because the pilot survey found the recall data on this topic too unreliable. It may well be the case that it was generally felt that gaining cultural data that would contribute significantly to policy formation was seen as too complex for such a long questionnaire, and that the real concern with culture was more bureaucratic. That is, that the interest was really to obtain data 'to broadly evaluate specific agency programs first, and broad policy second, and a generally-held view that some data are better than none' (Altman & Taylor 1996: 200). Indeed, a conclusion of the previous assessment was that the interest in culture was primarily to do with cultural maintenance (Peterson 1996: 151), referred to by the ABS as 'cultural retention' (2004c: 2).

In this light, it is interesting to consider whether the nature of the 'culture' questions in the 2002 survey reflect this. While the principal culture questions fall within the 'Culture' section of the CAI questionnaire, there are a few other 'culture' questions elsewhere in the survey.

The first four questions in the culture section of the CAI questionnaire are grouped under the heading 'language'. Two of these are about languages spoken and two about whether the interviewee has difficulty being understood in service delivery situations. These are quintessentially culture questions and—in tandem with the 1994 survey—allow for some estimate of the maintenance of Aboriginal languages and for policy recommendations in respect of service delivery, from education to employment. The results of this part of the survey are discussed separately by Frances Morphy and Inge Kral in the next chapter.

The next seven questions, which deal with 'cultural participation/involvement in social activities' are divided into two parts. For these and all of the questions where it made sense, people could choose more than one response. The 'cultural activities' a person participates in are defined as: funerals, ceremonies, sports carnivals, 'festivals or carnivals involving arts, craft, music or dance', and being 'involved with any Aboriginal or Torres Strait Islander organisations'? (CAI Questionnaire: p.55). The first question in relation to these asks whether people had been involved in any or none of these; the second asks whether the person had done any of the following, including as part of their job: made any arts or crafts, performed any music, dance or theatre or written or told any stories. The third question was whether they had been paid for any of these activities and the fourth asks which of the activities they had been paid for. These all provide interesting information about what people are actually doing, although the way that the results are presented in ABS (2004c) is not particularly helpful. There, funerals and ceremonies are collapsed with sports carnivals and the rest, which means events that, it can be assumed, Aboriginal people get themselves to unaided, are lumped in with festivals and events that are subsidised and organised by State agencies and committees, etc. The former are an index of independent cultural retention, versus the latter which are, of course, cultural, in the sense that they are what people are doing as part of their everyday lives now, but more a measure of what might be called contemporary state-aided social capital.

The next three questions were about kinds of involvement with sport in the last 12 months, with the possibilities being: as player or participant, as coach, instructor or teacher, as referee, umpire or official, as committee member or administrator, or other. The third question asks whether, in the last three months, the person has participated in any of the following activities: recreational group or cultural group activities; community or special interest group activities; church or religious activities; went out to a café, restaurant or bar; took part in sport or physical activities; attended sporting event as a spectator; visited library, museum or art gallery; attended movies, theatre or concert; visited park, botanic gardens, zoo or theme park; or none of these. Rather than being concerned with cultural retention, the focus of these questions seems to be on establishing some of the things people do with their non-work time. While the possibilities offered in respect of sport make good sense, the range of possibilities referred to in the third question do not all seem to be of equal significance, nor the relevance of some, clear. Church attendance does seem important, not least because it is frequently associated, in theory, with a limitation on certain kinds of activity, but mixing café attendance with going to a bar does not seem helpful. As to the list of mainstream cultural activities, the purpose of the questions is again unclear and their possible policy implications obscure.

The next four questions are grouped under the heading 'cultural identification' and include whether the person identifies with a tribal group, a language group, or a clan; whether they recognise a homeland or traditional country and whether they are currently living there; and, finally, whether they are allowed to live there. The identity questions are significant given native title and the changes taking place in the census. Those relating to homelands are, however, ambiguous, since one usage of this term is as an equivalent of outstation, whereas the usage here appears to relate to living on Aboriginal land to which one feels one has a right. Clarifying this distinction would be helpful.

The last three questions on the CAI questionnaire, for which permission was first asked to talk about the subject, are under the heading 'removal from family'. The person is asked whether they were taken from their natural family by a mission, the government or welfare; whether any of their relatives were taken away by the same groups; and which of their relatives were taken away from their natural families. Seven classes of close relative are named, and there are three alternatives of 'other', 'don't know', and 'don't want to answer'. The purpose of these questions and their usefulness in cross-correlating with other issues has been well demonstrated (Borland & Hunter 2000; Hunter 2001).

Changes to survey questions

A considerable number of changes have been made to the questions asked under the section 'Culture'. In respect of the 'Language module' (CAI Questionnaire: p.54), the questions are similar but better phrased. In respect of the 'Cultural participation/involvement in social activities' module (CAI Questionnaire: p.55), the first question is changed by the inclusion of a distinct category for 'Sports carnival'. In 1994 the list of activities was followed by a question about whether and what obstacles prevented people from going to the various activities, with six possibilities and 'Other' listed. The equivalent question is now found in a shortened form in the 'Transport' module (CAI Questionnaire: p.31). The 1994 question about a place to meet for cultural activities has been dropped. In its place are a whole set of new questions about making craft, cultural performance, and whether people were paid for them; about sport and the role taken in it (CAI Questionnaire: p.56—Q02IISA), and about social activities (CAI Questionnaire: p.57-Q03IISA). The questions under the 'Cultural identification' module (CAI Questionnaire: p.58) have been considerably simplified, especially in relation to homelands/traditional country. Now there are just three questions, in place of ten. The question in the 1994 survey on elders has also been removed. Finally, the questions in the 'Removal from family' module (CAI Questionnaire: p.59) have been slightly changed: the initial question remains the same but in place of asking about who brought the removed person up, two questions are asked about whether other close relatives were removed.

Overall, the changes are an improvement. They provide more information on the actual social and sporting activities people undertake, and wasted questions (such as those on elders, on the obstacles to people going to activities and the questions about a place for cultural activities) have been removed. In respect of the various kinds of activities asked about, it seems that better questions could be found to replace those related to libraries, theatres and zoos.

Findings on cultural participation/involvement in social activities

A comparison of the findings between 1994 and 2002 is the principal way in which to give the figures some meaning (see Table 20.1). Broadly speaking, there is little overall change. Despite a small decline in the numbers participating in funerals, attendance at funerals remains at a very high level. An increase in the participation in ceremonies could be seen as heartening, if it is true, but it seems counter-intuitive. On the other hand, the increase in participation in festivals and carnivals, which include sport, seems entirely plausible.

Table 20.1. Cultural participation/ involvement in social activities in 1994 and 2002[a]

	1994	2002
	%	%
Funerals	53.2	46.6
Ceremonies	19.3	23.5
Festivals/carnivals involving arts, craft, music, dance & sport	41.7	45.9
Been involved with any ATSI organisation	23.7	26.1

a. The 1994 figures use the original weights on a population of 13 plus, while the 2002 tables are for a population of 15 plus. For the purposes of this chapter, it is assumed that this difference can be ignored, as it will not substantively affect the distributions.
Source: ABS (2004c: 31)

It would be interesting to divide these figures between remote and non-remote for both periods but that has not been possible to date. However, for 2002 that division has been supplied (see Table 20.2). The differences are marked in respect of funerals, ceremonies and sports carnivals, broadly involving more than double the percentage of people in remote Australia. Interestingly, the proportion of people involved with ATSI organisations is virtually the same, indicating how relations with government are a central part of contemporary Aboriginal life everywhere.

Table 20.2. Cultural participation/involvement in social activities by remoteness, 2002

	Remote	Non-remote	Australia total
	%	%	%
Funerals	74.1	36.3	46.6
Ceremonies	45.0	15.5	23.5
Sports carnivals	52.8	21.2	29.8
Festival/carnival involving arts, craft, music or dance	41.7	33.5	35.7
Involved with ATSI organisations	24.9	26.5	26.1

Source: ABS (2004c: 38)

If the survey is going to be useful in measuring cultural retention through the participation in cultural and social activities, it is important these categories are not changed again.

Table 20.3. Paid and unpaid participation in cultural activities, 2002

	Paid	Unpaid
Arts and crafts	13 993	31 672
Dance or theatre	7510	15 958
Writing or telling stories	7670	27 773
Cultural activities	21 703	55 620

Source: Customised cross-tabulations from the 2002 NATSISS CURF

The 1994 survey failed to ask anything about the arts and crafts industry, so it would seem that the new questions about paid and unpaid participation in cultural activities are to fill that gap (see Table 20.3). It is not clear, however, what to make of the numbers in the unpaid column. Nor is it clear what to make of the commodification of these activities. Art and craft is, one can assume, production for sale and rather different from the 'dance and theatre' category, and the 'writing or telling stories' which presumably relate to the subsidising of public performances, teaching in schools and the like. The last category is not at all helpful, as it is too generalised. But if it means that cultural activities such as funerals and ceremonies are being subsidised, then it could be a concern, because it would indicate bureaucratic intrusion into areas of personal, social and cultural life which people should be able to sustain themselves. If they are unable to do so, it has to be asked what role and purpose the state has in trying to maintain these practices on Aboriginal people's behalf.

Table 20.4. Participation in sporting activity by remoteness, 2002

	Remote	Non-remote	Australia total	Australia total (number)
	%	%	%	
Player or participant	49.2	44.8	46.0	129 864
Coach instructor or teacher	10.1	7.5	8.2	23 240
Referee, umpire or official	9.4	5.0	6.2	17 559
Committee member or administrator	12.3	3.2	5.7	16 089
Other	9.8	1.0	3.4	9609

Source: ABS (2004c: 38)

The detail provided on participation in sport is good, and overtime will provide a measure of the significance of this category of activity in Aboriginal life (see Table 20.4). At present, sport is enormously important, with nearly half the population claiming to be actively involved and over half involved as a spectator.

Table 20.5 shows that data for other activities in the last three months include the nebulous and unhelpful category of 'Went out to a café, restaurant or bar', which lumps alcohol consumption with eating and coffee drinking. What is needed to improve the social and policy significance of the information on participation in sport is the addition of some measure of intensity.

There are some important differences in remote/non-remote participation, particularly in respect of church or religious activities, with 40.7 per cent of remote people participating in such activities while only 17.3 per cent of non-remote people do so.

Table 20.5. Activities participated in during the three months before the 2002 NATSISS

	Australia total
	%
Recreational group or cultural group activities	27.7
Community or special interest group activities	19.5
Church or religious activities	23.7
Went out to a café, restaurant or bar	57.1
Took part in sport or physical activities	37.8
Attended sporting event as a spectator	48.1
Visited library, museum or art gallery	23.1
Attended movies, theatre or concert	32.6
Visited park, botanic gardens, zoo or theme park	32.7
Fishing or hunting in a group	13.9
Not involved in social activities	10.0

Source: ABS (2004c: 38) and customised cross-tabulations from the 2002 NATSISS CURF

Cultural identification

The question of identification with a tribal group, language group or clan produced a surprising result (see Table 20.6). The 1994 NATSIS was right at the beginning of the impact of native title. I predicted that with the impact of native title claims, this kind of cultural identification would grow because, particularly in settled Australia, native title claims are usually built around such identifications, and claims are filed in the name of such groups (Peterson 1996). However, remarkably, there has been a decline of around four percentage points between 1994 and 2002 in the propensity to identify with a tribal group. This may mean that by 2002, native title no longer gripped the imagination of many Aboriginal people.

Table 20.6. Identification with a tribal group, a language group or a clan 2002

	Number	Per cent
Yes	152 806	54.1
No	123 340	43.7
Don't know	6058	2.1

Source: Customised cross-tabulations from the 2002 NATSISS CURF

The most interesting figure in respect of homelands (an ambiguous term) is the number of people who do not identify with a homeland, which is up by 5 per cent from the 1994 Census. If this is correct, it is perhaps an indirect measure, of some kind, of the population of Aboriginal descent that is integrated into the wider society, and linked to the changes in identification so marked in the census. It is not surprising, however, that only 21.9 per cent of people live on their own homeland, since just under one-third of the population lives in cities and over two-thirds in settled Australia (see Table 20.7). Since 1994, there has been a decline in about 7 per cent of the population living on their own homeland, a fact which probably reflects real changes going on across the continent.

Table 20.7. Relations with homeland/traditional country, by remoteness in 2002

	Remote	Non-remote	Australia total
	%	%	%
Recognises homelands/traditional country	85.8	63.4	69.6
Does not recognise homelands/traditional country	14.2	36.6	30.4
Living there now	38.0	15.8	21.9
Not allowed to visit traditional country	0.6	0.5	0.5

Source: ABS (2004c: Table 12)

The decline in people denied visiting access to their country from 49 per cent of remote Indigenous people to virtually zero is remarkable, and suggests that either the original 1994 figure or this figure is wrong.

Cultural and family responsibilities

References to 'family responsibilities or considerations' are found at two points through the 2002 NATSISS questionnaire for non-remote areas (CAI Questionnaire: pp.20, 24). They appear in relation to why people finished school: '20—Other personal/family reasons', or why they are not looking for work (CAI Questionnaire: p.24). More specifically, they appear in the 'Cultural responsibilities' module (CAI Questionnaire: p.20). Here, cultural responsibilities are defined as including, 'such things as telling traditional stories, being involved in ceremonies and attending events such as funerals or festivals'. The question asked was: 'Because you work, is it possible to meet all your cultural responsibilities?'. Pressure to fulfill cultural responsibilities is also listed as a possible source of problems (stresses) for the interviewee or their family and friends (CAI Questionnaire: p.49). The definition of 'cultural responsibilities' used is problematic because it includes both voluntary and involuntary responsibilities. Telling traditional stories or attending festivals would usually be rather more optional than attending a funeral for a close relative.

Conclusion

The figures relating to the declining proportion of respondents with a tribal/clan affiliation and to those identifying with a homeland, raise the question of the extent to which the NATSIS results are affected by the recent dramatic increases in the proportion of people self-identifying as Indigenous. It seems likely that this may be having the effect of adding to that section of the Indigenous population who are closest in their characteristics to the general population. If this is correct, it is surprising because one might expect that one of the reasons for the recent increase in self-identification was that people were seeking to be included in native title claims and, therefore, likely to take on a tribal/clan identification and/or identify with a homeland as part of that. The fact that this does not seem to be the case suggests that those commentators who have argued that the motivation of many who recently self-identify as Indigenous is to take advantage of the alleged material benefits of being Indigenous are wrong, and this underlines the complexity of the identity issues.

The questions in the 2002 survey are, overall, a considerable improvement over those included in 1994. They are more detailed and specific and a number of wasted questions have been removed. Nevertheless, there is an opportunity to improve the questions further and to introduce other questions. In particular, more care needs to be given to separating different kinds of activity, such as those that are organised independently of the state from those that are subsidised, and those which are basically optional from those which are not.

Since 1994 the difficult issue of the role of 'culture' in sometimes aggravating some of the social problems people have been facing has been cautiously opened up for public debate (Atkinson 1990; Brady 2004; Martin 1993; Pearson 2001; Peterson 1999: 856–59; Sutton 2001). The sort of cultural issues addressed in this literature are not, however, easily examined by survey, since they relate to issues such as aspects of sharing practices, child rearing, and ways of relating, which may have implications for areas such as health, domestic violence and material wellbeing. While such matters may not be easily addressed directly by survey, the possibility exists that there may be proxy ways in which these matters can be approached. However, this again may be difficult, and even if correlations were to be shown between the proxy and particular negative outcomes, the policy implications and consequences might be obscure. There are, however, other questions that could be usefully asked from a policy point of view.

In the previous commentary on the 1994 survey (Peterson 1996) it was suggested that the technique adopted by Moisseeff (1994) in her interesting community survey of Port Augusta could be applied here. It would help researchers learn about important issues such as people's priorities, provided the results could be abstracted at a community level. She asked people about their main priorities for the community, offering them 12 possibilities which, apart from providing concrete evidence of people's priorities, produced interesting differences between the weight and ranking given to priorities between men and women. Further, rather than assuming what is significant in terms of cultural maintenance, such as attending ceremonies etc, it would be better to let the interviewees nominate what they see as important in this respect. It would be of greater policy significance if the survey questions were more specific, focusing on information needs and facilities that directly affect people's daily life.

In a more extended analysis, it would be important to cross-tabulate a number of the results from the culture questions with other sections of the survey. Thus, for example, it would be interesting to see if sports participation, living on homelands or church attendance correlates with measures of community wellbeing, age and gender, as might be expected. Some direct policy implication could flow from any such positive correlations.

21. Language

Inge Kral and Frances Morphy

It is well understood that Australia's Indigenous languages are endangered, with even the strongest languages having only some few thousand speakers (McConvell & Thieberger 2003; Schmidt 1990). The NATSISS can provide a process whereby data on language use and rates of language loss are gathered as evidence for the implementation and support of language maintenance programs. In this paper, we discuss the application of NATSISS as such an instrument.

This paper is divided into three main sections. In the first section, the questions on language that were asked in the 2002 NATSISS are discussed. Then the 2002 NATSISS evidence on the status and viability of the Indigenous languages of Australia is reviewed. The final section focuses on the issue of whether Indigenous language speakers differ significantly from other Indigenous people in terms of their education and labour force status.

The language questions in 2002 NATSISS

The 2002 NATSISS questionnaires asked the following language-related questions:

Q01LANG: Which language do you mainly speak at home?

1. English
2. Aboriginal language
3. Torres Strait Islander language
4. Other language

Those who answered 2 or 3 were then sequenced past the next question (presumably on the assumption that the answer was self-evident):

Q02LANG: Do you speak any Aboriginal or Torres Strait Islander languages?

1. Yes
2. Yes, some words only
3. No

The next question was asked of everyone:

Q03LANG: When you go to a service or office where only English is spoken, do you have problems with:

1. Understanding people there?
2. People there understanding you?
3. Neither

The final question in the language module varied in its form between the 'non-remote' and the 'remote' questionnaires. On the 'non-remote', the question was asked of those who had answered either 1 and/or 2 to the previous question. Its form was:

Q04LANG: Do you ever need someone to go with you to help you understand?

On the remote area questionnaire, the final question was asked of those who had answered 3 to the previous question. Its form was:

Q04LANG: Is that because you take somebody to help you understand?

This difference is interesting: it is assumed that a person in a non-remote area will only answer 'neither' if it is really true that they have no difficulty in being understood or in making themselves understood, whereas in remote areas the assumption is that a 'neither' answer does not necessarily rule out the possibility of communication difficulties.

There is one other question in which language featured:

Q01CULT: Do you identify with a tribal group, a language group, or a clan?

1. Yes
2. No
3. Don't know

It is notable that many more people, in all areas of Australia, answered yes to this question than claimed to speak an Indigenous language (IL).

In the 2002 NATSISS glossary (ABS 2004c: 73) Aboriginal and Torres Strait Islander languages are defined as those in the Australian Indigenous languages group of the *Australian Standard Classification of Languages* (ABS 1997). This list includes the Australian creole languages Kriol and Torres Strait Creole. Oceanic pidgins and creoles and 'Aboriginal English' are explicitly excluded in the glossary. Bill Arthur has drawn our attention to language data on Torres Strait Islander language use which calls into question the manner in which this definition was applied in the field. For Torres Strait Islanders who did not speak English at home, particularly those living in Torres Strait, the 'other' category is suspiciously high, and the 'Aboriginal or Torres Strait Islander Language' category is suspiciously low. It looks very much as if Torres Strait Creole was usually categorised as 'other' rather than as a Torres Strait language. It is not possible to check whether something similar happened with Kriol, because its speakers do not form a distinct sub-set of the population in the way that Torres Strait Islanders do. The suspicion is that Kriol was sometimes classified as 'Aboriginal English' (see discussion in section three of the paper). This is likely, given the difficulty some Indigenous people have with self-definition in terms of language 'mainly' spoken. It might also reflect a lack of sociolinguistic knowledge among interviewers. Since Kriol is an Aboriginal language, distinct from English, we must conclude that the number of speakers of Indigenous languages in the Northern Territory and Western Australia is underestimated in NATSISS. Ironically, the two creole languages of Australia are probably, in terms of numbers of speakers, the two largest Indigenous languages of the continent.

The first point to be made is that the questions asked do not correlate well with those that were asked about language use in the 2001 Census, making it difficult to compare the two data sets in any meaningful way. We return to this issue below.

The posing of two questions in the 2002 NATSISS—one on language use at home (asked first) and one about speaking an Indigenous language or languages—is a good idea in principle, for reasons that will be discussed at some length below. However, the alternatives allowed for in question 02 do not necessarily allow for accurate responses. There are many degrees of language competence between full, fluent command of a language and knowing only a few words, and many Indigenous people are likely to fall somewhere on the continuum between the two extremes. It would be better to distinguish fluent from partial command of a language, and between the ability to speak and the ability to 'understand' a language, in addition to the final option of knowing only a few words, as the data give no indication of the numbers of people who may understand an Indigenous language well without actually being able to speak it. If the Indigenous language is not used at home, it might also be useful to have a question regarding with whom speakers use the Indigenous language (see McConvell & Thieberger 2001).

Although questions 01 and 02 appear to be quite straightforward, in remote communities it is not unusual to encounter people who would have difficulty identifying whether they speak either English or an Indigenous language. This is particularly true in situations where the language is shifting and people regularly code-switch depending on who they are talking to at home or in the broader community. For example, even at home, people may code-switch between a strongly English-based dialect when talking to some relatives, then change register and use a more complex version of the Indigenous language when talking to someone else. Language status, identity and context are other factors which predetermine how people would answer questions 01 and 02, irrespective of how fully they speak either language. In addition, the answers may differ according to where and with whom the questions are asked.

These comments are predicated on the assumption that the 2002 NATSISS questions about language are designed to gather sociocultural data on language use and rates of language loss or maintenance. However, it could be conjectured that these questions are designed mainly with a different set of issues in mind, and clues to this are the final questions on communication difficulties. There is an inherent tendency to create problems out of Indigenous language use as a barrier to communication with mainstream agencies, rather than viewing it as a (valuable) component of a person's cultural identity.

Questions 03 and 04 raise further complexities. The underlying assumptions in the questions are too broad. In many service or office contexts, people may have sufficient spoken Standard Australian English (SAE) to comprehend and participate in the everyday 'over the counter' spoken exchanges and would therefore answer 'neither'. The question does not, however, allow for a complex explanation of the range of difficulties that may be encountered when the spoken English becomes more bureaucratic, lexically dense or context-specific, depending on the situation. The questions also do not distinguish between problems with spoken English and problems with literacy. It is quite typical for Aboriginal people from remote communities to have a high level of competence in listening and speaking in English as a second language and be able to participate in complex spoken interactions, yet be unable to read or fill in forms independently. It is more likely that people in this situation would need someone to accompany them to assist with any potential reading and writing difficulties, as much as with understanding spoken English. They may also need help with accessing touch screen information, such as at a Centrelink office.

Is there a future for Indigenous languages?

In his comments on the language questions in the 1994 NATSIS, Nic Peterson considered that the figures 'hold out some hope for the future' of Indigenous languages (Peterson 1996: 153). Since there seems to be no significant difference (see Table 21.1) in the numbers of people reporting that they speak an Indigenous language in the 2002 NATSISS, it is tempting to conclude that this comment still holds. However, the questions asked in both 1994 and 2002 allow for some more detailed scrutiny of people's language use, and lead to a less sanguine conclusion.

Table 21.1. Those who speak an Indigenous language by State, 1994 and 2002

State/ Territory	1994 %	2002 %
Northern Territory	74	76.6
South Australia	23	31.4
Western Australia	21	27.3
Queensland	15	15.2
Total	21	21.1

Source: The 1994 data are from Peterson (1996: 152) and the 2002 data are from ABS (2004c: Table 2)

A language remains viable in the long term only so long as children continue to learn it as their first language. [1] Children will only learn a language as their first

[1] The absolute number of speakers is not necessarily significant, provided that other social factors work in favour of a language's continued existence: in pre-colonial times (as in Papua New Guinea, where

language if it is the language mainly used at home by the adults and older children in the household. Both the 1994 NATSIS and the 2002 NATSISS, importantly, asked not only whether people spoke an Indigenous language but also what language was used mainly at home—see the endangerment index proposed in the State of Indigenous Languages (SOIL) report (McConvell & Thieberger 2001). Table 21.2 tabulates the 2002 NATSISS results by State.

Table 21.2. Those whose main language at home is an Indigenous language, by state, 2002

	State/Territory								
	NT	SA	WA	QLD	ACT	VIC	NSW	TAS	Total
	%	%	%	%	%	%	%	%	%
Speaks an Indigenous language	76.6	31.4	27.3	15.2	10.7	7.8	3.2 [a]	1.2 [a]	21.1
Aboriginal language is main language spoken at home	63.0	12.1 [b]	11.4	1.7 [a]	1.2 [b]	0.5 [b]	0.1 [b]	c	10.9
TSI language is main language spoken at home	0.2 [b]	0.3 [b]	0.1 [b]	3.8 [a]	c	0.7 [b]	0.1 [b]	c	1.2

a. Estimate has a relative standard error of 25% to 50% and should be used with caution.
b. Estimate has a standard error greater than 50% and is considered too unreliable for general use.
c. Nil or rounded to zero (including null cells).
Source: ABS (2004c: Table 2)

Several points can be made about Table 21.2. The first is that data on language use in those areas where very few people speak an Indigenous language is subject to high standard errors and is essentially unusable. In the rest of this discussion, the less fine-grained division between remote and non-remote will mainly be used as the basis for comparison. However, even bearing in mind this proviso, two things are clear. Firstly, it is only in very remote Australia (predominantly the NT and remote SA, WA and QLD) that a significant number of people speak an Indigenous language. This is a reflection of past colonial circumstances across the country, and is not unexpected. The second and more worrying fact is that levels of Indigenous language use in the home are significantly lower in all cases except in the Northern Territory. Even there, 13 per cent of those who speak an Indigenous language are not using it as the main language in the home, and are therefore not passing on that language as a first language to the younger generations. According to the ABS (2004c: 31), the numbers of people who use an Indigenous language at home has not changed significantly since the 1994

one-quarter of the world's languages are spoken) many Indigenous Australian languages would have had no more than 200 speakers (see e.g. Blake 1981: 43).

NATSIS. However, if these figures—both in 1994 and 2002—are a true reflection of language use at home, we can expect to see the numbers of people who claim to speak an Indigenous language beginning to fall in subsequent surveys, as the cohort now aged below 15 years begins to figure significantly in the data.

The 2002 NATSISS (and its predecessor, the 1994 NATSIS) did not collect data at the level of actual languages used, unlike the 2001 Census. Unfortunately, the census question on language does not produce data that is strictly comparable with the data collected in the 2002 NATSISS. NATSISS asks two main questions: whether a person speaks an Indigenous language, and what the main language is that is spoken at home. The census asks only one question that varies according to whether the Special Indigenous form was used or not, and only asks about language use at home. The word 'main' is missing from the census language questions, so that arguably the responses to this question and the 2002 NATSISS question on language use in the home could elicit different answers from individuals living in households where more than one language is commonly spoken.

The census also asks the respondent to give the name of the language spoken at home. The 2002 NATSISS does not—presumably because this fine-grained level of detail would not produce statistically robust results. Because the census questions differ from those of NATSISS, it is unfortunately not possible to use census data on individual languages to determine whether it is all languages that are undergoing death by slow attrition, or whether some languages are thriving, relatively speaking, while others are on their way out. The latter scenario is the more probable. The NATSISS data, because they are not fine-grained, tend to give a superficial impression of gradual decline overall, whereas in reality selective language death probably continues apace. This has obvious policy implications at a local level, but NATSISS data cannot help to pinpoint the areas of difference at that level.

The NATSISS questions provide two other windows on language use and potential changes in use. One is the possibility of correlating the language data with the age data. The results are given in Table 21.3.

Table 21.3. Knowledge and use of Indigenous language by age group

	Age					
	15–24	25–34	35–54	45–54	55+	Total
	%	%	%	%	%	%
Speaks an Indigenous language	18.2	22.3	21.8	19.5	26.1	21.1
Aboriginal language is main language spoken at home	10.0	12.1	10.0	9.7	13.4	10.9
TSI language is main language spoken at home	1.1[a]	0.8[a]	1.6[a]	1.0[a]	1.3[b]	1.2

a. Estimate has a relative standard error of 25% to 50% and should be used with caution.
b. Estimate has a standard error greater than 50% and is considered too unreliable for general use.
Source: ABS (2004c: Table 3)

This is the national picture, and it would have been desirable to break this table down by remoteness as well. At the national level, it is not possible to conclude much from this Table, except that those aged 55+ are more likely to speak an Indigenous language than those in younger cohorts, and that they are also somewhat more likely to be living in a household where the main language spoken is an Indigenous language. As far as it goes, this seems to fit with the other data on the attrition of Indigenous languages.

The other measure of possible language change is given by the additional question where people could report that they 'speak some Indigenous words only'. As noted above, the bipartite division between 'speaks a language' and 'speaks some words only' is a somewhat crude measure of language capacity—there are many intermediate stages between full fluency and the retention of a few words. Despite that caveat, the comparative results from these two questions are suggestive. Table 21.4 shows this comparison, broken down into remote versus non-remote areas.

Table 21.4. Language fluency, by remoteness category

	Remote	Non-remote	Total
	%	%	%
Speaks an Indigenous language	54.2	8.6	21.1
Speaks some Indigenous words only	17.3	23.8	22.0
Does not speak an Indigenous language	28.5	67.6	56.9

Source: ABS (2004c: Table 12)

The remote and non-remote columns display, essentially, two stages in the process of language loss. In non-remote Australia, Indigenous languages are largely defunct, especially if the numbers of people who speak one as a main language at home is a guide to the long-term viability of a language (see Table

21.2). However, a substantial proportion of people (nearly one-quarter) still know some words from one or more Indigenous languages, and in all probability use them in their variety of 'Aboriginal English' or SAE. These words may remain in use for some time to come as a badge of Indigenous identity. Over two-thirds of Indigenous people in non-remote areas do not even claim to know any words from an Indigenous language (they may know some, but may not be aware of their origin). In remote areas, over half of Indigenous people still claim to speak an Indigenous language, a substantial number (17.3%) say they speak some words only, and a more substantial number (28.5%) claim not to speak an Indigenous language. Had NATSISS questions been asked 100 years ago in non-remote Australia, the results might well have been similar to those for remote areas today, and there is a prospect that in 100 years from now the picture for remote Australia will be similar to that for non-remote Australia today.

Are speakers of Indigenous languages different from other Indigenous people?

We have already seen one way in which Indigenous language speakers differ from other Indigenous Australians: they are concentrated overwhelmingly in remote Australia. Are they indistinguishable from other remote Indigenous people, or do they form a distinctive sub-set of that population? In the following comparisons, we treat all Indigenous language speakers as if they were remote dwellers and compare them to the remote population as a whole.

Level of schooling

To be fully interpretable, the populations shown in Table 21.5 would have to be broken down by age. But if that were done, the raw numbers in each cell would make the data statistically unreliable. It would appear that Indigenous language speakers are significantly less likely than the remote population as a whole to complete high school (or even to attend school, since the first column includes those who never went to school).

Table 21.5. Educational attainment: Indigenous language versus remote areas

	Year 9 or below	Year 10/11	Year 12
	%	%	%
Indigenous language speakers	29.8	17.6	15.5
Remote total	46.1	28.2	8.5

Source: Customised cross-tabulations from the 2002 NATSISS CURF

There is a somewhat unexpected reversal shown in the final column, where Indigenous language speakers appear to be more likely to finish Year 12 than the remote population as a whole. Quantifying educational attainment data in remote and very remote Australia is complex and tells us little, unless it is broken down by age and region and also accompanied by sociocultural information pertaining to that region. For example, compulsory schooling came relatively late to many remote Aboriginal regions, so there has been minimal transmission of the culture of formal schooling, and in some places there is still little access to secondary education, making comparisons with other data questionable. Experience has shown that some Aboriginal people in remote Australia have difficulty identifying what year level they completed, as completed credentials are rare and the English nomenclatures may be meaningless for a few reasons. Firstly, the use of the terms Year 9, Year 10, etc may not be commonly recognised, particularly by people who went to high school when the descriptors differed (Form 3, Form 4, etc). Secondly, many people who went to school as teenagers may have been to school up to say age 15, but their actual academic level was most likely much lower than their mainstream age counterparts. Also, in many places teenagers are often grouped together in collective 'post-primary' classes and sometimes leave school not knowing what year level they have attained. See Morphy (2002: 47) for comments on difficulties with the 'education' questions in the 2001 Census in an outstation community in the Northern Territory, and Kral (forthcoming) for self-definition of educational attainment in very remote communities.

A final point from our experience in gathering such data in remote communities is that many people have finished school around age 12 and it is not uncommon for people to answer '12' and for interviewers to erroneously interpret this as Year 12. This alerts us to the linguistic and cultural complexities embedded in the survey itself that cannot be ameliorated simply by using interpreters.

Labour force status

In this section we make a distinction between very remote areas where CDEP is the main employer and remote areas where non-CDEP mainstream employment is available (see Table 21.6). In both remote and very remote areas, Indigenous language speakers are more likely to be on CDEP than those who do not speak an Indigenous language. In remote areas, those who speak only some Indigenous words are intermediate between the other two categories, but in very remote areas they are even more likely to be on CDEP than those who speak an Indigenous language. We return to that slight puzzle below. Overall, people in remote Australia are much less likely to be on CDEP (16.9%) than they are in very remote Australia (42.2%). In both areas, speakers of Indigenous languages are less likely to have a non-CDEP job than are their counterparts who speak only some words or who do not speak an Indigenous language. Rates of official unemployment are generally lower in very remote Australia than in remote Australia, and the lowest rate of unemployment is among Indigenous language speakers in very remote areas. Finally, relatively more Indigenous language speakers in both types of area are 'not in the labour force' compared to their counterparts in the other two categories.

This somewhat complex picture probably results from the interaction of several factors. CDEP clearly has a more prominent role in very remote Australia than in remote Australia, and this is almost certainly partly due to the dearth of non-CDEP jobs. It also seems to take up more of the unemployment 'slack' in very remote areas. Altman, Gray and Levitus (2005: 9), in their discussion of the profile of CDEP participants, note that 'CDEP participants are more likely to speak an Indigenous language than the mainstream employed', and that this holds in all areas, not just in remote and very remote areas. They equate Indigenous language speakers with the category 'more traditionally-oriented' people who have 'strong maintenance of customary practices', and note that CDEP seems to be popular with this category of people. Elsewhere, Altman (2005a) and Arthur (2002) both attribute this to the flexibility of CDEP, which allows people to continue to maintain non-work related aspects of their lives, particularly their ceremonial obligations and their subsistence hunting and gathering activities.

Table 21.6. Whether speaks an Indigenous language (remote versus very remote areas), by labour force status

	Remote			Very remote		
	Speaks an Indigenous language	Speaks only some Indigenous words	Not an Indigenous language speaker	Speaks an Indigenous language	Speaks only some Indigenous words	Not an Indigenous language speaker
Employed: CDEP (%)	25.5	18.4	10.9	41.9	48.6	38.0
Employed: non-CDEP (%)	17.0	36.2	39.0	10.7	20.2*	24.2**
Total employed (%)	42.5	54.6	49.9	52.6	68.8	62.1
Unemployed (%)	10.2	13.6*	8.5	3.8	4.4	6.0**
Total in the labour force (%)	52.7	68.2	58.4	56.4	73.3	68.1
Not in the labour force (%)	47.3	31.8	41.6	43.6	26.7	31.9

Source: Customised cross-tabulations from the 2002 NATSISS CURF

The category 'not an Indigenous language speaker' in very remote areas shows a somewhat unexpectedly large proportion of people on CDEP. We suspect that this is because of the failure to distinguish Kriol speakers consistently as speakers of an Indigenous language. As a result, in very remote areas, the category 'not an Indigenous language speaker' may in fact conceal two very distinct sub-populations. These are local Kriol or Aboriginal English speakers who are on CDEP and speakers of more SAE, some of whom may well be from interstate or have returned to traditional homelands after growing up elsewhere. In some very remote communities CDEP is virtually the only employment option, so the CDEP category embraces all 'workers' irrespective of language spoken or prior educational qualifications.

Conclusion

The Indigenous Australian population is, statistically speaking, quite small. It is characterised by its diversity, which has multiple origins. Firstly, it was always diverse. Secondly, it experienced the colonisation of the continent in very different ways and at different times. Thirdly, the variable effects of social engineering (by missionaries, government officials, industry, successive changes in state policy, and so on) have had significant and sometimes different local impacts. What we have, in effect, is multiple sub-populations. The data on language, at least, suggest that treating the current Indigenous population as a homogeneous entity that is amenable to standard sampling techniques—particularly when the sample chosen is rather small—is perhaps an intrinsically flawed endeavour. This is particularly so when the methodological requirements for sensitivity to the linguistic and cultural complexities may be unattainable. This paper seeks also to show that much of the decontextualised information attained using a survey such as NATSISS is only minimally useful when disassociated from the particular social, cultural and historical factors of each context. A sample survey like NATSISS can throw up questions for further detailed research, but it cannot of itself produce data that is meaningful at a level where it could be the basis for the coherent development of policy. More in-depth and comprehensive information about the status of Indigenous languages should therefore be sought in the Australian Institute of Aboriginal and Torres Strait Islander Studies (AIATSIS)/Federation of Aboriginal and Torres Strait Islander Languages (FATSIL) National Indigenous Language Survey (AIATSIS/FATSIL 2005).

22. Torres Strait Islanders and the national survey model

Bill Arthur and John Hughes

This is the fourth event related to the NATSISS-type surveys that I have participated in. The three others have been a pre-NATSIS workshop in 1992 (Arthur 1992), a post-NATSIS workshop in 1996 (Arthur 1996) and a publication prepared by me and published jointly by the ABS in 1997 (ABS/CAEPR 1997). [1] That previous work followed my principal brief at the CAEPR which was to increase the commitment to provide policy-relevant data and information on Torres Strait Islanders (Islanders). In this role, much of what I said in the former work noted first that Islanders were not generally given much priority in standard ABS publications, and that there were certain conditions specific to Islanders that should be noted and recorded in any social surveys. In this paper I will review the earlier work to estimate the quality and coverage of data on Islanders in the 2002 NATSISS.

The Islander homeland is Torres Strait, a remote archipelago of small islands located on the political and cultural borderland between Melanesian Papua New Guinea (PNG) and Aboriginal Australia. Under Queensland legislation, Islanders were not allowed to move out of the Strait until the 1950s, after which many shifted to the mainland with a degree of enthusiasm. Many moved to improve their access to education, employment and to the generally higher level of services on the mainland. This movement, from a one-spot source, has continued and has been likened to a 'Diaspora'. As shown in Figure 22.1, initially, most Islanders moved to mainland Queensland, [2] and although they later spread to other States and Territories, the greatest numbers are still found in Queensland. [3] Those in Torres Strait live in the small regional town of Thursday Island and its

---

[1] The 1992 and 1996 workshops were both hosted by CAEPR at the ANU (Altman 1992; Altman & Taylor 1996a).

[2] For policy purposes, the region and communities that make up Torres Strait are those within the purview of the Torres Strait Regional Authority. This includes two communities on the tip of Cape York Peninsula (Seisia and Bamaga). These communities largely consist of Islanders and their descendants who were relocated there from islands in the Strait after World War II (Arthur 1990). References made here to Islanders within Torres Strait include those in Seisia and Bamaga, while references made to Islanders on the mainland means all of the others. This, for the reasons of policy noted above, is despite the fact that Seisia and Bamaga are geographically on Cape York (part of the mainland of Queensland and Australia).

[3] Figure 1 also shows that the mainland population increased dramatically following the move out of Torres Strait. It is thought that this can be explained partly by numbers of Islanders forming unions and having children with non-Islanders. Assuming these children identify as Torres Strait Islanders, such 'out-marriage' mathematically increases population growth (Sanders & Arthur 1997). In addition, there is some concern about the accuracy of more recent census data due to the possibility of other island people identifying in censuses as Torres Strait Islanders (see ABS/CAEPR 1997: 29–31).

surrounding small island communities, while Islanders outside Torres Strait reside mostly in urban coastal centres, just like the vast majority of Australia's population. This aspect of the population distribution has implications for socioeconomic status: when compared with mainland urban centres, the formal labour market in the Strait is rather restricted.

Figure 22.1. Torres Strait Islander population, 1880–2001: a contextual note

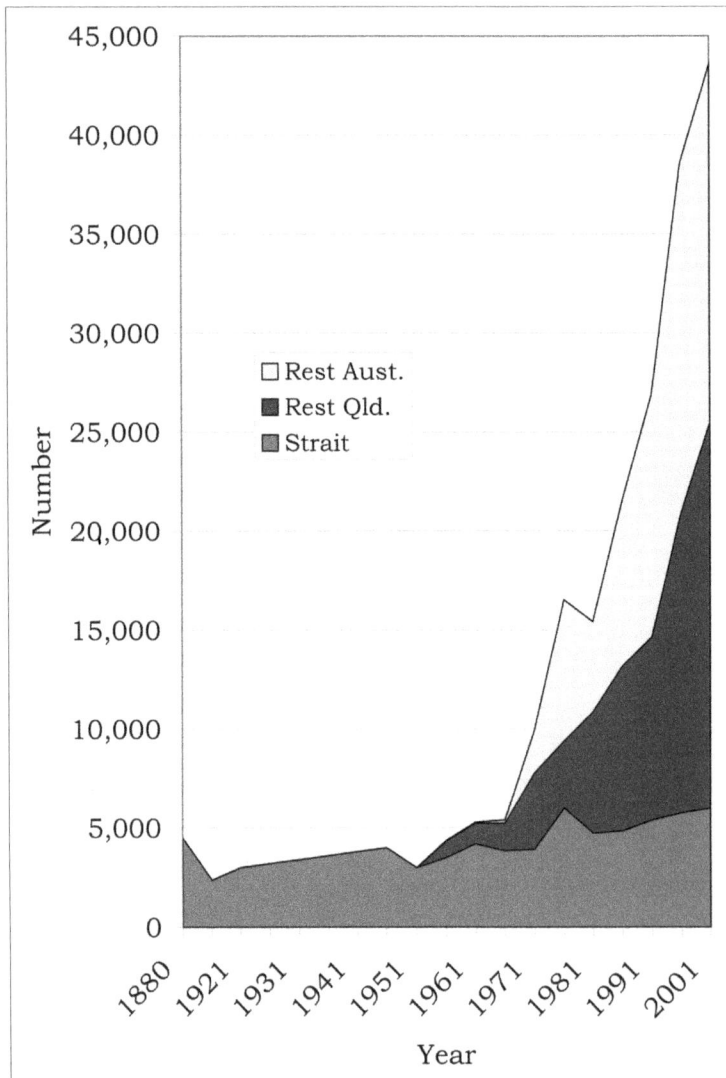

Source: Census of population and housing, various years.

Other work, largely (but not only) based on census data, has revealed several policy-related factors regarding Islanders, including that:

- only in Queensland are Islanders given any particular or Islander-specific attention in that State government's programs (Arthur 1998)
- Islanders in Torres Strait have a different socioeconomic status from those in the rest of Queensland. This is largely because, as noted above, they live in different kinds of places—a remote archipelago and urban centres (Arthur & Taylor 1994), and
- Islanders seem to have had a different socioeconomic status from Aboriginal people at a national level (Taylor & Gaminiratne 1992).

The historical distribution of Islanders, together with the policy and socioeconomic factors noted above, suggest that it would be valuable to obtain policy-relevant data on Islanders for particular groupings. Using data from the 2001 Census, these groupings and their size are shown in Tables 22.1 and 22.2. The groupings are Islanders in Torres Strait, Islanders on the mainland of Queensland, and Islanders in other States. We should note that these populations are quite small: there are only 6000 Islanders in Torres Strait, 19 000 on the Queensland mainland and 18 000 in all of the other States/Territories, and this makes them rather hard to survey adequately. Also, whereas in Torres Strait, Islanders are the majority, in all other places they are a small part of the Indigenous population. Overall, as we can see, Islanders are only some 11 per cent of the total Indigenous population.

Table 22.1. Relevant Islander population groups, 2001

Population groupings	Number	Per cent of total Islander population
Islanders in Torres Strait	6000	14
Islanders on the mainland of Queensland	19 450	44
Islanders in other States and Territories	18 124	42
All Islanders	43 574	100

Source: 2001 Census

Table 22.2. Relevant Islander and Aboriginal population groups, 2001

Population groupings	Number	Per cent of total Australian Indigenous population
All Islanders	43 574	11
All Aboriginal people	366 429	89
All Indigenous people	410 003	100

Source: 2001 Census

I will now comment on some aspects of the 1992 pre-NATSIS workshop, the 1996 post-NATSIS workshop, and the 2002 NATSISS. Bearing in mind what I have said about the significance of the distribution of Islanders, this will be my principal interest.

The 1992 pre-NATSIS workshop

In 1992 at the pre-NATSIS workshop, I proposed that factors of culture, socioeconomic status and history legitimised the view that Islanders are a distinct group that should be viewed separately from Aboriginal people (Arthur 1992). I noted, however, that the ABS did not generally produce standard publications that were dedicated solely to Islanders, or that allowed us to compare Islanders with Aboriginal people (Arthur 1992: 64). I noted that Islanders in Torres Strait obtained their income from a mix of sources: commercial fishing, welfare, and subsistence (what Jon Altman later characterised as a 'hybrid economy' (2002)) and that it would be good if NATSIS could find out more about each element of this income (Arthur 1992: 62). I also proposed that, unlike the situation in mainland Aboriginal communities, in Torres Strait the dinghy was probably a more important work tool than the car and so any survey should attempt to find out more dinghy running costs, rather than car costs (Arthur 1992: 62). Given what we knew about language use in Torres Strait, I suggested that proficiency in oral communication should be determined there with some reference to the use of Creole (Arthur 1992: 64–5). However, as something of a 'catch-all', I noted that it was likely that most of what we wanted to know or needed to know about Islanders could probably be gleaned from the five-yearly censuses (Arthur 1992: 60) and, indeed, this remains very much my position today (see below).

The 1996 post-NATSIS workshop and publication

The 1992 workshop was followed by a post-NATSIS workshop in 1996. At this workshop I noted that the 1994 NATSIS had provided some new data across many of the survey categories, such as those for health, family and culture, law and justice (Arthur 1996: 170). However, I also said that the 1994 standard publications did not provide any data specifically for Islanders. Rather, it gave data on Indigenous people in Torres Strait. On the other hand, as the vast majority of Indigenous people in the Strait are, in fact, Islanders, this represented a de-facto measure of Islanders in Torres Strait and was therefore quite adequate (Arthur 1996: 166–7).

To improve the Islander-specific findings from the 1994 survey, the ABS and I collaborated on a publication dedicated solely to data on Islanders. However, because of the sample size, this work could only deal with Islanders in Torres Strait and in the rest of Queensland (Arthur 1996: 166–7). This publication looked at several of the NATSIS categories (see ABS/CAEPR 1997).

Also, the 1994 NATSIS did not provide data on Islanders in States other than Queensland (Arthur 1996: 170) Therefore, a comparison between the characteristics of the Islander population in Torres Strait and that in the rest of Australia could not be made. Nor did the data allow one to compare fully all Aboriginal people with all Islanders.

The 2002 NATSISS

The 2002 NATSISS provided the following data. For the standard survey categories, the ABS (2004c: 20) allows us to compare the total Australian Aboriginal population with the total Torres Strait Islander population. This allows the national comparison of Islanders and Aboriginal people that I requested in the earlier work. The data are not provided at the State and Territory level, but I am less certain that this level of analysis is really required. As noted above, in many cases the census data are adequate.

For slightly fewer categories in the 2002 NATSISS, ABS (2004c: 50) allows us to compare:

- Islanders in the Strait with Islanders in the rest of Queensland, and
- Islanders in the Strait with Islanders in all of the rest of Australia (that is, all of the mainland).

This provides the data requested earlier and allows an adequate, if not a good, comparison of the Torres Strait and the national Islander populations. The 2002 NATSISS does not provide results for Islanders in State and Territories other than Queensland, but again these may not really be necessary and in most cases census data will probably suffice for policy purposes.

Some results from 2002 NATSISS

Figure 22.2. Selected characteristics for Islanders living in the Torres Strait and rest of Australia

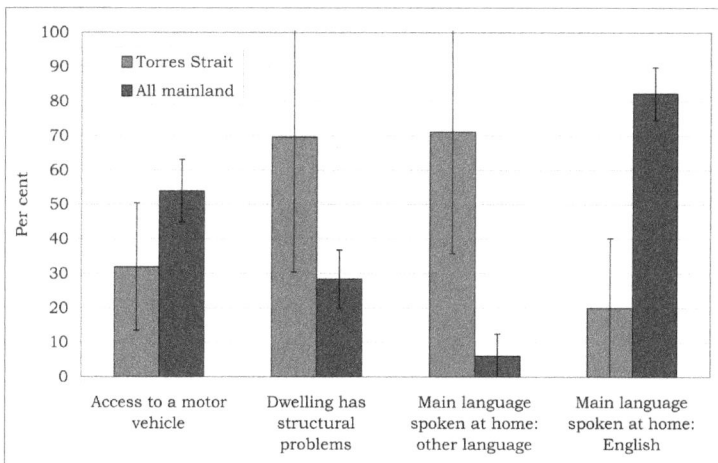

Source: ABS (2004c: Table 23).

Figure 22.2 shows several results from the 2002 NATSISS comparing Islanders in Torres Strait with those in the rest of Australia. We see that on the mainland, the main language spoken at home was English while in the Strait it was an 'other' language, in this case Creole. As we are aware that Islanders are engaging with mainstream society on the mainland, the different use of language in the two locations is exactly what we would expect. As regards rental housing, houses in Torres Strait were more likely to have structural problems than were those on the mainland and, again, this is what we would expect as most housing in the Strait is community rental housing located on small islands. This is generally of a lower standard than public rental housing on the mainland. Regarding access to a motor vehicle, although the data in Figure 22.2 are probably correct, in Torres Strait, as I noted 1992, the more important data would be about access to a dinghy and the 2002 result on motor cars may not be very relevant. On the other hand, while some more data on access to dinghies might be relevant, it is probably quite easy to obtain for other sorts of surveys that focus directly on Torres Strait.

Figure 22.3. Selected comparisons between Aboriginal Australians and Torres Strait Islanders

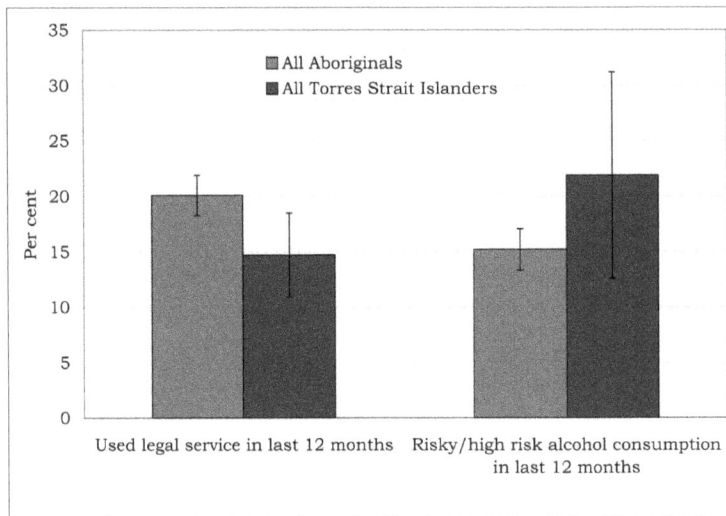

Source: ABS (2004c: Table 1).

Figure 22.3 shows several results from the 2002 NATSISS comparing all Islanders with all Aboriginal people nationally. The data here suggest that Islanders are 'riskier' drinkers than Aboriginal people. This seems counter-intuitive given what we already know about Aboriginal and Torres Strait Islander drinking habits (see Brady & Chikritzhs in this volume, and Arthur 1996) and I am unable to explain this apparent anomaly.

The data also show that Islanders use the legal service less than Aboriginal people. It is worth noting here that although the imperative for the 1994 NATSIS came from recommendations of the Royal Commission into Aboriginal Deaths in Custody, the Commission's reports did not list any Islanders as having died in custody. This was despite the fact that overall, Queensland had the second highest number of deaths in custody of any State or Territory (Arthur 1992: 64). I can also add that an Islander working in a Queensland legal service office told me that he had not dealt with a single Islander client in his time there. Therefore, the NATSIS data may well reflect the situation on the ground.

Table 22.3 summarises the various survey type events discussed above, indicating the extent to which data requested have been supplied.

Table 22.3. The survey events and data requested: 1992 to 2002

Data requested in 1992 for 1994 NATSIS	Data provided in 1994 NATSISS	Data requested in 1996 post-NATSISS	Data provided in 2002 NATSISS	Comments
Dedicated Islander data	No	Dedicated Islander data	Yes	Adequate
Data on income sources	No		No, but too hard for a national survey—better by another technique	Not relevant for a national survey
Data on dinghies	No		Probably not, but possibly better from other surveys	Adequate
Data on Creole	Yes		Yes	Adequate
		Data for other States	No, but probably data from censuses will suffice	Adequate
		Data for the mainland	Yes	Adequate

Conclusion

This paper has not been a full study of the 2002 NATSISS data. However, a brief analysis of the data indicates that they reveal little that is unexpected. The type of data collected and the geography of the collection and its publication—the geography of the statistics—seem quite adequate. There are still no data for Islanders in States other than Queensland, but are these really required? There are not many Islanders, and they are a small part of the Australian Indigenous population, making them quite hard to pick up in a national Indigenous survey. Other work has analysed the socioeconomic status of Islanders in other States from census data and it is likely that, as predicted in 1992, these will be prove adequate for policy purposes. Where census data are inadequate, it is also likely that another form of survey would better capture the features of such a small population. In addition, it worth recalling that the rationale for the NATSIS type surveys was the Royal Commission into Aboriginal Deaths in Custody and this rationale may have had less relevance for Islanders than for Aboriginal people. Therefore, it is possibly time to reassess the purpose of the NATSISS type of survey against current issues. In doing this, some attention should be given to demarcating the potential for the censuses to provide adequate data. In considering the potential for NATSISS-type surveys to address current policy issues, a relevant priority of questions might be: In what subject areas is the census adequate? Which of the remaining subject areas can be addressed by a national survey? By which means can the balance of the information needed (if any) be collected?

23. Social justice and human rights: using Indigenous socioeconomic data in policy development

Tom Calma

The perspective that I intend to bring to this discussion is a human rights one. I want to reflect on the importance and utility of Indigenous socioeconomic data in contributing to *improved enjoyment of human rights* by Indigenous peoples in Australia.

Generally speaking, in Australia we have not converted our international human rights obligations into domestic law and practice very well. This is particularly the case in relation to economic, social and cultural rights, such as the right to health, housing, education and so forth. As a result, human rights standards do not enjoy prominence as a tool in planning policy or in holding governments accountable in Australia.

Despite this, there are significant developments occurring in the human rights system worldwide which demonstrate the suitability of a human rights framework for advancing policy development and which ultimately can guide improvements in Indigenous socioeconomic outcomes.

What I intend to do today is to identify some recent developments at the international level relating to human rights and outline a human rights framework for addressing Indigenous socioeconomic status. I will then consider this in light of recent developments in Indigenous policy making in Australia, and make some comments about:

- the adequacy of existing processes for assessing Indigenous socioeconomic outcomes
- how these processes relate to Australia's human rights obligations and how adequate they are in holding governments accountable for the commitments they have made through the COAG, as well as through the *new arrangements* at the federal level, and
- future directions for data collection and the identification of research priorities for analysing Indigenous socioeconomic outcomes from a human rights and social justice perspective.

Recent international developments—the human rights context

I want to begin at the international level. Recent years have seen significant discussion at the United Nations (UN) on how to better integrate human rights into its everyday activities. There are currently three main, inter-related developments.

Firstly, there is the focus on the need to reform the operations of the UN itself. Debate on this issue is currently focused around the report by Kofi Annan, the Secretary-General of the United Nations, which is titled, *In Larger Freedom: Towards Development, Security and Human Rights for All.*

This report proposes some radical change to the structures of the UN, from the Security Council through to replacing the Commission on Human Rights. The report will form the basis of the summit of world leaders taking place in New York in September this year. The Summit will assess progress five years down the track in implementing the UN Millennium Declaration and in meeting the targets set in the declaration, known as the Millennium Development Goals.

The report of the Secretary-General is focused on achieving better integration of the purposes of the UN as laid down in the UN Charter—namely, peace and security, development, and human rights. As the Secretary-General's report (Annan 2005: para 17) states:

> Humanity will not enjoy security without development, it will not enjoy development without security, and it will not enjoy either without respect for human rights. Unless all these causes are advanced, none shall succeed.

The report proposes a speeding up of a process that has been underway for at least the past decade; of integrating human rights into the everyday operations of the UN. The process is being termed 'mainstreaming'. It envisages a significantly expanded role for the Office of the High Commissioner for Human Rights, as well as making human rights issues central across all areas of UN activity, with a particular focus on the rule of law, democratic processes and addressing extreme poverty.

The *second* development is the increased focus and commitment of the UN on poverty eradication. This is exemplified by the Millennium Development Goals [1] and the commitments of all governments to achieve these goals by the year 2015.

---

[1] The goals are: eradicate extreme poverty and hunger; achieve universal primary education; promote gender equality and empower women; reduce child mortality; improve maternal health; combat HIV/AIDS, malaria and other diseases; ensure environmental sustainability; and develop a global partnership for development.

The focus of these goals is very much centred on *developing nations*. The usual context in which the involvement of countries like Australia is discussed is in relation to international aid, technical assistance and debt relief. But the implications of this focus on poverty eradication clearly relate to the situation of Indigenous peoples in Australia.

As part of the UN reform process, a large part of the focus on integrating human rights into UN activities more generally has been on improving the human rights system. In her report to the UN General Assembly for the upcoming summit in September 2005, the High Commissioner for Human Rights (Louise Arbour) has set out how she envisages the human rights system can be used to address the challenge of poverty eradication. Her comments are directly relevant to the challenges in improving Indigenous socioeconomic outcomes in Australia. Arbour (2005) states:

> We will focus on promoting rights-sensitive poverty reduction strategies and the application of rights-based approaches to development, and advancing the right to development. In doing so, we will emphasise the free, active and meaningful participation of rights holders, the accountability of duty bearers, non-discrimination at all levels and the political and economic empowerment of those that development seeks to lift out of poverty. In addition, we will use the human rights framework to buttress and solidify pledges made by the richer countries.

Significant work has already commenced with the purpose of elaborating the linkages between human rights, poverty eradication and development. Past social justice and native title reports have highlighted this work, such as the extensive focus on human rights by the United Nations Development Programme (UNDP) (including through its annual Human Development Reports), increased focus on the right to development, and also through the drafting of guidelines on human rights and poverty eradication by the High Commissioner for Human Rights and the UNDP. I will refer to this work in more detail shortly.

The *third* development internationally, which complements these issues, is that we are also facing an increased level of consideration of issues specifically relating to Indigenous people in the UN system. This is very much a result of the inaugural four years of work of the Permanent Forum on Indigenous Issues. This forum, to which Mick Dodson was elected as the representative of the Indigenous peoples of the Pacific region last year, has been very successful in highlighting the plight of Indigenous peoples at the highest levels of the UN.

At its *fourth* session earlier this year, for example, the forum considered the application to Indigenous peoples of the Millennium Development Goals relating to the provision of universal primary education and the eradication of extreme poverty. In its report to the Economic and Social Council, the forum stated that

'Indigenous and tribal people are lagging behind other parts of the population in the achievement of the [Millennium Development Goals]…in most, if not all, the countries in which they live' and called for greater attention to the situation of Indigenous peoples (Permanent Forum on Indigenous Issues 2005: paras 4–6).

Indigenous peoples from so-called 'first world' countries such as the United States of America, Canada and Australia have consistently voiced their concerns at the Permanent Forum about the extent of poverty and marginalisation in their communities and the need for increased international attention to this situation. In doing so, they have challenged the traditional focus of UN activity being solely on the developing world.

The Permanent Forum has also led developments in the UN to improve systems for collecting and disaggregating data relating to Indigenous peoples at the international level. In January 2004 they convened an international workshop on data collection and disaggregation for Indigenous peoples (Permanent Forum on Indigenous Issues 2004). This workshop identified the need for full participation of Indigenous peoples in all stages of data collection and analysis as an essential component of emerging participatory development practice.

So what does all of this mean?

The main point to take from this overview is that we are seeing a significantly increased focus in the UN system on the situation of Indigenous peoples and on poverty eradication and development as human rights issues. The normative frameworks of the human rights system are being increasingly applied to these issues.

As a result, there is increased attention to how governments go about addressing poverty within their borders, as well as how they participate in international cooperation efforts, and increased scrutiny of government's performance on these issues. For Australia, this has the following implications:

- Firstly, we can expect that the 'Achilles heel' that is the poor performance of successive Australian governments in addressing Indigenous marginalisation will receive increasing attention internationally and this attention will broaden in its focus as human rights approaches to poverty and development become 'mainstreamed' across the international system.
- Secondly, we can expect greater scrutiny of our policy approaches to addressing poverty and development among Indigenous peoples as a matter of compliance with our human rights obligations.

A human rights approach to assessing Indigenous socioeconomic outcomes

So, then, what does a human rights approach to assessing Indigenous socioeconomic outcomes entail?

A rights-based approach includes the following elements:

- it emphasises the accountability of governments for socioeconomic outcomes among different sectors of civil society by treating these outcomes as a matter of legal obligation, to be assessed against the norms established through the human rights system
- it emphasises the process for achieving improvements in these outcomes, with the free, active and meaningful participation of affected groups being critical. There is emerging—through international law—a broader acceptance of this under the heading of free, prior and informed consent
- it establishes fundamental principles to guide policy development (e.g. that disadvantaged groups should not be discriminated against and should be treated equally), and
- it requires governments, working in partnership with affected groups, to demonstrate that they are approaching these issues in a targeted manner, and are accountable to the achievement of defined goals within a defined timeframe.

This latter point is known as the 'progressive realization' principle. This principle is set out in Article 2 of the International Covenant on Economic, Social and Cultural Rights. This is one of the main international human rights treaties to which Australia is a party.

Article 2 (1) of the covenant requires Australia: to take steps '…to the maximum of its available resources, with a view to achieving *progressively* the full realization of the rights recognised in the present Covenant by all *appropriate* means' (emphasis added).

The committee overseeing the covenant has commented that while the obligation 'to take steps' means that the full realisation of relevant rights may be achieved progressively, the taking of such steps cannot be delayed and, further, those steps should be deliberate, concrete and targeted as clearly as possible towards meeting the obligations recognised in the covenant (Committee on Economic Social and Cultural Rights 2004: para 2).

The High Commissioner for Human Rights has described this principle and its relevance to policy making as follows:

> Since the realization of most human rights is at least partly constrained by the availability of scarce resources, and since this constraint cannot be eliminated overnight, the international human rights law explicitly allows for progressive realization of rights… While the idea of progressive achievement is common to all approaches to policy making, the distinctiveness of the human rights approach is that it imposes certain

conditions on the behaviour of the State so that it cannot use progressive realization as an excuse for deferring or relaxing its efforts. [2]

Accordingly,

> The idea of progressive realization has two major strategic implications. First, it allows for a time dimension in the strategy for human rights fulfillment by recognizing that the full realization of human rights may have to occur in a progressive manner over a period of time. Second, it allows for setting priorities among different rights at any point in time since the constraint of resources may not permit a strategy to pursue all rights simultaneously with equal vigour (High Commissioner for Human Rights 2004: 22).

This approach requires that governments identify appropriate indicators, in relation to which they should set ambitious but achievable benchmarks, so that the rate of progress can be monitored and, if progress is slow, corrective action taken. Setting benchmarks enables government and other parties to reach agreement about what rate of progress would be adequate. Such benchmarks should be:

- specific, time-bound and verifiable
- set with the participation of the people whose rights are affected, to agree on what is an adequate rate of progress and to prevent the target from being set too low, and
- reassessed independently at their target date, with accountability for performance (UNDP 2000: 101).

My predecessor, as Social Justice Commissioner, has elaborated on this rights-based approach in the context of addressing Indigenous disadvantage. In particular, he identified five integrated requirements that need to be met to integrate a human rights approach into redressing Indigenous disadvantage and to provide sufficient government accountability. These requirements are:

- making an unqualified national commitment to redressing Indigenous disadvantage

---

[2] See Hunt, Osmani & Novak (2004: paras 19–20). These conditions are: 'First, the State must take immediate action to fulfill any rights that are not seriously dependent on resource availability. Second, it must prioritize its fiscal operations so that resources can be diverted from relatively non-essential uses to those that are essential for the fulfilment of rights that are important for poverty reduction. Third, to the extent that fulfilment of certain rights will have to be deferred, the State must develop, in a participatory manner, a time-bound plan of action for their progressive realization. The plan will include a set of intermediate as well as long-term targets, based on appropriate indicators, so that it is possible to monitor the success or failure of progressive realization. Finally, the State will be called to account if the monitoring process reveals less than full commitment on its part to realize the targets'.

- facilitating the collection of sufficient data to support decision-making and reporting, and developing appropriate mechanisms for the independent monitoring and evaluation of progress towards redressing Indigenous disadvantage
- adopting appropriate benchmarks to redress Indigenous disadvantage, negotiated with Indigenous peoples, State and Territory governments and other service delivery agencies, with clear timeframes for achievement of both long-term and short-term goals
- providing national leadership to facilitate increased coordination between governments, reduced duplication and overlap between services, and
- ensuring the full participation of Indigenous organisations and communities in the design and delivery of services (UNDP 2000: 93).

The Council for Aboriginal Reconciliation's (CAR's) final report to Parliament, *Australia's Challenge* also commends this approach. Recommendation 1 calls on:

> ...the Council of Australian Governments to agree to implement and monitor a national framework for all governments and ATSIC to work to overcome Indigenous disadvantage through setting benchmarks that are measurable, have timelines, are agreed with Indigenous peoples and are publicly reported (CAR 2000).

To date, this recommendation has not been fully implemented.

Recent developments in Indigenous policy in Australia

So, let me now turn to the situation in Australia and apply this approach. We have just seen the first twelve months of what are generally referred to as the *'new arrangements'* for the administration of Indigenous affairs at the federal level. This has involved:

- the abolition of ATSIC
- the transfer of Indigenous-specific programs to mainstream government departments and their continuation alongside mainstream programs
- the creation of a central OIPC with regional ICCs which are intended to deliver services on a whole-of-government basis
- the introduction of direct negotiation with Indigenous communities or sections of communities through SRAs
- the proposed introduction of regional agreement-making processes, to be known as Regional Participation Agreements, and
- stronger leadership within the mainstream of government with their oversight by a ministerial taskforce and a working group of secretaries of government departments.

I have discussed these new arrangements at length in the *Social Justice Report 2004* and also indicated in that report how my office will continue to monitor them over the coming years. In my forthcoming *Social Justice Report*, which should be released in the latter part of this year, I am focusing on the SRA process, issues relating to representation and participation of Indigenous peoples in these new arrangements post-ATSIC, and the links between a rights-based approach to Indigenous health and the new arrangements.

The new arrangements are supplemented by commitments at the inter-governmental level through the COAG. Through successive communiqués on reconciliation and Indigenous affairs since 2000, COAG has committed to:

- eight trials of whole-of-government activity in Indigenous communities
- the development of a whole-of-government reporting framework on Indigenous disadvantage with biennial reports on progress, and
- a series of principles for service delivery to Indigenous communities.

The new arrangements and the COAG commitments are also now beginning to be fused together with the signing of bilateral agreements on Indigenous affairs between the Commonwealth and States or Territories. At this time, there is only one agreement formally in place, with the Northern Territory. However, it is expected that other agreements will follow in the coming months. These agreements, I understand, will build on the COAG commitments and see the relevant State or Territory agree to work with the Commonwealth on its new arrangements and in identified priority areas.

Overall, these changes are large in scope and ambitious. I remain of the view that there is still a lack of information about the new approach among Indigenous peoples and communities, and significant challenges to ensure that Indigenous communities can participate fully in these processes, and also do so equitably. It is clear that it will take several years before the new arrangements are fully in place.

Implications for data collection and research

Any assessment of the new arrangements needs to acknowledge their evolving nature. In doing so, it must be said that they raise many challenges for policy makers, Indigenous communities and researchers in how we assess Indigenous socioeconomic outcomes.

Up front, it must be said that the new arrangements involve the making of significant commitments to Indigenous peoples and communities. They are based on the idea of 'matched up' government or whole of government activity. In theory, this can deliver substantial benefits to Indigenous people. These benefits may include improving the accessibility and use of mainstream services and, accordingly, potentially opening up wider sources of funding. Or they might include simplifying administrative processes for communities in their dealings with government, to provide more sustained funding processes and better target actual need in communities.

The *rhetoric* of the new arrangements addresses many of the significant concerns that have been raised in successive reviews and reports, including that of the Commonwealth Grants Commission on Indigenous funding. These new arrangements are also matched by substantial commitments through COAG to address Indigenous disadvantage.

But what is trying to be achieved is hard work and it also has to be acknowledged *that it could go wrong*. While there is still no formal evaluation of the COAG trials, for example, we know that these trials cannot possibly be replicated nationally due to their resource-intensive nature. Whole-of-government activity is also notoriously difficult and potentially very expensive. If not done well, it could result in greater administrative costs and processes, cost and blame-shifting between jurisdictions and within government, a blurring of responsibilities, and a lack of transparency about process and outcomes.

The new arrangements also involve the rejection of using representative Indigenous structures to interface with Indigenous peoples and communities and proposes a model of direct engagement by government through agreement-making processes. This is not the preferred model of many Indigenous people and it is untested to date. We also do not know how this aspect of the new arrangements will work, and if it is not implemented from a community development/capacity development perspective, then sustainable and tangible outcomes may not be achieved.

So the 3.1 billion dollar question is: how will we know whether the new arrangements are delivering improved outcomes or not?

Outcomes for the period since the new arrangements have been in place will not show up in data collections and analysis for at least another two to four years. We are unlikely to see analysis of the 2006 Census until 2007 or 2008, and analysis of the next NATSISS until at least 2008. The latest report on overcoming disadvantage by the SCRGSP that was released in July this year reflects on data that pre-dates the new arrangements on most indicators. So it will not be until 2007 that we see any data which is compiled in accordance with the commitments of COAG and reported in a holistic manner that relates to the new arrangements.

It also remains to be seen how governments at all levels will link their programs and activities to achieving improvements in the headline indicators and strategic change indicators contained in the COAG reporting framework on Indigenous disadvantage.

There is also a disaggregation issue. It is not easy to manipulate current data to identify regional trends and variations. Given the reliance of the new arrangements on regional approaches, with coordination through regional ICC offices, the proposed use of Regional Participation Agreements for structuring regional representation and priority setting, and the continuation of COAG trials in select regions, being able to disaggregate to this level is very important for being able to establish the success or otherwise of the new approach.

And in terms of the intended focus of the new arrangements on increasing the accessibility of mainstream services for Indigenous people and communities, there is currently not even any way to identify what mainstream funding is being expended on Indigenous issues. If we cannot even say what services are or are not being utilised then we certainly cannot assess whether the services provided are culturally appropriate, accessible, available or adapted.

By the way, these latter terms—appropriate, available, accessible and adapted—have specific meanings within a human rights framework which have been elaborated in relation to the right to the highest attainable standard of health, housing, education and so forth. [3] They can be used to structure analysis on the adequacy of programs to address economic, social and cultural rights. When you add these concepts to that of progressive realisation and benchmarking, you can see how a human rights approach can offer a powerful, structured and targeted approach to these issues.

What these factors indicate to me is that it is not going to be easy for us to establish what works and what does not in the new arrangements, nor to identify the impact of these processes on Indigenous socioeconomic outcomes. This will particularly be so at the regional and sub-regional level.

A related issue here is how SRAs will link to existing data collection. This is not clear at present, although the SRA approach is a staggered one with initial agreements intended to be single issue/simple agreements and forthcoming agreements intended to be more holistic and integrated in their focus.

Now, in saying this, I am identifying the challenges that lie ahead and hoping to promote informed discussion about ways to address this. What I have said is not just for the sake of it.

There is movement to address some of these issues from the government. One of the most interesting processes currently underway in relation to these issues is the Australian Government Indigenous Management Information System (AGIMIS) project. The OIPC describes this project as follows:

> The main objective of the AGIMIS Project is to develop an Indigenous management information system. This will assist to support the long-term policy, program implementation and reporting requirements of the 'joined

---

[3] See, for example, the General Comments of the Committee on Economic, Social and Cultural Rights (2004).

up', whole of government approach to Indigenous funding, program performance monitoring and reporting.

The AGIMIS Project will collect data and provide reports to monitor investment by Government, initially on Indigenous-specific activity and at a later stage on mainstream services accessed by Indigenous people. The information will allow input to the measurement of overall outcomes and the assessment of effectiveness and efficiency of programs. [4]

As I understand this project, it involves establishing a database that can ultimately identify what programs are active in what communities, what services they are providing, and so on. It will draw data down from existing systems maintained by individual agencies and departments. And, ultimately, it will include a mix of information about Indigenous-specific services—the initial focus of the project—and, later on, mainstream services.

The prime challenge for this project is consistency and compatibility of data. The *potential*, however, is a powerful tool for identifying the nature and scope of government activity on a local and regional basis. It could be a useful tool for advancing the proposals first made by the ABS and the CGC for ranking Indigenous need on a regional basis.

Clearly, a significant challenge will be allowing for an interface between this data on government activity and available data collections on Indigenous socioeconomic outcomes. I am not sure how widely OIPC are consulting in developing this tool—whether there is any discussion with Indigenous communities about it or whether it is being kept as internal to government. But there are the various concept papers and reports of the project on the OIPC web site. This project offers much potential in addressing some of the need that exists for better data.

In terms of disaggregation, there is a further process underway that may provide some ways forward. In this year's budget, the government announced a new 'Healthy for life' program totalling $113.6m over the next five years. This initiative involves, among other things, the establishment of a number of 'healthy for life' sites providing primary health care interventions. Each site will be subject to a formal evaluation process and has benchmarks set for the life of the program. These include halving incidence of low-birth-weight babies within five years. My office will be interested to see how people collect the data to establish whether these benchmarks are met.

I also note that the budget included a measure titled 'Indigenous communities—developing a twenty-to-thirty-year vision'. This is intended to 'assist Indigenous communities to develop their long-term vision and aspirations

---

[4] See: http://www.oipc.gov.au/AGIMIS/default.asp

and identify what is required to achieve these goals'. The funding for this process is negligible—$1 million for each of the next two years. I have no further information about the project. I do not know if this involves a detailed needs assessment and projection of community need in the context of population projections for individual communities, such as the work done by John Taylor in Wadeye (Taylor 2004b). This is something my office will also be interested to watch.

Conclusion

In conclusion, I know that I have probably raised more issues than I have provided answers for today. In particular, I have highlighted the importance of a benchmarking approach and an approach that is built on engagement and participation of Indigenous people at all stages of the process. My point is that we are currently a long way from this situation. We do not have a system for saying, when all variables and people's desires are taken into account, what is a reasonable and acceptable rate of progress. We do not have a system which identifies what is an acceptable Indigenous socioeconomic outcome and, hence, whether government is meeting its responsibilities well.

I will be providing some ways forward on this in my forthcoming *Social Justice Report* in terms of setting forward a twenty-year agenda for addressing health inequality. I hope this agenda will be the focus of debate and attention over the next twelve months.

But clearly, data collection is critical as a tool for accountability and for enhancing Indigenous participation. We need to work with government to improve the existing systems and to build on the initiatives they have in place. I hope that my contribution has emphasised how important these issues are and the priority that government should place on them.

But we also have to look beyond government. In a climate where the new arrangements are not fully locked into place and where issues such as data collection and monitoring and evaluation processes remain very much under development or consideration, it is incumbent on researchers, Non-Government Organisations and service deliverers to ask what we can do to supplement government processes and document and evidence the success or otherwise of these processes. There has never been a more important time for this research to occur.

24. Influencing Indigenous policy making with statistics

Jon Altman and Boyd Hunter

Providing a brief concluding chapter to a very comprehensive and data-rich volume is a challenging task. This also proved to be the case in making some concluding comments to the conference on which this volume is based. In this final chapter, we seek to combine comments made in summing up the conference with some broad-brush attempts to encapsulate key issues that have been raised, both by conference delegates and by readers when this volume was peer reviewed.

Tim Rowse raises the role of theory in scholarly analysis in his opening chapter. Without a clear theoretical framework or disciplinary perspective it is difficult to interpret statistical data. While few of the chapters in this monograph focus on theoretical issues explicitly, the majority are clearly theoretically grounded. For example, a theory of poverty is used by economists to estimate the wellbeing of households. Hence, information on levels of income is qualified by information on costs and on preferences implicit in household expenditure surveys. Hunter's chapter on income, financial stress, and social exclusion uses such theory to derive measures of equivalence scales and poverty. Clearly, underlying preferences and cost structures are not necessarily the same for all sub-population (or cultural) groups, and therefore the imposition of any particular equivalence scales can, and should, be contested. Hunter's chapter provides an example of how such contestation can be undertaken.

From an academic perspective, statistics should never just be presented atheoretically—indeed, it is a maxim of the dominant western scholarship paradigm that while sound empirical research is informed by theory, conversely, sound theory is informed by empirical research. Consequently, one cannot be divorced from the other. A complicating factor, however, is that data can be consistent with different theories either within one discipline, or informed by different disciplinary perspectives. In such circumstances, it is essential to acknowledge such contestation and the impacts that it might have on the interpretation of statistical evidence.

It is easy to make a case, especially in the emotive area of Indigenous affairs, that 'something needs to be done'. It is far more difficult to provide evidence about 'what policies will be most effective'. In addition to incorporation of theoretical models in any evaluation of the evidence, analysts need to be acutely aware of the reliability and shortcomings of any data they use. The important scene setting work by Biddle and Hunter in Chapter 3 demonstrates that one must be careful in how the data are used: of the size of sampling errors, the likely nature of non-sampling error, the availability of standard errors, and the misspecification of questions that do not assist in answering any policy question coherently. Given this scope for questioning the reliability of data, it is important to validate any information from a questionnaire-based social survey instrument against any existing alternate survey or administrative data sets. Aware of this need, many chapters in this monograph seek to do this. Hopefully, the discussion by Biddle and Hunter alerts those with sufficient expertise to conduct their own validation exercises if they have residual concerns about the quality of any data.

Hence, while data from the 2002 NATSISS can be used for sophisticated analysis, it is important that qualifications about data quality are clearly stipulated. The data from the 2002 NATSISS can be used for policy formulation and broad evaluation. However, such undertakings are rarely straightforward; they are generally complicated and require appropriate contextualisation, both of theoretical perspective and data quality issues. Unfortunately, popular discourse, especially in the print and electronic media, often misuses information when analysts look for a sensationalistic angle in their story. Or, alternatively, statistics are amendable to selective use to support a predetermined position. The chapter on substance use by Brady and Chikritz is illustrative of how dangerous it can be to use data of doubtful integrity without appropriate heavy qualification.

What is the quality of current information about Indigenous wellbeing? The previous CAEPR monographs that focused on national surveys for Indigenous Australians were provided to conference participants to ensure that the discussion did not become too focused on the latest survey instruments and outputs, without placing them in a historical context (Altman 1992; Altman & Taylor 1996b). However, this is difficult, for it is clearly tempting (unless undertaking comparative analysis) to focus on what is the most contemporary.

An important issue that was raised in previous workshops in 1992 and 1996 (and resulting monographs) is the need for Indigenous identifiers in other ABS household surveys like the Labour Force Survey (LFS) and the Household Income and Expenditure Surveys. This issue did not garner much attention at the 2005 conference, but we believe it remains an important issue to consider and debate. We recognise the statistical sampling problems created by the relatively small size of the Indigenous population and the fundamentally different geographic distribution to that of other Australians. Nevertheless, there is still a case for regular augmentation in sampling frames to allow robust identification of Indigenous Australians from these mainstream surveys. The ABS has recognised this need in the annual augmentation of the LFS, and the recent release of experimental labour force statistics for the period 2002–04. These statistics are accompanied by suitable qualifications (and standard error tables) about the sensitivity of the data owing to sampling error and the need to be cautious when using these experimental data for comparative analysis over both time and space (ABS 2006). As the ABS is all too aware, it is extremely difficult to convey the need for such caution to the mainstream media. There are other situations where data have not been released by the ABS to date, apparently because of quality concerns. On one hand, it is undeniable that the ABS has statistical expertise to make such judgments about data quality. On the other hand, it might be preferable if such judgments were scrutinised via the peer review process rather than by in-house processes within the ABS (see discussion in Hunter & Taylor 2001).

An associated issue is what policy questions are being addressed with the collection of social statistics. This is important because social policy questions need to be cast as a priori and theoretically-informed social science hypotheses. Typically, this concern has arisen ex-post-facto, after surveys have been completed (see chapters in Altman & Taylor 1996b). Sometimes this has resulted from decisions on the questions to be asked being based on a process involving consultation with the diverse stakeholders represented on the survey reference group. Clearly, in the development of the 2008 NATSISS, there is room for recognising that there is a clear difference between subject expertise (both theoretical and empirical expertise) and statistical expertise. The ABS has a great deal of the latter, but probably less of the former. Social science expertise, including that of Indigenous practitioners of the social sciences, is needed to decide on the questions to be asked. It might be more productive to resolve such first order issues before taking them for approval to a survey reference group. And it might also be productive for the ABS to enhance the input of social sciences expertise at the outset and not just as one set of stakeholders on reference groups.

The issues of evaluation and the need for data for community and regional development purposes under the new arrangements in Indigenous affairs arose on many occasions during the conference. In order to conduct policy and program evaluations, there is a need for access to administrative data that appears to be increasingly shrouded in confidentiality and hence inaccessible for independent assessments. Several contributors to this volume emphasised the need for better access to administrative data sets, especially in relation to labour market programs, with specific examples raised in discussions focusing on the very limited access to Centrelink information on job-seekers and Department of Employment and Workplace Relations information on CDEP scheme participation.

The reduced availability of NATSISS data at the regional level is of potential concern, since such data could be useful for evaluating the efficacy of new administrative arrangements, especially as these emphasise local processes. While the geographic information in the 2002 NATSISS is—in some ways—more satisfactory than in the 1994 survey, this information was provided by remoteness classification and State, rather than the smaller regional aggregations reported earlier (e.g. ATSIC regions). Data will be needed at regional and community levels to match the jurisdictions of proposed Regional Partnership Agreements and the scores of SRAs that have now been completed. However, it is far from clear whether a survey instrument is the best means of acquiring the necessary data for these purposes, given that the SRAs are so community-focused. And yet there is a growing and worrying mismatch between the level at which data are available and the level at which they are increasingly needed. It is imperative that data are not just produced at broad remote/non-remote geographical level. Such broad levels of aggregation are unlikely to reflect the lived reality of Indigenous people and they are disconnected from the focus in the new arrangements on the community and regional level. On one hand, an instrument like the census does provide information at the community level, but it is only conducted every five years and lacks Indigenous-specific questions. On the other hand, there is a new emerging need for task-specific data collections, including qualitative data about processes, if the new arrangements in Indigenous affairs are ever to be rigorously evaluated. It is not clear to us what analogous processes to academic peer review exist in the policy community to put in place the transparency and accountability referred to in the COAG (2004) principles covering delivery of services to Indigenous people across Australia.

The 2002 NATSISS does provide some advances on previous data collection, but it still has limitations. The lack of an adequate disaggregated geography has just been noted. But to allow a higher degree of disaggregation, the NATSISS sample may need to be augmented. However, as the ABS made clear at the conference, there are obvious costs and benefits associated with such an option that need to be recognised. And while the contributions to this monograph contain many specific suggestions for improving particular questions in the 2008 NATSISS, these clearly need to be balanced against the needs for consistency in inter-temporal analysis, that is, between the 1994 NATSIS and the 2002 NATSISS.

It is important not to adopt terms like 'evidence-based policy formulation' uncritically. Almost by definition, any such policy development tends to place more emphasis on outcomes than process. Several chapters in this volume have noted an increasing emphasis on outcomes at the expense of process in recent policy initiatives. This development is not universally endorsed by the authors here. The artificial distinction between these two aspects of policy has been highlighted in debate about the relative merits of practical reconciliation (which focuses on outcomes) and symbolic reconciliation (which focuses on processes or rights). Altman and Hunter (2003) argue this distinction is unclear, unhelpful, and may militate against effective policy formation, especially if it alienates Indigenous people. Any socially-oriented policy is unlikely to work unless it secures the firm cooperation of the people whose welfare it seeks to enhance.

One instrumental concern with outcomes-based policy is that where outcomes are poorly defined, we may get a distorted set of policies based on hitting targets rather than actually delivering benefits. An example is the CDEP scheme and its role in delivering employment equality to Indigenous Australians. Assuming employment is the desired policy outcome, the CDEP scheme can be regarded as either a huge success or a huge failure depending on whether or not people in the CDEP scheme are defined as employed. The real issue is whether CDEP participation delivers the kinds of benefits normally associated with (market-based) employment. Alternatively, if the CDEP scheme is not just about employment, then all the other social, cultural and locational benefits associated with it need to be clearly articulated and evaluated.

An evidence-based policy approach in Indigenous affairs requires transparent critical scrutiny of the analyses of both researchers and policy makers. Peer review is one way to ensure that the data presented provides, in some sense, evidence for a particular proposition. While the focus of this monograph has been on data quality and the information content in the 2002 NATSISS, it is also important that analysts and policy-makers be held accountable. This monograph is a modest contribution to this process. As academics and policy commentators, we need to be constantly on the lookout to challenge loose research (and associated loose language and loose claims of success or failure), especially if it allows gratuitous appeals to populist or emotive sentiments.

One of the potential strengths of an edited volume is that it does not impose any one view on individual contributors. While this monograph is in many ways inter-disciplinary, the peer review process encourages, and indeed requires, robust criticism of the content of various chapters. The conference provided the first opportunity for peer review, while subsequent independent academic review provided a second layer of quality control.

In the spirit of providing a critical reflection on the inter-play between theory and empirical evidence, it is of interest to focus on two chapters as exemplars of this issue, but from opposite ends of the spectrum. Will Sanders is a political scientist who is one of Australia's foremost experts on Aboriginal housing. In his analysis of housing need, Sanders uses a particular approach that focuses on populations and households and that also focuses on institutional arrangements (see also Sanders 2005). Sanders's chapter here is interesting because he takes an unorthodox position arguing that from his perspective, and for his research purposes, better data already exist in the five-yearly census than in the Indigenous-specific NATSISS. This is impossible to argue with; the proof is in the many important and sensible contributions that Sanders has made to debates about Indigenous housing over the years (see Sanders 1990, 1993, 2005). And yet it could be argued that Sanders overlooks a potential major benefit of the NATSISS: it provides information that would allow multi-factor explanations of Indigenous housing shortfalls beyond the institutional. We highlight the housing issue because of obvious relationships at the level of the individual between housing tenure and housing adequacy and a wide range of socioeconomic variables (including social factors, such as education, labour force status and health status). These are important areas for researchers and policy makers to examine using NATSISS rather than the census which provides an inadequate number of variables. Using NATSISS it would be possible to go beyond reliance on population estimates from the census as the basis for Indigenous housing analysis, but Sanders makes it clear that multi-variate analysis is not his particular preferred approach—he leaves that task for others, perhaps from different disciplinary perspectives.

The lack of multi-variate analysis in the housing chapter is shared by many other chapters in this monograph and is partly explained by the lack of space available to authors both at the conference and in this volume to develop a proper theoretical and empirical framework (that would differ substantially between the various subject areas). [1] Survey data that include a sufficiently large number of Indigenous people can allow researchers to explore subtle behavioural relationships when informed by suitable theoretical models and appropriate multi-variate techniques. The data available from the 2002 NATSISS presents an important opportunity to conduct new analyses to inform policy makers and test theoretical propositions/hypotheses.

Bob Gregory's chapter, conversely, illustrates why one should be wary about making too many claims for theoretical propositions. Gregory uses a conventional neoclassical economics framework to argue that economic policy must be informed by an understanding of supply and demand side issues. In reality, though, it is often difficult to separately identify whether supply or demand or what mix of supply and demand are generating the outcomes—in Gregory's particular case, in the labour market. That is, while it is important to be cognizant of relevant theoretical issues, it is also vital not to make too many claims on available data. Notwithstanding this, Gregory is probably right to speculate that policies are likely to be ineffective if one ignores the distinction between supply and demand just because the available data do not allow one to separately identify the relevant issues for Indigenous Australians. While Gregory is more concerned with 'asking the right questions' than providing definitive evidence for any one proposition, this is not entirely the case in public debate where he seems to privilege supply of Indigenous labour over demand for labour and to postulate policy positions that highlight the need for labour migration when employment outcomes are not guaranteed (Gregory 2005).

---

[1] Even if the multi-variate analysis is not always conducted in this volume, several authors allude to the importance of controlling for confounding the bi-variate analysis in their respective chapters in this volume (e.g. Gray & Chapman; Radoll).

One of the continual tensions for the ABS is the extent to which it should include and frame questions in future NATSISS data collections that will either allow direct comparisons with non-Indigenous Australians or include questions that have the greatest relevance for Indigenous people themselves. To the extent that one leans towards the latter, it will be necessary to frame questions in the most culturally appropriate way—a process which ideally should be determined, in terms of appropriate phrasing of the question, with extensive input from Indigenous people themselves. However, the diversity in cultural practice across Indigenous Australia means that even the most sensitive data collection techniques will be unreliable to a certain extent (i.e. have a certain level of non-sampling error). The inherent unreliability of survey data on cultural variables was identified in Nic Peterson's chapter, especially the difficulty in interpreting the changes in culture and identity between 1994 and 2002. While cultural data are exceptionally sensitive to the methodology and general context of a survey, one should not underestimate their importance as they are essential background for many of the theoretical perspectives used to interpret data for Indigenous Australians. This may be an intractable problem: despite the best intentions of the ABS, surveys may be the wrong instrument to collect cultural data and more field-based ethnographic techniques are likely to better capture diversity.

If the policy vision for the future of Indigenous Australia is 'equality' in the sense of Indigenous people achieving the same outcomes as non-Indigenous peoples, then greater emphasis should be placed on collecting data which will allow direct comparisons with non-Indigenous Australians. Whether questions are framed in a culturally sensitive way would be a secondary concern to that of achieving direct comparability with existing general population surveys. This also pre-supposes that ultimately Indigenous Australians will share the same aspirations as non-Indigenous Australians, ultimately a culturally determined presupposition.

If, on the other hand, the policy vision is cultural plurality, self-determination and choice, then comparability with general population data would be of secondary concern. Much discussion in this monograph can be seen in this light, including frequent queries being raised about different techniques and questions used in remote and non-remote areas (or CAs/NCAs). Such differences presumably arose in a survey of Indigenous Australians from a range of factors, including:

- recognition of different aspirations and degrees of integration with non-Indigenous society
- issues of cultural appropriateness in survey design and conduct
- treatment of CDEP jobs as employment in most contexts
- laments about the absence of adequate data on the 'customary economy', and

- inadequate recognition of hunting, gathering and fishing activities as legitimate economic activities (see chapters by Biddle & Hunter; Altman, Buchanan & Biddle; and Taylor & Kinfu).

It is noteworthy that the tension we are highlighting here has been embedded in both policy and disciplinary approaches for many years now. On the policy front, Altman and Sanders (1991) highlighted the tension between equality and equity in the now superseded Aboriginal Employment Development Policy (AEDP). The AEDP sought statistical equality between Indigenous and non-Indigenous Australians by the year 2000. The measurement of attainment of this policy goal would need comparable statistical information on Indigenous and other Australians, with the five-yearly census being the obvious instrument. However, the AEDP also sought to recognise cultural diversity, connoted by the term 'equity' (or fairness). Measuring this would not only require statistics about cultural differences, but would also require trade-offs with the goal of statistical equality. For example, if on equity grounds people chose and were supported to live at remote localities distant from mainstream opportunities, their chances of equality would be heavily circumscribed.

On the disciplinary front, Altman and Rowse (2005: 159–77) have recently highlighted a similar tension within the social sciences differentially interpreting the goals of Indigenous affairs policy as either of socioeconomic equality or the facilitation of choice and self-determination. Using anthropology and economics as exemplars, they have shown how anthropology dwells on cultural difference and presumes that difference to be a social good, whereas economics dwells on socioeconomic inequality and presumes that it is the negative legacy of historic exclusion, racism and neglect. There is clearly a tension between the two ideals of 'equality' and 'plurality' in a liberal democracy like Australia, and associated challenges to statistics that seek to measure 'equality' in some extremely remote, difficult and structurally constrained circumstances where many Indigenous people choose to live.

Perhaps the most common sentiments expressed by the authors of chapters in this monograph were hopes for enhancements to the data through greater comparability (particularly between the 2002 NATSISS and future national surveys), merging or supplementing the survey data with administrative data collections, and collection of true longitudinal or panel data. While strong arguments can be made for all of these propositions, it is important to acknowledge that all such strategies are not without cost. Matching ABS statistics with departmental records over time, or establishing a 'true' longitudinal data set (focusing on the same individuals over time), may not be an optimal use of scarce resources—both approaches are expensive and the resulting data introduce their own sets of limitations for researchers. [2] However, tracking individuals across time can potentially offer insights into causal relationships and dynamics of social interactions, both of which are almost impossible to examine with cross-sectional data. Consequently, it is vital to consider how longitudinal data are best collected for Indigenous Australians. The experience of the Longitudinal Study of Indigenous Children, which is currently being designed by FaCS and should be in the field in 2007, will itself provide data on how future longitudinal data might be collected for Indigenous Australians.

This monograph covers a wide range of research topics, and this obviously reflects the very wide range of issues about which the 2002 NATSISS collected statistical information. There is an inevitable trade-off between the depth to which any issue is explored and the number of different issues canvassed. This can be seen as reflecting either positively or negatively on the survey. Almost invariably, writers in each subject area pointed to both positive aspects of the 2002 data and to items they felt could have been included or expanded upon for the purposes of their own research.

This publication contains numerous recommendations, and ultimately it will not be feasible for the ABS to adopt all, or even most, of them. Indeed, the main aim of this monograph is to stimulate public debate about Indigenous issues among policy makers, academics, and other users of ABS data. While, as noted at the outset, detailed data from the 2002 NATSISS only became available last year, it is important that debate about these important statistics is disseminated as widely as possible. It is unfortunately the case that the numbers that can attend a two-day conference in Canberra will always be limited. Hopefully, this publication with its diverse perspectives and subjects, mainly drawn from the 2002 NATSISS, should make a contribution to the quality of analyses of Indigenous affairs issues and the overall standard of Australian public debate. It is our hope that the contents of this monograph might improve the overall standard of public discourse on Indigenous issues by highlighting some of the statistics that are currently available, when and for what they can be used, and when we need to take extreme care in how we use them.

---

[2] However, Gray & Hunter illustrate that synthetic panel data can be constructed to partially redress the lack of 'true' longitudinal data for Indigenous individuals (Gray & Hunter 2002; Hunter & Gray 2001).

References

Aboriginal Drug and Alcohol Council (ADAC) 2004. *Responding to the Needs of Indigenous People Who Inject Drugs*, ADAC, Adelaide.

ABS 1987. *Internal Migration Australia*, cat. no. 3408.0, ABS, Canberra.

——1995. *National Aboriginal and Torres Strait Islander Survey 1994: Detailed Findings*, cat. no. 4190.0, ABS, Canberra.

——1996a. *National Aboriginal and Torres Strait Islander Survey 1994: All ATSIC Regional Councils*, cat. no. 4196.0.001–36, ABS, Canberra.

——1996b. *National Aboriginal and Torres Strait Islander Survey: An Evaluation of the Survey*, cat. no. 4184.0, ABS, Canberra.

——1997. *Australian Standard Classification of Languages*, cat. no. 1267.0, ABS, Canberra.

——1999. *1995 National Health Survey: Aboriginal and Torres Strait Islander Results*, cat. no. 4806.0, ABS, Canberra.

——2000a. *The Health and Welfare of Australia's Aboriginal and Torres Strait Islander Peoples, 2000*, cat. no. 4704.0, ABS, Canberra.

——2000b. *Housing and Infrastructure in Aboriginal and Torres Strait Islander Communities: Australia 1999*, cat. no. 4710.0, ABS, Canberra.

——2002a. *2001 National Health Survey: Aboriginal and Torres Strait Islander Results*, ABS cat. no. 4715.0, ABS, Canberra.

——2002b. *National Health Survey 2001, Summary Results*, cat. no. 4364.0, ABS, Canberra.

——2003a. *2002 Deaths, Australia*, cat. no. 3302.0, ABS, Canberra.

——2003b. *General Social Survey 2002: Summary of Results, Australia 2002*, cat. no. 4159.0, ABS, Canberra.

——2003c. *Household and Income Distribution, Australia: 2000–01*, cat. no. 6523.0, ABS, Canberra.

——2004a. *Australian Social Trends*, cat. no. 4102.0, ABS, Canberra.

——2004b. *Deaths, Australia*, cat. no. 3302.0, ABS, Canberra.

——2004c. *National Aboriginal and Torres Strait Islander Social Survey 2002*, cat. no. 4714.0, ABS, Canberra.

——2004d. *National Aboriginal and Torres Strait Islander Social Survey (NAT-SISS) 2002: Output Data Items*, cat. no. 4714.0.55.001, ABS, Canberra.

——2004e. *Prisoners in Australia*, cat. no. 4517.0, ABS, Canberra.

——2005a. *Australian Social Trends*, cat. no. 4102.0, ABS, Canberra.

——2005b. *National Aboriginal and Torres Strait Islander Social Survey Confidentialised Unit Record File Technical Paper*, cat. no. 4720.0, ABS, Canberra.

——2005c. *National Aboriginal and Torres Strait Islander Social Survey: Expanded Confidentialised Unit Record File*, cat. no. 4720.0, ABS, Canberra.

——2006. *Labour Force Characteristics of Aboriginal and Torres Strait Islander Australians: Experimental Estimates from the Labour Force Survey*, cat. no. 6287.0, ABS, Canberra.

ABS/AIHW 2005. *The Health and Welfare of Australia's Aboriginal and Torres Strait Islander Peoples, 2005*, cat. no. 4704.0, ABS, Canberra.

ABS/CAEPR 1997. *National Aboriginal and Torres Strait Islander Survey 1994: Torres Strait Islanders in Queensland*, cat. no. 4179.3, ABS, Canberra.

AIATSIS/FATSIL 2005. *National Indigenous Languages Survey Report 2005*, Commonwealth of Australia, Canberra.

Altman, J.C. (ed.) 1992. *A National Survey of Indigenous Australians: Options and Implications*, CAEPR Research Monograph No. 3, CAEPR, ANU, Canberra.

——1996. 'Aboriginal economic development and land rights in the Northern Territory: past performance, current issues and strategic options', *CAEPR Discussion Paper No. 126*, ANU, Canberra.

——1999. 'The proposed restructure of the financial framework of the Land Rights Act: a critique of Reeves', in J. C. Altman, F. Morphy and T. Rowse (eds), *Land Rights at Risk? Evaluations of the Reeves Report*, CAEPR Research Monograph No. 14, CAEPR, ANU, Canberra.

——2002. 'Aboriginal economy and social process: The Indigenous hybrid economy and its sustainable development potential', *Arena*, 56 (January): 38–9.

——2003a. 'The eastern Kuninjku harvesting economy in the mid-wet season: 1980 and 2003 compared', unpublished field report to the ARC Key Centre for Tropical Wildlife Management, Charles Darwin University, Darwin.

——2003b. 'Economic development and participation for remote Indigenous communities: best practice, evident barriers, and innovative solutions in the hybrid economy', presentation to Ministerial Council for Aboriginal and Torres Strait Islander Affairs (MCATSIA), Sydney, 28 November.

——2004. 'Economic development and Indigenous Australia: contestations over property, institutions and ideology', *Australian Journal of Agricultural and Resource Economics*, 48 (3): 513–34.

———2005a. 'CDEP 2005: A new home and new objectives for a very old program?' *CAEPR Topical Issue 2005/07*, CAEPR, ANU, available online at <http://www.anu.edu.au/caepr/topical.php>.

———2005b. 'Development options on Aboriginal land: sustainable Indigenous hybrid economies in the twenty-first century', in L. Taylor, G. K. Ward, G. Henderson, R. Davis, et al. (ed.), *The Power of Knowledge, The Resonance of Tradition*, Aboriginal Studies Press, Canberra.

——— and Allen, L.M. 1992. 'Indigenous participation in the informal economy: statistical and policy implications', in J.C. Altman (ed.), *A National Survey of Indigenous Australians: Options and Implications*, CAEPR Research Monograph No. 2, CAEPR, ANU, Canberra.

———, Biddle, N. and Hunter, B.H. 2005. 'A historical perspective on Indigenous socioeconomic outcomes, 1971–2001', *Australian Economic History Review*, 45 (3): 273–95.

———, ——— and Levitus, R. 2005. 'Policy issues for the Community Development Employment Projects Scheme in rural and remote Australia', *CAEPR Discussion Paper No. 271*, CAEPR, ANU, Canberra.

——— and Hinkson, M. 2005. 'The social universe of Kuninjku trucks', submitted to the *Journal of Material Culture*.

——— and Hunter, B.H. 1998. 'Indigenous poverty', in R. Fincher and J. Nieuwenhuysen (eds), *Australian Poverty: Then and Now*, Melbourne University Press, Melbourne.

——— and ——— 2003. 'Evaluating Indigenous socioeconomic outcomes in the reconciliation decade, 1991–2001', *Economic Papers*, 22 (4): 1–16.

——— and Nieuwenhuysen, J. 1979. *The Economic Status of Australian Aborigines*, Cambridge University Press, Cambridge.

——— and Rowse, T. 2005. 'Indigenous affairs', in P. Saunders and J. Walter (eds), *Ideas and Influence: Social Science and Public Policy in Australia*, UNSW Press, Sydney.

——— and Sanders, W. 1991. 'The CDEP scheme: administrative and policy issues', *Australian Journal of Public Administration*, 50 (4): 515–25.

——— and Taylor, J. (eds) 1996a. *The 1994 National Aboriginal and Torres Strait Islander Survey: Findings and Future Prospects*, CAEPR Research Monograph No. 11, CAEPR, ANU, Canberra.

——— and Taylor, J. 1996b. *The 1994 National Aboriginal and Torres Strait Islander Survey: Findings and Future Prospects*, CAEPR Research Monograph No. 11, CAEPR, ANU, Canberra.

Anderson, I. and Sibthorpe, B. 1996. 'The NATSIS and policy and planning in Aboriginal health', in J.C. Altman and J. Taylor (eds), *The 1994 National Aboriginal and Torres Strait Islander Survey: Findings and Future Prospects*, Research Monograph No. 11, CAEPR, ANU, Canberra.

Andrews, K. 2005. *Welfare to Work Reforms*, Australian Council of Social Service Conference, Parliament House, Canberra, 29 September 2005.

Annan, K. (Secretary-General of the United Nations) 2005. 'In larger freedom: Towards development, security and human rights for all', *UN Doc: A/59/2005*, UN, New York.

Arbour, L. (High Commissioner for Human Rights) 2005. 'In larger freedom: Towards development, security and human rights for all—addendum: plan of action', *UN Doc: A/59/2005/Add.3*, UN, Geneva.

Arthur, W. 1990. *Torres Strait Development Study 1989*, Australian Institute of Aboriginal Studies, Canberra.

——1992. 'The provision of statistics about Torres Strait Islanders', in J.C. Altman (ed.), *A National Survey of Indigenous Australians: Options and Implications*, CAEPR Research Monograph No. 3, CAEPR, ANU, Canberra.

——1996. 'Torres Strait Islanders', in J.C. Altman and J. Taylor (eds), *The 1994 National Aboriginal and Torres Strait Islander Survey: Findings and Future Prospects*, CAEPR Research Monograph No. 11, CAEPR, ANU, Canberra.

——1998. 'Access to government programs and services for mainland Torres Strait Islanders', *CAEPR Discussion Paper No. 151*, CAEPR, ANU, Canberra.

——2002. 'Autonomy and the Community Development Employment Projects Scheme', *CAEPR Discussion Paper No. 232*, CAEPR, ANU, Canberra.

—— and Taylor, J. 1994. 'The comparative economic status of Torres Strait Islanders in Torres Strait and mainland Australia', *CAEPR Discussion Paper No. 72*, CAEPR, ANU, Canberra.

Atkinson, J. 1990. 'Violence against Aboriginal women: reconstitution of community law—the way forward', *Aboriginal Law Bulletin*, 2 (46): 6–9.

Australian Construction Services (ACS) 1993. *1992 National Housing and Community Infrastructure Needs Survey: Final Report Stage 1 Australia, States and Territories*, A report to the Aboriginal and Torres Strait Islander Commission, ACS, Canberra.

Australian Government 1987. *Aboriginal Employment Development Policy Statement: Policy Paper No. 1*, AGPS, Canberra.

Australian Institute of Criminology (AIC) 1996. *National Police Custody Survey 1995, Preliminary Report*, AIC, Canberra.

——2004. *Australian Crime: Facts and Figures*, AIC, Canberra.

Australian Institute of Health and Welfare (AIHW) 2002. *2001 National Drug Strategy Household Survey, Detailed Findings*, AIHW cat. no. PHE 66, AIHW, Canberra.

——2005a. *Mortality*, AIHW, Canberra, viewed 4 August 2005, <http://www.aihw.gov.au/Indigenous/health/mortality.cfm>

——2005b. *Statistics on Drug Use in Australia 2004*, AIHW cat. no. PHE 62, AIHW, Canberra.

Baker, J. 2001. 'The scope for reducing Indigenous imprisonment rates', *Crime and Justice Bulletin*, 55 (March 2001): 1–10.

Bell, M. 1992. *Internal Migration in Australia 1981–1986*, AGPS, Canberra.

——2001. 'Understanding circulation in Australia', *Journal of Population Research*, 18 (1): 1–18.

—— and Maher, C. 1995. *Internal Migration in Australia 1986–1991: Overview Report*, AGPS, Canberra.

Biddle, N. 2005. 'Health benefits of education in Australia: Indigenous/non-Indigenous comparisons', paper presented at the Australian Social Policy Conference 2005, University of New South Wales, 20–22 July, Sydney.

—— and Hunter, B.H. 2004. 'Methodological issues for analysis of NATSISS 2002', Presented in the CAEPR seminar series on 10 November 2004, ANU, Canberra.

Blake, B.J. 1981. *Australian Aboriginal Languages*, Angus & Robertson, Sydney.

Booth, A. and Carroll, N. 2005. 'The health status of Indigenous and non-Indigenous Australians', *CEPR Discussion Paper No. 486*, CEPR, ANU, Canberra.

Borland, J. and Hunter, B.H. 2000. 'Does crime affect employment status? The case of Indigenous Australians', *Economica*, 67 (1): 123–44.

Borrie, W. 1975. *Population and Australia: A Demographic Analysis and Projection (Volume Two)*, AGPS, Canberra.

Brady, M. 2004. *Indigenous Australia and Alcohol Policy. Meeting Difference With Indifference*, University of New South Wales Press, Sydney.

——2005. 'Making use of medics: Overcoming cultural constraints in alcohol interventions', in L. Taylor, G. Ward, G. Henderson, R. Davis, et al. (eds), *The Power of Knowledge: Resonance of Tradition*, Aboriginal Studies Press, Canberra.

——, Sibthorpe, B., Bailie, R., Ball, S., et al. 2002. 'The feasibility and acceptability of introducing brief intervention for alcohol misuse in an urban Aboriginal medical service', *Drug and Alcohol Review*, 21: 375–80.

Brandon, P. 2004. 'Identifying the diversity of Australian children's living arrangements: a research note', *Journal of Sociology*, 40 (2): 179–92.

Brice, G.A. 2000. *Australian Indigenous Road Safety*, Transport SA, Adelaide.

Broadhurst, R.G., Ferrante, A., Loh, N., Reidpath, D., et al. 1994. *Aboriginal Contact with the Criminal Justice System in Western Australia: A Statistical Profile*, Crime Research Centre, The University of Western Australia, Perth.

Broom, L. and Jones, F. L. 1973. *A Blanket a Year*, ANU Press, Canberra.

Cameron, A.C. and Trivedi, P.K. 2001. *Regression Analysis of Count Data*, Cambridge University Press, Cambridge.

Cameron, J. and Gibson-Graham, J.K. 2003. 'Feminising the economy: metaphors, strategies, politics', *Gender, Place and Culture*, 10 (2): 145–57.

Canadian Royal Commission on Aboriginal Peoples 1996. *Report of the Royal Commission on Aboriginal Peoples*, Group Communication, Ottawa.

Cane, S. and Stanley, O. 1985. *Land Use and Resources in Desert Homelands*, North Australia Research Unit, ANU, Darwin.

Carcach, C.A. and Mukherjee, S.K. 1996. 'Law, justice, indigenous Australians and the NATSIS: policy relevance and statistical seeds', in Altman, J.C. and Taylor, J. (eds), *The 1994 National Aboriginal and Torres Strait Islander Survey: Findings and Future Prospects*, CAEPR Research Monograph No. 11, CAEPR, ANU, Canberra.

CDHSH 1996. *National Drug Strategy Household Survey: Urban Aboriginal and Torres Strait Islander Peoples Supplement, 1994*, AGPS, Canberra.

Commonwealth Grants Commission 2001. 'Report on the Indigenous Funding Inquiry', Commonwealth Grants Commission, Vol. 1–3, Canberra.

Chikritzhs, T., Catalano, P., Stockwell, T., Donath, S., et al. 2003. *Australian Alcohol Indicators, 1990–2001: Patterns of Alcohol Use and Related Harms for Australian States and Territories*, National Drug Research Institute, Curtin University of Technology, Perth.

——, Jonas, H., Heale, P., Dietze, P., et al. 2000. 'Alcohol-caused deaths and hospitalisations in Australia, 1990–1997', *National Alcohol Indicators, Technical Report No.1*, National Drug Research Institute, Curtin University of Technology, Perth.

—— and Pascal, R. 2004. 'Trends in youth alcohol consumption and related harms in Australian jurisdictions, 1990–2002', *National Alcohol Indicat-*

ors, Bulletin No. 6, National Drug Research Institute, Curtin University of Technology, Perth.

Citro, C.S. and Michael, R.P. 1995. 'Adjusting poverty thresholds', in C.S. Citro and R.P. Michael (eds), *Measuring Poverty and New Approach*, National Academic Press, Washington.

Council of Australian Governments (COAG) 2004. 'National framework of principles for delivering services to Indigenous Australians', *COAG Communique*, 24th June 2004, COAG, Canberra.

Committee on Economic Social and Cultural Rights (CESCR) 2004. 'General comment: the nature of States parties obligations', *UN Doc: E/1991/23, 14/12/90*, UN, New York.

Commonwealth of Australia 1991. *Royal Commission into Aboriginal Deaths in Custody*, vol. 2, (Commissioner E. Johnston), AGPS, Canberra.

Cooke, P. 1994. *Planning a Future at Cape York Outstations*, Cape York Land Council, Carins.

Coombs, H.C., Dexter, B.G. and Hiatt, L.R. 1982. 'The outstation movement in Aboriginal Australia', in E. Leacock and R. Lee (eds), *Politics and History in Band Societies*, Cambridge University Press, Cambridge.

Council for Aboriginal Reconciliation (CAR) 2000. *Australia's Challenge*, CAR, Canberra.

Cross, S. and Madson, L. 1997. 'Models of self: self-construals and gender', *Psychological Bulletin*, 122: 5–37.

Crossley, T. and Kennedy, S. 2002. 'The reliability of self-assessed health status', *Journal of Health Economics*, 21: 643–58.

Cunneen, C. and McDonald, D. 1997. *Keeping Aboriginal and Torres Strait Islander People out of Custody: An Evaluation of the Implementation of the Recommendations of the Royal Commission into Aboriginal Deaths in Custody*, ATSIC, Canberra.

Daly, A.E. 1995. *Aboriginal and Torres Strait Islander People in the Australian Labour Market*, cat. no. 6253.0, ABS, Canberra.

——1996. 'Post-secondary qualifications and training for Indigenous Australians', in J.C. Altman and J. Taylor (eds), *The 1994 National Aboriginal and Torres Strait Islander Survey: Findings and Future Prospects*, Research Monograph No. 11, CAEPR, ANU, Canberra.

——2005. 'Bridging the digital divide: the role of community online access centres in Indigenous communities', *CAEPR Discussion Paper No. 273*, CAEPR, ANU, Canberra.

Department of Communications, Information Technology and the Arts (DCITA) 1999. *Networking the Nation*, online at http://www.dcita.gov.au/, DCITA, Canberra.

——2002. *Telecommunications Action Plan for Remote Indigenous Communities*, DCITA, Canberra.

Department of Employment and Workplace Relations (DEWR) 2003. 'Job Network evaluation stage two: progress report', Evaluation and Programme Performance Branch, Labour Market Policy Group, DEWR, Canberra.

Economic and Social Council 2005. *Permanent Forum on Indigenous Issues: Report on the Fourth Session (16–27 May 2005)*, United Nations, New York.

Fairbairn, T. 1985. *Island Economies: Studies from the South Pacific*, Institute of Pacific Studies, University of the South Pacific, Suva.

Fink, C. and Kenny, C. J. 2003. *W(h)ither the Digital Divide?* The World Bank, available online at http://freeculture2.soc.american.edu/uploads/359/W_h_ither_DD__Jan_.pdf, Washington DC.

Fisk, E.K. 1985. *The Aboriginal Economy in Town and Country*, George Allen and Unwin, Sydney.

Fogarty, W. 2005. ''You got any truck?' Vehicles, and decentralised, mobile service provision in remote Indigenous Australia', *CAEPR Working Paper No. 30*, CAEPR, ANU, Canberra.

Fox, J. and Tracy, P. 1986. *Randomized Response: A Method for Sensitive Surveys*, Sage, Beverly Hills, CA.

Gale, F. 1972. *Urban Aborigines*, ANU Press, Canberra.

—— and Wundersitz, J. 1982. *Adelaide Aborigines. A Case Study of Urban Life 1966–1981*, Development Studies Centre, ANU, Canberra.

Gerrard, G. 1989. 'Everyone will be jealous for the Mutika', *Mankind*, 19 (2): 95–111.

Gibson-Graham, J.K. 1996. *The End of Capitalism (As We Knew It): A Feminist Critique of Political Economy*, Blackwell Publishers, Cambridge, Mass.

——2005. 'Surplus possibilities: post-development and community economies', *Singapore Journal of Tropical Geography*, 26 (1): 4–26.

——n.d. 'A diverse economy: rethinking economy and economic representation', <http://www.communityeconomies.org>, accessed 19 July 2005.

Gibson, K. 1999. 'Community economies: economic politics outside the binary frame', paper presented at the Rethinking Economy Conference, ANU, August.

Gray, A. 1988. 'Aboriginal child survival: an analysis of the results from the 1986 Census of Population and Housing', *ABS Occasional Paper*, cat. no. 4126.0, ABS, Canberra.

——2004. 'The formation of contemporary Aboriginal settlement patterns in Australia: government policies and programmes', in J. Taylor and M. Bell (eds), *Population Mobility and Indigenous Peoples in Australasia and North America*, Routledge, London and New York.

—— and Tesfaghiorghis, H. 1991. 'Social indicators of the Aboriginal population of Australia', *CAEPR Discussion Paper No. 18*, CAEPR, ANU, Canberra.

Gray, D., Saggers, S., Atkinson, D. and Strempel, P. 2004. 'Substance misuse and primary health care among Indigenous Australians', *Aboriginal and Torres Strait Islander Primary Health Care Review, Consultant Report No. 7*, Commonwealth of Australia, Canberra.

——, Sputore, B., Stearne, A., Bourbon, D., et al. 2002. 'Indigenous drug and alcohol projects 1999–2000', *ANCD Research Paper 4*, Australian National Council on Drugs, Canberra.

Gray, M.C., Altman, J.C. and Halasz, N. 2005. 'The economic value of wild resources to the Indigenous community of the Wallis Lake catchment', *CAEPR Discussion Paper No. 272*, CAEPR, ANU, Canberra.

—— and Hunter, B.H. 2002. 'A synthetic cohort analysis of labour market outcomes of Indigenous and non-Indigenous Australians, 1986–96', *Australian Economic Review*, 35 (4): 391–404.

—— and —— 2005. 'The labour market dynamics of Indigenous Australians', *Journal of Sociology*, 41 (4): 389–408.

—— and Taylor, J. 2004. *Health Expenditure, Income and Health Status Among Indigenous and Other Australians*, CAEPR Research Monograph No. 21, ANU E-Press, Canberra.

Gregory, R.G. 2005. 'Between a rock and a hard place: economic policy and the employment outlook for Indigenous Australians', in D. Austin-Broos and G. Macdonald (eds), *Culture, Economy and Governance in Aboriginal Australia*, University of Sydney Press, Sydney.

Gurmu, S. and Trivedi, P. 1996. 'Excess zeros in count models for recreational trips', *Journal of Business and Economic Statistics*, 14 (4): 469–77.

Hamilton, A. 1987. 'Coming and going: Aboriginal mobility in North-West South Australia, 1970–71', *Records of the South Australia Museum*, 20: 47–57.

Harding, A., Lloyd, R. and Greenwell, H. 2001. *Financial Disadvantage in Australia 1990 to 2000*, The Smith Family, Sydney.

Hemstrom, O. 2001. 'Per capita alcohol consumption and ischaemic heart disease mortality', *Addiction*, 96 (1): S93–S112.

Henry, G.T. 1990. *Practical Sampling*, Sage Publications, Newbury Park.

Higgins, L., Cooper-Stanbury, M. and Williams, P. 2000. *Statistics on Drug Use in Australia, 1998*, AIHW cat. no. PHE 16, AIHW, Canberra.

High Commissioner for Human Rights 2004. *Human Rights and Poverty Reduction—A Conceptual Framework*, UN, Geneva.

Holcombe, S. 2004. 'The sentimental community: a site of belonging. A case study from Central Australia', *The Australian Journal of Anthropology*, 15 (2): 163–84.

Hughes, H. 2005a. *The Economics of Australian Apartheid: The Causes of Indigenous Deprivation*, Presented to the 34th Conference of Economists, University of Melbourne, 26–28 September, Melbourne.

——2005b. 'The economics of Indigenous deprivation and proposals for reform', *Issues Analysis No. 63*, The Centre for Independent Studies, St Leonards, NSW.

—— and Warin, J. 2005. 'A new deal for Aborigines and Torres Strait Islanders in remote communities', *Issue Analysis No 54*, The Centre for Independent Studies, St Leonards, NSW.

Hunt, P., Osmani, S. and Nowak, M. 2004. *Summary of the Draft Guidelines on a Human Rights Approach to Poverty Reduction*, OHCHR, Geneva.

Hunter, B.H. 1996. 'Indigenous Australians in the labour market: the NATSIS and beyond', in J.C. Altman and J. Taylor (eds), *The 1994 National Aboriginal and Torres Strait Islander Survey: Findings and Future Prospects*, CAEPR Research Monograph No. 11, CAEPR, ANU, Canberra.

——1998. The effect of high rates of arrest on educational attainment among indigenous Australians, Presented to the Annual Meeting of the American Economist Association, 3–5 January, Chicago.

——2001. *Factors Underlying Indigenous Arrest Rates*, New South Wales Bureau of Crime Statistics and Research, Sydney.

——2004a. *Indigenous Australians in the Contemporary Labour Market*, cat. no. 2052.0, ABS, Canberra.

——2004b. 'Taming the social capital hydra?' *Learning Communities: International Journal of Learning in Social Contexts*, 2004 (2): 19–35.

——2005. 'Changes in the economic, health and social status of Indigenous Australians in remote and settled Australia, 1994–2002', Presented to the 34th Conference of Economists, University of Melbourne, 26–28 September, Melbourne.

——, Arthur, W.S. and Morphy, F. 2005. 'Social justice', in W.S. Arthur and F. Morphy (eds), *Macquarie Atlas of Indigenous Australia*, Macquarie Library Pty Ltd, Sydney.

—— and Borland, J. 1999. 'The effect of the high rate of Indigenous arrest on employment prospects', *Crime and Justice Bulletin*, 45 (June): 1–8.

—— and Gray, M.C. 2001. 'Analysing recent changes in Indigenous and non-Indigenous Australians income: a synthetic panel approach', *Australian Economic Review*, 34 (2): 135–54.

——, —— and Jones, R. 2000. 'An analysis of data from the longitudinal survey of ATSI job seekers: Labour market participation patterns and pathways to employment', Report to the Department of Employment, Workplace Relations and Small Business, CAEPR, ANU, Canberra, archived at http://pandora.nla.gov.au/pan/39721/20040128/www.workplace.gov.au/WP/Content/Files/WP/EmploymentPublications/topic1.pdf

——, Kennedy, S. and Biddle, N. 2004. 'Indigenous and other Australian poverty: revisiting the importance of equivalence scales', *Economic Record*, 80 (251): 411–22.

——, Kinfu, Y. and Taylor, J. 2003. 'The future of Indigenous work: forecasts of labour force status to 2011', *CAEPR Discussion Paper No. 251*, CAEPR, ANU, Canberra.

—— and Schwab, R.G. 1998. 'The determinants of Indigenous educational outcomes', *CAEPR Discussion Paper No. 160*, CAEPR, ANU, Canberra.

—— and Smith, D.E. 2002. 'Surveying mobile populations: lessons from recent longitudinal surveys of Indigenous Australians', *Australian Economic Review*, 35 (3): 261–75.

—— and Taylor, J. 2001. 'The reliability of Indigenous employment estimates', *Agenda*, 8 (2): 113–28.

—— and —— 2004. 'Indigenous employment forecasts: implications for reconciliation', *Agenda: A Journal of Policy Analysis and Reform*, 11 (2): 179–92.

Hunter, E. 1993. *Aboriginal Health and History. Power and Prejudice in Remote Australia*, Cambridge University Press, Cambridge.

Ironmonger, D. 1996. 'Counting outputs, capital inputs and caring labour; Estimating gross household product', *Feminist Economics*, 2 (3): 37–64.

Jones, F.L. 1970. *The Structure and Growth of Australia's Aboriginal population*, ANU Press, Canberra.

——1972. 'Fertility patterns among Aboriginal Australians', *Human Biology in Oceania*, 1 (4): 245–54.

Jones, R. 1994. *The Housing Need of Indigenous Australians, 1991*, CAEPR Research Monograph No. 8, CAEPR, ANU, Canberra.

——1999. *Indigenous Housing 1996 Census Analysis*, Aboriginal and Torres Strait Islander Commission, Canberra.

Kawachi, I., Kennedy, B.P. and Wilkinson, R.G. 1999. *The Society and Population Health Reader: Income Inequality and Health*, New Press, New York.

Kinfu, Y. 2005. 'Spatial mobility among Indigenous Australians: patterns and determinants', *Working Papers in Demography No. 97*, Demography Program, RSSS, ANU, Canberra.

—— and Taylor, J. 2002. 'Estimating the components of Indigenous population change: 1996–2001', *CAEPR Discussion Paper No. 240*, CAEPR, ANU, Canberra.

—— and Taylor, J. 2005. 'On the components of Indigenous population change', *Australian Geographer*, 36 (2): 233–55.

Kolig, E. 1989. 'The mobility of Aboriginal religion', in M. Charlesworth, H. Morphy, D. Bell and K. Maddock (eds), *Religion in Aboriginal Australia: An Anthology*, University of Queensland Press, Brisbane.

Kral, I. (forthcoming). An ethnography of literacy in a remote Aboriginal community, PhD Thesis (Anthropology), ANU, Canberra.

Lake, P. 1989. 'Alcohol and cigarette use by urban Aboriginal people', *Aboriginal Health Information Bulletin*, 11 (May): 20–2.

Langton, M. 2002. *A New Deal? Indigenous Development and the Politics of Recovery*, University of Sydney, Sydney.

Larson, A. 1996. 'What injectors say about drug use: preliminary findings from a survey of Indigenous injecting drug users', *IDU Working Paper No. 2*, Australian Centre for International and Tropical Health and Nutrition, University of Queensland, Brisbane.

Larson, R. and Richards, M. 1994. *Divergent Realities: The Emotional Lives of Mothers, Fathers, and Adolescents*, Basic Books, New York.

Lawrence, R. 1991. 'Motorised transport in remote Australia', *Australian Aboriginal Studies*, 1991 (2): 62–6.

Lloyd, R. and Bill, A. 2004. *Australia Online: How Australians are Using Computers and the Internet*, cat. no. 2056.0, ABS, Canberra.

—— and Helwigg, O. 2000. 'Barriers to the uptake of new technologies', *Discussion Paper No. 53*, NATSEM, Canberra.

Long, J.S. 1997. *Regression Models for Categorical and Limited Dependent Variables*, Sage Publications, Thousand Oaks, CA.

Long, L. 1992. 'Changing residence: Comparative perspectives on its relationship to age, sex and marital status', *Population Studies*, 46: 141–58.

Martin, D.F. 1993. Autonomy and relatedness: An Ethnography of Wik People of Aurukun Western Cape York Peninsula, PhD thesis, Anthropology Department, ANU, Canberra.

—— 1998. 'The supply of alcohol in remote Aboriginal communities: potential policy directions from Cape York', *CAEPR Discussion Paper No. 162*, CAEPR, ANU, Canberra.

——, Morphy, F., Sanders, W.G. and Taylor, J. 2004. *Making Sense of the Census: Observations of the 2001 Enumeration in Remote Aboriginal Australia*, CAEPR Research Monograph No. 22, 2nd Edition, ANU E-Press, Canberra.

McConvell, P. and Thieberger, N. 2001. 'State of Indigenous languages in Australia—2001', *Australia State of the Environment Second Technical Paper Series*, Department of the Environment and Heritage, (Natural and Cultural Heritage), Canberra. Available online at http://www.deh.gov.au/soe/techpapers/languages/indicator1d.html

—— and —— 2003. 'Language data assessment at the national level: learning from the State of the Environment process in Australia', in J. Blythe and M. Brown (eds), *Maintaining the Links: Language, Identity and the Land*, Foundation for Endangered Languages, Bath.

Moisseeff, M. 1994. *Davenport Community Profile 1993: Aspects of an Aboriginal village in South Australia*, AIATSIS, Canberra.

Moller, J., Dolinis, J. and Cripps, R. 1996. *Aboriginal Injury-Related Hospitalisation 1991–92: A Comparative Overview Report*, prepared by the AIHW National Injury Surveillance Unit, Research Centre for Injury Studies, Canberra.

Moodie, P.M. 1973. *Aboriginal Health*, ANU Press, Canberra.

Morgan, M., Strelein, L. and Weir, J. 2004. 'Indigenous rights to water in the Murray Darling Basin: in support of the Indigenous final report to the Living Murray Initiative', *AIATSIS Research Discussion Paper No. 14*, AIATSIS, Canberra.

Morphy, F. 2004a. 'Indigenous household structures and ABS definitions of the family: What happens when systems collide, and does it matter?' *CAEPR Working Paper No. 26*, CAEPR, ANU, Canberra.

——2004b. 'When systems collide: The 2001 Census at a Northern Territory outstation', in D.F. Martin, F. Morphy, W.G. Sanders and J. Taylor, *Making Sense of the Census: Observations of the 2001 Enumeration in Re-*

mote Aboriginal Australia, Research Monograph No. 22, 2nd Edition, ANU E-Press, Canberra.

Myers, F. 1989. 'Burning the truck and holding the country: Pintupi forms of property and ideology', in E.N. Wilmsen (ed.), *We Are Here: Politics of Aboriginal Land Tenure*, University of California Press, Berkeley.

Nathan, P. and Japanangka, D.L. 1983. *Settle Down Country: Pmere Arlaltyewele*, Central Australian Aboriginal Congress. Kibble Books, Alice Springs.

National Centre for Social Applications of Geographic Information Systems (NCSAGIS) 2003. *Indigenous Housing Need: Homelessness, Overcrowding and Affordability 2001 Census Analysis*, Report to the Aboriginal and Torres Strait Islander Commission, Canberra.

National Health and Medical Research Council (NHMRC) 2001. *Australian Alcohol Guidelines: Health Risks and Benefits*, Commonwealth of Australia, Canberra.

Neutze, M., Sanders, W. and Jones, R. 2000. 'Estimating Indigenous housing need for public funding allocation: A multi-measure approach', *CAEPR Discussion Paper No. 197*, CAEPR, ANU, Canberra.

Ngaanyatjarra Pitjantjatara Yankunytjatjara Women's Council 1990. *Minyma Tjuta Tjunguringkula Kunpuringanyi (Women Growing Strong Together)*, Pitjantjatara Council Inc, Alice Springs.

Nussbaum, M. 2001. *Women and Human Development: the Capabilities Approach*, Cambridge University Press, Cambridge.

Olson, M. 1965. *The Logic of Collective Action: Public Goods and the Theory of Groups*, Harvard University Press, Cambridge.

Payne, H. 1984. 'Residency and ritual rights', in J. Kassler and J. Stubbington (eds), *Problems and Solutions: Occasional Essays in Musicology Presented to Alice M. Moyle*, Hale and Iremonger, Sydney.

Pearson, N. 2000. *Our Right to Take Responsibility*, Noel Pearson and Associates, Cairns.

——2001. 'On the human right to misery, mass incarceration and early death', *Arena Magazine*, 56: 22–31.

——2005. 'Working towards peace and prosperity', *The Australian*, 26 October: 16.

Perkins, J., Sansom-Fisher, R., Blunden, S., Lunnay, D., et al. 1994. 'The prevalence of drug use in urban Aboriginal communities', *Addiction*, 89: 1319–31.

Permanent Forum on Indigenous Issues 2004. 'Report of the workshop on data collection and disaggregation for Indigenous peoples', *UN Doc: E/C.19/2004/2*, UN, New York.

——2005. 'Report on the fourth session (16–27 May 2005), Economic and Social Council, Official Records, Supplement No.23', *UN Doc: E/C.19/2005/9*, UN, New York.

Peterson, N. 1996. 'Cultural issues', in J.C. Altman and J. Taylor (eds), *The 1994 National Aboriginal and Torres Strait Islander Survey: Findings and Future Prospects*, CAEPR Research Monograph No. 11, CAEPR, ANU, Canberra.

——1999. 'Hunter-gatherers in first world nation states: bring anthropology home', *Bulletin of the National Museum of Ethnology*, 23 (4): 847–61.

——2000. 'An expanding Aboriginal domain: mobility and the initiation journey', *Oceania*, 70 (3): 205–18.

Pollack, D.P. 2001. 'Indigenous land in Australia: a quantitative assessment of Indigenous land holdings in 2000', *CAEPR Discussion Paper No. 221*, CAEPR, ANU, Canberra.

Prensky, M. 2001. 'Digital natives, digital immigrants', *On the Horizon*, 9 (5): 1–6.

Preston, S. and Palloni, A. 1986. 'Fine-tuning brass-type mortality estimates with data on ages of surviving children', *Population Bulletin*, UN, New York, 13: 72–91.

Ramsay, M. 1996. 'The relationship between alcohol and crime', *Research Bulletin*, Home Office Research and Statistics Directorate, London, 38: 37–43.

Rehm, J., Greenfield, T., Walsh, G., Xie, X., et al. 1999. 'Assessment methods for alcohol consumption, prevalence of high-risk drinking and harm: a sensitivity analysis', *International Journal of Epidemiology*, 28: 219–24.

Riedmann, A. 1993. *Science that Colonizes: A Critique of Fertility Studies in Africa*, Temple University Press, Philadelphia.

Rogers, A. and Castro, L.J. 1981. 'Model migration schedules', *Research Report RR-81-30*, International Institute for Applied Systems, Laxenburg, Austria.

——, Racquillet, R. and Castro, L. J. 1978. 'Model migration schedules and their applications', *Environment and Planning A*, 10 (5): 475–502.

Room, R. 2001. 'New findings in alcohol epidemiology', in N. Rehm, R. Room and G. Edwards (eds), *Alcohol in the European Region—Consumption, Harm and Policies*, WHO Regional Office for Europe, Copenhagen.

Rowley, C.D. 1966. 'Some questions of causation in relation to Aboriginal affairs', in I.G. Sharp and C.M. Tatz (eds), *Aborigines in the Economy: Employment, Wages and Training*, Jacaranda Press, Brisbane.

——1970a. *The Destruction of Aboriginal Society*, ANU Press, Canberra.

——1970b. *Outcasts in White Australia*, ANU Press, Canberra.

——1970c. *The Remote Aborigines*, ANU Press, Canberra.

Saggers, S. and Gray, D. 1998. *Dealing with Alcohol: Indigenous Usage in Australia, New Zealand and Canada*, Cambridge University Press, Melbourne.

Sandall, R. 2000. *The Culture Cult: Designer Tribalism and Other Essays*, Westview Press, Boulder, CO.

Sanders, W. 1990. 'Reconstructing Aboriginal housing policy for remote areas: how much room for manoeuvre?' *Australian Journal of Public Administration*, 49 (1): 38–50.

——1993. 'Aboriginal housing', in C. Paris (ed.), *Housing Australia*, MacMillian, Melbourne.

——1996. 'Housing data from the NATSIS: can it assist with program evaluation?' in J.C. Altman and J. Taylor (eds), *The 1994 National Aboriginal and Torres Strait Islander Survey: Findings and Future Prospects*, CAEPR Research Monograph No. 11, CAEPR, ANU, Canberra.

——2005. 'Housing tenure and Indigenous Australians in remote and settled areas', *CAEPR Discussion Paper No. 275*, CAEPR, ANU, Canberra.

——and Arthur, W.S. 1997. 'A Torres Strait Islanders Commission? Possibilities and issues', *CAEPR Discussion Paper No. 132*, CAEPR, ANU, Canberra.

Sarossy, G. 1996. 'Findings from the NATSIS evaluation', in J.C. Altman and J. Taylor (eds), *The 1994 National Aboriginal and Torres Strait Islander Survey: Findings and Future Prospects*, CAEPR Research Monograph No. 11, CAEPR, ANU, Canberra.

Saunders, P. 2002. *Poor Statistics: Getting the Facts Right About Poverty in Australia: Issue Analysis No. 23*, Centre for Independent Studies, Centre for Independent Studies, Sydney.

——2005. *The Poverty Wars: Reconnecting Research with Reality*, UNSW Press, Sydney.

Schmidt, A. 1990. *The Loss of Australia's Aboriginal Language Heritage*, AIATSIS, Canberra.

Schwab, R.G. 1996. 'Indigenous participation in schooling: a preliminary assessment of the NATSIS findings', in J.C. Altman and J. Taylor (eds), *The 1994 National Aboriginal and Torres Strait Islander Survey: Findings and*

Future Prospects, CAEPR Research Monograh No. 11, CAEPR, ANU, Canberra.

Shoobridge, J. 1997. *The Health and Psychological Consequences of Injecting Drug Use in an Aboriginal Community*, National Centre for Education and Training on Addiction, Adelaide.

Sibthorpe, B., Anderson, I. and Cunningham, J. 2001. 'Self-assessed health among Indigenous Australians: How valid is a global question?' *American Journal of Public Health*, 91 (10): 1660–3.

Sims, G. 1992. 'A national survey of the Aboriginal and Torres Strait Islander population: the proposal', in J.C. Altman (ed.) *A National Survey of Indigenous Australians: Options and Implications,* CAEPR Research Monograph No. 3, CAEPR, ANU, Canberra.

Smith, B.R. 2000a. Between places: Aboriginal decentralisation, mobility and territoriality in the Region of Coen, Cape York Peninsular (Qld, Australia), PhD thesis, Department of Anthropology, London School of Economics, London.

——2004. 'The social underpinnings of an 'outstation movement' in Cape York Peninsula, Australia', in J. Taylor and M. Bell (eds), *Population Mobility and Indigenous Peoples in Australasia and North America*, Routledge, London and New York.

Smith, D.E. 2000b. *Indigenous Families and the Welfare System: Two Community Case Studies*, CAEPR Research Monograph No. 17, CAEPR, ANU, Canberra.

——and Roach, L.M. 1996. 'Indigenous voluntary work: NATSIS empirical evidence, policy relevance and future data issues', in J.C. Altman and J. Taylor (eds), *The 1994 National Aboriginal and Torres Strait Islander Survey: Findings and Future Prospects*, CAEPR Research Monograph No. 11, CAEPR, ANU, Canberra.

Smith, L. 1982. 'Aboriginal health and Aboriginal health statistics', *Aboriginal Health Project Bulletin*, 1: 14–24.

——1980. *The Aboriginal Population of Australia*, Aborigines in Australian Society 14, ANU Press, Canberra.

Stanner, W.E.H. 1970. 'Foreword', in H.P. Schapper (ed.), *Aboriginal Advancement to Integration*, ANU Press, Canberra.

Starkey, P., Ellis, S., Hine, J. and Ternell, A. 2002. *Improving Rural Mobility: Options for Developing Motorized and Non-Motorized Transport in Rural Areas*, The World Bank, Washington, D.C.

Statistics Canada 2003. *Aboriginal Peoples Survey 2001: Concepts and Methods Guide*, Statistics Canada, Ottawa.

——2005. *Statistics: Power from data*, http://www.statcan.ca/english/edu/power/toc/contents.htm (last viewed 5 August 2005), Statistics Canada, Ottowa.

Steering Committee for the Review of Government Service Provision 2003. *Overcoming Indigenous Disadvantage: Key Indicators 2003 Report*, Productivity Commission, Melbourne.

——2005. *Overcoming Indigenous Disadvantage: Key Indicators 2005 Report*, Productivity Commission, Melbourne.

Stockwell, T., Donath, S., Cooper-Stanbury, M., Chikritzhs, T., et al. 2004. 'Under-reporting of alcohol consumption in household surveys: a comparison of quantity-frequency, graduated-frequency and recent recall', *Addiction*, 99 (8): 1024–33.

Stotz, G. 2001. 'The colonizing vehicle', in D. Miller (ed.), *Car Cultures*, Oxford Press, New York.

Sutton, P. 2001. 'The politics of suffering: Indigenous policy in Australia since the seventies', *Anthropological Forum*, 11 (2): 125–73.

Taffe, S. 2005. 'The role of FCAATSI in the 1967 referendum: mythmaking about citizenship or political strategy?', in T. Rowse (ed.), *Contesting Assimilation*, API-Network, Perth.

Taylor, C. 2004a. *Modern Social Imaginaries*, Duke University Press, Durham and London.

Taylor, J. 1996. 'Surveying mobile populations: lost opportunity and future needs', in J.C. Altman and J. Taylor (eds), *The 1994 National Aboriginal and Torres Strait Islander Survey: Findings and Future Prospects*, CAEPR Research Monograph No. 11, CAEPR, Canberra.

——2002. 'The spatial context of Indigenous service delivery', *CAEPR Working Paper No. 16*, CAEPR, ANU, Canberra.

——2004b. *Social Indicators for Aboriginal Governance: Insights from the Thamarrurr Region, Northern Territory*, CAEPR Research Monograph No. 24, ANU E Press, Canberra.

—— and Bell, M. 1996a. 'Mobility among Indigenous Australians', in P.L. Newton and M. Bell (eds), *Population Shift: Mobility and Change in Australia*, AGPS, Canberra.

—— and Bell, M. 1996b. 'The mobility status of Indigenous Australians', in P.L. Newton and M. Bell (eds), *Population Shift: Mobility and Change in Australia*, AGPS, Canberra.

—— and Bell, M. 2004. 'Continuity and change in Indigenous Australian population mobility', in J. Taylor and M. Bell (eds), *Population Mobility and Indigenous Peoples in Australasia and North America*, Routledge, London and New York.

—— and Bell, M. 2005. 'Measuring circulation among Indigenous Australians', Paper presented to the International Association for Official Statistics Conference, Te Papa Tongarewa, Museum of New Zealand, 14–15 April, Wellington.

—— and Gaminiratne, K.H.W. 1992. 'A comparison of the socioeconomic characteristics of Aboriginal and Torres Strait Islander people', *CAEPR Discussion Paper No. 35*, CAEPR, ANU, Canberra.

—— and Hunter, B.H. 1998. *The Job Still Ahead: Economic Costs of Continuing Indigenous Employment Disparity*, Office of Public Affairs, ATSIC, Canberra.

—— and Scambary, B. 2005. *Indigenous People and the Pilbara Mining Boom: A Baseline for Regional Participation*, CAEPR Research Monograph No. 25, ANU E Press, Canberra.

—— and Stanley, O. 2005. 'The opportunity costs of the status quo in the Thamarrurr Region', *CAEPR Working Paper No. 28*, CAEPR, ANU, Canberra.

Tharawal Aboriginal Corporation 1994. 'Aboriginal health in southwest Sydney', *Report of the Tharawal Aboriginal Family Environment Health Study*, Tharawal Aboriginal Corporation, Sydney.

Thoits, P. 1995. 'Stress, coping, and social support processes: Where are we? What next?' *Journal of Health and Social Behaviour*, (Extra Issue): 53–79.

Toohey, P. 2000. 'Sly grog', *The Australian*, 30 September 2000, www.theaustralian.com.au.

Torres Strait Health Workshop Working Party 1993. *Torres Strait Health Strategy*, (no publisher noted), Thursday Island.

Tsumori, K., Saunders, P. and Hughes, H. 2002. 'Poor arguments: A response to the Smith Family Report of Poverty in Australia', *Issue Analysis No. 21*, Centre for Independent Studies, Sydney.

Tuck, M. 1989. 'Drinking and disorder: a study of non-metropolitan violence', *Home Office Research Study 108*, Home Office, London.

United Nations 1983. *Manual X: Indirect Techniques for Demographic Estimation*, UN, New York.

——1985. *Socio-economic Differentials in Child Mortality in Developing Countries*, Sales No. E85.XIII.T, UN, New York.

United Nations Development Programme 2000. *Human Development Report*, UN, Geneva.

Wadiwel, D. 2005. 'Transport disadvantage in Aboriginal communities', *NCOSS News*, 32 (5): 7.

Whiteford, P. 1985. 'A family's needs: Equivalence scales, poverty and social security', *Research Paper No. 27*, Department of Social Security, Development Division, Canberra.

WHO 2000. 'International guide for monitoring alcohol consumption and alcohol related harm', in T. Stockwell and T. Chikritzhs (eds), *WHO/MSD/MSB/00.5*, WHO, Geneva.

Williams, P. 2001. 'Deaths in custody: 10 years on from the Royal Commission', *Trends and Issues in Crime and Criminal Justice No.203*, Australian Institute of Criminology, Canberra.

Wolter, K. M. 1985. *Introduction to Variance Estimation*, Springer-Verlag, New York.

Young, D. 2001. 'The Life and death of cars: Private vehicles on the Pitjantjatara Lands, South Australia', in D. Miller (ed.), *Car Cultures*, Oxford Press, New York.

Young, E. A. and Doohan, K. 1989. *Mobility for Survival: A Process Analysis of Population Movement in Central Australia*, North Australia Research Unit, Darwin.

Zubrick, S. R., Lawrence, D. M., Silburn, S. R., Blair, E., et al. 2004. *The Western Australian Aboriginal Child Health Survey: The Health of Aboriginal Children & Young People*, Telethon Institute for Child Health Research, Perth.

Notes on Contributors

Jon Altman

Jon Altman has a disciplinary background in economics and anthropology. He has been involved in research on Indigenous Australians since the late 1970s. In 1977 he participated in the research project The Economic Status of Australian Aborigines that was among the first to use statistical information from the 1971 Census to document socioeconomic difference between Indigenous and other Australians, as well as diversity within the Indigenous population. In 1990 he became the inaugural director of CAEPR at ANU, where he is still located. In 1992 he convened a workshop prior to NATSIS 1994 and in 1996 co-convened a workshop with John Taylor that analysed outcomes from that survey. Jon divides his research effort between a focus on national economic and policy issues and a specific regional focus on western Arnhem Land and the Kuninjku community with whom he has worked for 26 years.

Bill Arthur

Bill Arthur began researching Indigenous affairs in the 1980s. Much of his initial work was land-related and was carried out for Indigenous organisations such as the Kimberley Land Council. Since 1990 his research has focused more in issues of economic development for Torres Strait Islanders. During the preparation of this paper he was a Visiting Research Fellow at the CAEPR.

Larissa Behrendt

Larissa Behrendt is a Professor of Law and Indigenous Studies and Director of Ngiya, the National Institute of Indigenous Law, Policy and Practice. She has a Doctorate of Laws from Harvard University, and is admitted to the ACT Supreme Court as a Barrister-of-Law. Larissa has worked as a practicing lawyer in the areas of Aboriginal land claims and family law, has taught at the University of New South Wales and Australian National University Law Schools, and worked in Canada and at the United Nations with First Nations organisations. She is currently undertaking research on regional authority models, and has published on property law, indigenous rights, dispute resolution and Aboriginal women's issues. Her book *Achieving Social Justice: Indigenous Rights and Australia's Future* was published by The Federation Press in 2003.

Nicholas Biddle

Nicholas Biddle is currently a PhD student at the CAEPR, where he is researching education outcomes for Indigenous Australians as part of a linkage grant between CAEPR and the ABS. Nicholas has a Bachelor of Economics (Hons) from the University of Sydney and has completed the requirements for a Graduate Diploma in Education from Monash University. He is on study leave from the Australian Bureau of Statistics, where he has worked since 2001, focusing his research on neighbourhood income inequality, the measurement of Indigenous poverty, and health patterns of Australian migrants.

Dan Black

Dan Black is Director of the National Centre for Aboriginal and Torres Strait Islander Statistics in the Australian Bureau of Statistics. He has had a leadership role in ABS Indigenous statistics for over 20 years.

Maggie Brady

Maggie Brady is a social anthropologist with a longstanding research background in Indigenous health and substance abuse. She has held positions with Flinders University Medical School, the Northern Land Council, the Human Rights Commission, AIATSIS and ANU, and published extensively for both academic and community-based audiences. Her books include an account of drinking in Tennant Creek (*Where the Beer Truck Stopped*, 1988), an anthropological study of petrol sniffing (*Heavy Metal*, 1992), Indigenous accounts of drinking and abstention (*Giving Away the Grog*, 1995), a study of Indigenous alcohol policy and practice (*Indigenous Australia and Alcohol Policy*, 2004), and community-directed alcohol action books (*The Grog Book*, 1998, 2005; *Tackling Alcohol Problems — Strengthening Community Action in South Africa* 2005). Maggie is an ARC Postdoctoral Fellow at the CAEPR.

Geoff Buchanan

Geoff Buchanan is a Graduate Research Assistant with the CAEPR. Geoff has a multidisciplinary background covering environmental policy and economics, social policy and development, Indigenous Australian studies and commerce. His primary research focus is on Indigenous involvement in natural and cultural resource management.

Tom Calma

Tom Calma is Aboriginal and Torres Strait Islander Social Justice Commissioner and acting Race Discrimination Commissioner with the Human Rights and Equal Opportunity Commission. He is an Aboriginal elder from the Kungarakan tribal group and the Iwaidja tribal group whose traditional lands are south west of Darwin and on the Coburg Peninsula in Northern Territory, respectively. Tom has been involved in Indigenous affairs at a local, community, state, national and international level and worked in the public sector for over 30 years. He has extensive experience in public administration, particularly in Indigenous education programs and in developing employment and training programs for

Indigenous people from both a national policy and program perspective. He has served as a diplomat, as a Director of Aboriginal Hostels, in ATSIS, DEETYA, and as a Senior Adviser to the Minister of Immigration, Multicultural and Indigenous Affairs.

Bruce Chapman

Bruce Chapman is a Professor and Head of the Economics Program in the Research School of Social Sciences at ANU. He has a PhD from Yale University and is a labour and education economist having published over a hundred and forty articles in the areas of training, wage determination, student loans and higher education financing, unemployment, labour market program evaluation, the economics of crime, economic analyses of cricket and the role of income contingent loans in public policy. He has had extensive direct policy experience, including the motivation and design of HECS in 1988, as a senior economic advertiser to Prime Minister Paul Keating from 1994-1996, and as a consultant to the OECD, the World Bank, and the governments of around 12 countries (mostly in the area of student loans and university financing). In 2001 he was awarded the Order of Australia for 'contributions to the development of economics, labour market and social policy'. He is currently writing a book on the application of income contingent loans to a host of public policies, including for the financing of: tertiary student income support; elite athletes; low level criminal offences; drought relief; housing loans; and community investment projects.

Tanya Chikritzhs

Tanya Chikritzhs is a Research Fellow at the National Drug Research Institute. One of her main responsibilities is to co-ordinate the National Alcohol Indicators Project (NAIP), the aim of which is to track and report on trends in indicators of alcohol-related harm at national, state and local levels. The NAIP has recently been expanded to include specific alcohol indicators relevant to Indigenous communities with an emphasis on providing information which is relevant at local levels. Other recent research projects include an evaluation of the Northern Territory's Living With Alcohol Program and a current investigation into the efficacy of liquor restrictions throughout Australia and internationally.

Mick Dodson

Mick Dodson is a member of the Yawuru peoples, the traditional Aboriginal owners of land and waters in the Broome area of the southern Kimberley region of Western Australia. He is currently Director of the ANU's National Centre for Indigenous Studies. Professor Dodson was Australia's first Aboriginal and Torres Strait Islander Social Justice Commissioner. He completed a Bachelor of Jurisprudence and a Bachelor of Laws at Monash University, and has been awarded an honorary Doctor of Letters from the University of Technology Sydney and an honorary Doctor of Laws from the University of NSW.

From August 1988 to October 1990 Mick was Counsel assisting the Royal Commission into Aboriginal Deaths in Custody. He is a member and the current Chairman of the Australian Institute of Aboriginal and Torres Strait Islander Studies. He is also a board member of the Reconciliation Australia and Lingiari Foundations. Throughout his career, has been a prominent advocate on land rights and other issues affecting Aboriginal and Torres Strait Islanders.

Matthew Gray

Matthew Gray is the Deputy Director (Research) at the Australian Institute of Family Studies and was previously employed at CAEPR. He has undertaken research on a wide range of economic and social policy issues, including those relating to Indigenous Australians.

R. G. (Bob) Gregory

Bob Gregory is Professor of Economics at the Research School of Social Sciences, Australian National University. His current research interests focus mainly on Australian labour markets and welfare support. He has held academic appointments at the University of Melbourne, London School of Economics, Northwestern University, Harvard and the University of Chicago. He has also been a member of the Board of the Reserve Bank of Australia and the Institute of Family Studies, and a member of the Australian Science and Technology Council and Expert Panel of the Australian Research Council. In 1996 he was awarded the Order of Australia Medal.

Sarah Holcombe

Sarah Holcombe is now a Research Fellow at the CAEPR. She is currently working on two ARC projects. These are 'Indigenous Community Organisations and Miners: Partnering Sustainable Regional Development?' with the linkage partner Rio Tinto, and an 'Indigenous Community Governance' project with the linkage partner Reconciliation Australia. The research in the later project is also supported by the Desert Knowledge CRC and is part of the core project 'viable desert settlements'. Dr Holcombe's previous research focused on applied anthropology in the Northern Territory as a staff member for the Central and Northern Land Councils. Her PhD research in social anthropology was undertaken in the Central Australian Luritja community of Mt Liebig (Amunturrngu), on the processes by which this settlement evolved into an Indigenous community.

John Hughes

John Hughes is a publications editor at CAEPR. He has previously worked within the Australian Public Service as an Indigenous employment consultant and editor, and as a freelance author and publisher.

Boyd Hunter

Boyd Hunter is a Fellow at the CAEPR. He specialises in labour market analysis, racial discrimination, social economics, crime and justice statistics, neighbourhood inequality, poverty research, and coordinated the first longitudinal analysis of Indigenous job seekers for the Department of Employment and Workplace Relations. He contributes to many government committees and inquiries, including, the Design sub-Committee of the Longitudinal Study of Indigenous Children. Until recently, he was a Ronald Henderson Fellow and visiting Fellow with the New Zealand Treasury in Wellington. In 2003, he held a Australian Census Analytic Program Fellow that resulted in the ABS monograph, *Indigenous People in the Contemporary Australian Labour Market*.

Yohannes Kinfu

Yohannes Kinfu was a Research Fellow at CAEPR. His major research interests are in the area of statistical and mathematical modelling of population processes and dynamics, and he lectures on these areas in the Demography and Sociology Program at the ANU. He has worked for the UN-Regional Institute for Population Studies in Ghana and has held a visiting scientist position at the South African Medical Research Council, working on issues related to the measurement of mortality in Africa. In recent times, his research has focused on the demographic components of Indigenous population change in Australia. He has published on these issues in Australia and internationally.

Inge Kral

Inge Kral has worked in education in remote Aboriginal Australia for some 20 years across a range of sectors including Aboriginal community schools, adult education and training, and as an education consultant. Her areas of expertise include language and literacy (in Aboriginal languages and English), curriculum development and education policy. As a language and literacy consultant she has worked on education policy in the Ngaanyatjarra Lands in Western Australia and VET sector training and Aboriginal Health Workers in the Northern Territory. Recent publications include a (2004) NCVER Report *What is all that learning for? Indigenous adult literacy practices, training, community capacity and health* and the co-compilation of the *Ngaanyatjarra Picture Dictionary* (IAD Press 2005). Inge has an MA in applied linguistics through the University of Melbourne in which she researched the development literacy in Arrernte, a Central

Australian Aboriginal language. She is currently a PhD student at CAEPR. Her thesis is in anthropology, where she is conducting ethnographic research on social literacy practices in a remote Aboriginal community in the Western Desert.

Frances Morphy

Frances Morphy is a Research Fellow at CAEPR. She is an anthropologist and linguist whose research has focused mainly on the Top End of the Northern Territory, with interests in Australian Indigenous languages, the problem of inter-cultural translation, land rights and native title and the governance of Indigenous organisations. With other CAEPR colleagues she was an official observer of the 2001 Census enumeration in selected Indigenous communities in remote Australia. The results of this research were published in a CAEPR Research Monograph (Martin et al. 2002, *Making Sense of the Census: Observations of the 2001 Enumeration in Remote Aboriginal Australia*) and in a CAEPR Working Paper, 'Indigenous household structures and ABS definitions of the family: What happens when systems collide, and does it matter?' She was a member of the Australian Bureau of Statistics' 2006 Census Indigenous Enumeration Strategy Working Group. She is co-editor (with Bill Arthur) of the Macquarie Atlas of Indigenous Australia.

Nicolas Peterson

Nicolas Peterson lectures in anthropology in the School of Archaeology and Anthropology at ANU. He has carried out fieldwork in northeast Arnhem Land with Yolngu speakers on ecological issues and in central Australia with Warlpiri speakers on religious life and territorial organisation. He has a long standing interest in land and sea tenure and has worked on twelve major native title and land claims. Recently he has compiled and introduced *Donald Thomson in Arnhem Land* (2003, MUP), and co-edited *Photography's Other Histories* (2003, Duke) with Chris Pinney and *Citizenship and Indigenous Australians* (1998, Cambridge) with Will Saunders.

Peter Radoll

Peter Radoll is an Aboriginal doctoral candidate and an Associate Lecturer in Information Systems at ANU's Faculty of Economics and Commerce. Peter's PhD research focuses on Information Communication Technology use in remote, rural and urban Aboriginal Communities. His research has been presented to the Community Informatics Research Network at the Russian Academy of Sciences in Moscow. He currently holds several competitive scholarships.

Alistair Rogers

Alistair Rogers works in the Statistical Services Branch of the Australian Bureau of Statistics' Methodology Division. The branch provides mathematical statistical advice in support of the ABS survey program. Alistair led the development of the sample design framework underpinning the 2002 NATSISS, and more recently the 2004/5 National Aboriginal and Torres Strait Islander Health Survey.

Russell Ross

Russell Ross is an associate professor of economics at the University of Sydney. He has written extensively on the Australian labour market. His research interests centre on the under-utilisation of labour, with particular focus on labour market status of Indigenous Australians, participation of married women, labour market aspects of childcare availability, and labour market policy to improve outcomes for disadvantaged groups in the labour market. He was a member of the Employment Issues Technical Reference Group, advising the Australian Bureau of Statistics on the implementation and analysis of NATSIS 1994. He has authored reports for governments on the Indigenous labour market (NSW Government) and work-related childcare provision (Federal government, and the ABC). He and his co-author are currently in the process of preparing the third edition of their textbook *The Australian Labour Market*.

Tim Rowse

Tim Rowse is in the History Program, Research School of Social Sciences, at ANU. He has been researching Indigenous affairs since the early 1980s, publishing on public policy, public health and frontier history. From 1989 to 1994 he worked in the Menzies School of Health Research and produced *Remote Possibilities* (1992) and *Traditions for Health* (1996). In 2000-1, he wrote a critical synthesis of CAEPR's first ten years of work: *Indigenous Futures: Choice and Development for Aboriginal and Islander Australia* (2002). Working with CAEPR awakened his interest in the political determinants and political effects of the Indigenous statistical archive, and that has become his new research interest.

Will Sanders

Will Sanders is a Fellow at CAEPR, and has been researching Australian Indigenous affairs policy since 1981. He has worked in four departments of ANU during that time: the North Australia Research Unit in Darwin and the Urban Research Unit/ Program, the Department of Political Science and CAEPR in Canberra. He first published on housing policy for Indigenous people in 1990 in the *Australian Journal of Public Administration* and has been returning to the topic regularly since.

R. G. (Jerry) Schwab

Jerry Schwab is a Fellow at CAEPR. He is one of a very small number of anthropologists in Australia with research experience in the area of Indigenous education and training. He has been involved with educational research and development in both Australia and overseas (USA, United Arab Emirates and Egypt) since the mid-1980s. He has worked in the fields of literacy (especially early literacy), educational development (in schools and for academic staff in higher education) and program planning and evaluation. Since joining CAEPR in 1995, he has carried out primary and secondary research on issues as diverse as Aboriginal community-controlled schools, notions of educational 'failure' and 'success' among Indigenous students, Indigenous workforce development and Indigenous education outcomes at the primary, secondary and post-compulsory levels. He has long standing research interests in Indigenous school retention, the relationship between schools and communities and adult literacy. His current research is focused on land and resource management programs as an avenue for the educational and social re-engagement of Indigenous youth in remote regions.

Geoff Scott

Distinguished Professor Geoff Scott holds a research chair in public policy at the University of Technology, Sydney. He is a consultant to the NSW Aboriginal Land Council and a member of the board of the Aboriginal Housing Office.

John Taylor

John Taylor is a Senior Fellow at CAEPR and a member of the Australian Population Association. For the past twenty years his major research interests have focussed on the measurement and policy implications of demographic and economic change among Indigenous peoples. He has published widely on these issues in Australian and international books and journals, and his work has informed government, industry, and Indigenous agencies, most recently in relation to the COAG trial site at Wadeye. He is co-editor (with Martin Bell) of the seminal volume on Indigenous population mobility, *Indigenous Peoples and Population Mobility in Australasia and North America*, published by Routledge in London and New York.

Andrew Webster

Andrew Webster works in the National Centre for Aboriginal and Torres Strait Islander Statistics in the Australian Bureau of Statistics. He has managed the data output and dissemination program for the 2002 NATSISS. During the development phase of the survey, Andrew was out-posted to the then Aboriginal and Torres Strait Islander Commission, where he assisted with liaison between

ABS and ATSIC on survey content. Andrew has worked in the ABS for 12 years, in demography, family statistics and Indigenous statistics.

Ruth Weston

Ruth Weston is a Principal Research Fellow at the Australian Institute of Family Studies. Ruth has published extensively on such issues as quality of life, couple formation and stability, fertility decision-making, parent-adolescent relationships, and divorce.

CAEPR Research Monograph Series

1. *Aborigines in the Economy: A Select Annotated Bibliography of Policy Relevant Research 1985–90*, L. M. Allen, J. C. Altman, and E. Owen (with assistance from W. S. Arthur), 1991.

2. *Aboriginal Employment Equity by the Year 2000*, J. C. Altman (ed.), published for the Academy of Social Sciences in Australia, 1991.

3. *A National Survey of Indigenous Australians: Options and Implications*, J. C. Altman (ed.), 1992.

4. *Indigenous Australians in the Economy: Abstracts of Research, 1991–92*, L. M. Roach and K. A. Probst, 1993.

5. The Relative Economic Status of Indigenous Australians, 1986–91, J. Taylor, 1993.

6. *Regional Change in the Economic Status of Indigenous Australians, 1986–91*, J. Taylor, 1993.

7. *Mabo and Native Title: Origins and Institutional Implications*, W. Sanders (ed.), 1994.

8. *The Housing Need of Indigenous Australians, 1991*, R. Jones, 1994.

9. *Indigenous Australians in the Economy: Abstracts of Research, 1993–94*, L. M. Roach and H. J. Bek, 1995.

10. *The Native Title Era: Emerging Issues for Research, Policy, and Practice*, J. Finlayson and D. E. Smith (eds), 1995.

11. *The 1994 National Aboriginal and Torres Strait Islander Survey: Findings and Future Prospects*, J. C. Altman and J. Taylor (eds), 1996.

12. *Fighting Over Country: Anthropological Perspectives*, D. E. Smith and J. Finlayson (eds), 1997.

13. *Connections in Native Title: Genealogies, Kinship, and Groups*, J. D. Finlayson, B. Rigsby, and H. J. Bek (eds), 1999.

14. *Land Rights at Risk? Evaluations of the Reeves Report*, J. C. Altman, F. Morphy, and T. Rowse (eds), 1999.

15. *Unemployment Payments, the Activity Test, and Indigenous Australians: Understanding Breach Rates*, W. Sanders, 1999.

16. *Why Only One in Three? The Complex Reasons for Low Indigenous School Retention*, R. G. Schwab, 1999.

17. *Indigenous Families and the Welfare System: Two Community Case Studies*, D. E. Smith (ed.), 2000.

18. *Ngukurr at the Millennium: A Baseline Pro le for Social Impact Planning in South-East Arnhem Land*, J. Taylor, J. Bern, and K. A. Senior, 2000.

19. *Aboriginal Nutrition and the Nyirranggulung Health Strategy in Jawoyn Country*, J. Taylor and N. Westbury, 2000.

20. *The Indigenous Welfare Economy and the CDEP Scheme*, F. Morphy and W. Sanders (eds), 2001.

21. *Health Expenditure, Income and Health Status among Indigenous and Other Australians*, M. C. Gray, B. H. Hunter, and J. Taylor, 2002.

22. *Making Sense of the Census: Observations of the 2001 Enumeration in Remote Aboriginal Australia*, D. F. Martin, F. Morphy, W. G. Sanders and J. Taylor, 2002.

23. *Aboriginal Population Pro les for Development Planning in the Northern East Kimberley* J. Taylor, 2003.

24. *Social Indicators for Aboriginal Governance: Insights from the Thamarrurr Region*, Northern Territory, J. Taylor, 2004.

25. *Indigenous people and the Pilbara mining boom: A baseline for regional participation*, J. Taylor and B. Scambary, 2005.

For information on CAEPR Discussion Papers, Working Papers and Research Monographs (Nos 1-19) please contact:

> Publication Sales, Centre for Aboriginal Economic Policy Research, The Australian National University, Canberra, ACT, 0200
>
> Telephone: 02–6125 8211
> Facsimile: 02–6125 2789

Information on CAEPR abstracts and summaries of all CAEPR print publications and those published electronically can be found at the following WWW address:

> http://www.anu.edu.au/caepr/

www.ingramcontent.com/pod-product-compliance
Lightning Source LLC
Chambersburg PA
CBHW061241270326
41928CB00041B/3354